Ernst Schering Research Foundation Workshop 40
Recent Advances in Glucocorticoid Receptor Action

Springer
*Berlin
Heidelberg
New York
Barcelona
Hong Kong
London
Milan
Paris
Tokyo*

Ernst Schering Research Foundation
Workshop 40

Recent Advances in Glucocorticoid Receptor Action

A.C.B. Cato, H. Schäcke, K. Asadullah
Editors

With 82 Figures

Springer

Series Editors: G. Stock and M. Lessl

ISSN 0947-6075
ISBN 3-540-43229-9 Springer-Verlag Berlin Heidelberg New York

Library of Congress Cataloging-in-Publication Data

Recent advances in glucocorticoid receptor action / A. Cato, H. Schäcke, K. Asadullah, editors.
    p. cm. -- (Ernst Schering Research Foundation workshop ; 40)
  Includes bibliographical references and index.
  ISBN 3540432299 (alk. paper)
  1. Clucocorticoids--Receptors--Agonists--Congresses. 2.
Clucocorticoids--Receptors--Congresses. I. Cato, A. (Andrew), 1951- II. Schäcke, H.
(Heike), 1964- III. Asadullah, K. (Khusru), 1967- IV. Series.

This work is subject to copyright. All rights are reserved, whether the whole or part of the material is concerned, specifically the rights of translation, reprinting, reuse of illustrations, recitation, broadcasting, reproduction on microfilms or in any other way, and storage in data banks. Duplication of this publication or parts thereof is permitted only under the provisions of the German Copyright Law of September 9, 1965, in its current version, and permission for use must always be obtained from Springer-Verlag. Violations are liable for prosecution under the German Copyright Law.

Springer-Verlag Berlin Heidelberg New York
a member of BertelsmannSpringer Science+Business Media GmbH

http://www.springer.de

© Springer-Verlag Berlin Heidelberg 2002

The use of general descriptive names, registered names, trademarks, etc. in this publication does not imply, even in the absence of a specific statement, that such names are exempt from the relevant protective laws and regulations and therefore free for general use. Product liability: The publishers cannot guarantee the accuracy of any information about dosage and application contained in this book. In every individual case the user must check such information by consulting the relevant literature.

Typesetting: Data conversion by Springer-Verlag
Printing: Druckhaus Beltz, Hemsbach
Binding: J. Schäffer GmbH & Co. KG, Grünstadt
SPIN: 10866377   21/3130/AG–5 4 3 2 1 0 – Printed on acid-free paper

Preface

Since 1948, hydrocortisone (cortisol), the principal glucocorticoid (GC) of the human adrenal cortex has been successfully used at pharmacological concentrations for the suppression of clinical manifestations of rheumatoid arthritis. Numerous compounds with GC activity have also been developed and used.

Fifty years after their initial clinical use, GCs are still the most important and frequently prescribed class of anti-inflammatory drugs for various inflammatory disorders. They are administered either orally, parenterally (intravenous, intramuscular, intrathecal), or topically (cutaneous, intranasal, pulmonic, rectal). Despite the many beneficial effects of GCs, they also have their limitations and disadvantages that occur with varying prevalence on different organs and after different durations of therapy. These side-effects can range in severity from cosmetic (e.g. telangiectasias, hypertrichosis) to seriously disabling (e.g. induction of glaucoma, diabetes, osteoporosis) or even life-threatening disorders (e.g. gastric haemorrhage). These adverse effects of GCs seriously handicap their successful use as anti-inflammatory agents. There is therefore a strong need for the development of substances with the anti-inflammatory potency of classical GCs but with reduced side-effects.

Recent research has provided new insight into GC action offering unique opportunities for the development of novel therapeutic strategies based on new principles. To review these emerging possibilities, we organised a workshop on recent advances in GC-receptor action in Berlin, Germany from 31 October to 2 November 2001. This meeting was intended to bring together an outstanding group of scientists span-

The participants of the workshop

ning the broad scope of GC research and to have open discussions on future developments in the field of GCs action. This aim was successfully reached thanks to the generous support of the Ernst Schering Research Foundation and to the active contributions of the participants. At this workshop, we were able to cover clinical and basic aspects of GC action with talks and discussion from many internationally renowned clinicians and basic scientists. The proceedings of the workshop are contained in this volume and we wish to take this opportunity to thank the speakers and participants for their presentations and lively discussions. We trust that the readers will share with us our enthusiasm and continued excitement in the developments coming out of the field of GC research.

A.C.B. Cato, H. Schäcke, K. Asadullah

Contents

1. Glucocorticoids and Asthma
   P.J. Barnes 1

2. Immune Monitoring of Glucocorticoid Therapy
   P. Reinke, M. Bevilacqua, V. Tryon, J. Cheronis, H.-D. Volk   25

3. Topical Glucocorticoid Therapy in Dermatology
   W. Sterry, K. Asadullah 39

4. Bone Effects of Glucocorticoid Therapy
   K.H. Väänänen, P.L. Härkönen 55

5. Corticosteroids in Ophthalmology
   U. Pleyer, Z. Sherif 65

6. Special Problems in Glucocorticoid Treatment in Children
   U. Wahn 83

7. Functional Implications of Glucocorticoid Receptor Trafficking
   D.B. DeFranco 91

8. The Dynamics of Intranuclear Movement and Chromatin Remodeling by the Glucocorticoid Receptor
   G.L. Hager 111

9	Glucocorticoid Receptor Antagonism of AP-1 Activity by Inhibition of MAPK Family C. Caelles, A. Bruna, M. Morales, J.M. González-Sancho, M.V. González, B. Jiménez, A. Muñoz	131
10	Mast Cells as Targets for Glucocorticoids in the Treatment of Allergic Disorders O. Kassel, A.C.B. Cato	153
11	Cytosolic Glucocorticoid Receptor-Interacting Proteins A.-C. Wikström, C. Widén, A. Erlandsson, E. Hedman, J. Zilliacus	177
12	The Glucocorticoid Receptor β-Isoform: A Perspective on Its Relevance in Human Health and Disease M.J.M. Schaaf, J.A. Cidlowski	197
13	Cooperation of Nuclear Transcription Factors Regulated by Steroid and Peptide Hormones B. Groner, C. Shemanko	213
14	Induction and Repression of NF-κB-Driven Inflammatory Genes W. Vanden Berghe, K. De Bosscher, L. Vermeulen, G. De Wilde, G. Haegeman	233
15	DNA-Dependent Cofactor Selectivity of the Glucocorticoid Receptor A. Dostert, T. Heinzel	279
16	The Anti-inflammatory Action of Glucocorticoid Hormones P. Herrlich, M. Göttlicher	297
17	Analysis of Glucocorticoid Receptor Function in the Mouse by Gene Targeting C. Kellendonk, F. Tronche, H.M. Reichardt, A. Bauer, E. Greiner, W. Schmid, G. Schütz	305

18	Glucocorticoid-Inducible Genes That Regulate T-Cell Function *P.R. Mittelstadt, J. Galon, D. Franchimont, J.J. O'Shea, J.D. Ashwell* 319
19	Structural Analysis of the GR Ligand-Binding Domain *U. Egner* 341
20	SEGRAs: A Novel Class of Anti-inflammatory Compounds *H. Schäcke, H. Hennekes, A. Schottelius, S. Jaroch, M. Lehmann, N. Schmees, H. Rehwinkel, K. Asadullah* . . . 357

Subject Index 373

Previous Volumes Published in This Series 377

List of Editors and Contributors

Editors

K. Asadullah
Research Business Area Dermatology, Schering AG, Müllerstr. 178,
13342 Berlin, Germany
e-mail: Khusru.Asadullah@schering.de

A.C.B. Cato
Forschungszentrum Karlsruhe, Institute of Toxicology and Genetics,
P.O. Box 3640, 76021 Karlsruhe, Germany
e-mail: andrew.cato@itg.fzk.de

H. Schäcke
Research Business Area Dermatology, Schering AG, Müllerstr. 178,
13342 Berlin, Germany
e-mail: Heike.Schaecke@schering.de

Contributors

J. Ashwell
National Cancer Institute, National Institutes of Health, Laboratory of Immune Cell Biology, Rm. 1B-40, Bdg. 10, Bethesda, MD 20892, USA
e-mail: jda@pop.nci.nih.gov

P.J. Barnes
Department of Thoracic Medicine, Imperial College School of Medicine,
National Hearts and Lung Institute, Dovehouse Street, London, SW3 6LY, UK
e-mail: p.j.barnes@ic.ac.uk

A. Bauer
Deutsches Krebsforschungszentrum Heidelberg, Abteilung Molekularbiologie der Zelle I, Im Neuenheimer Feld 280, 69120 Heidelberg, Germany
e-mail: anton.bauer@univie.ac.at

M. Bevilacqua
Source Precision Medicine Inc., 2425 North 55th Street, Boulder, CO 80301, USA
e-mail: mbevilacqua@sourcepharma.com

A. Bruna
Departament de Bioquímica i Biologia Molecular, Div. IV, Faculatat de Farmàcia, Universitat de Barcelona, Avenida Diagonal, 643, 08028 Barcelona, Spain
e-mail: abruna@farmacia.far.ub.es

C. Caelles
Departament de Bioquímica i Biologia Molecular, Div. IV, Faculatat de Farmàcia, Universitat de Barcelona, Avenida Diagonal, 643, 08028 Barcelona, Spain
e-mail: caelles@farmacia.far.ub.es

J. Cheronis
Source Precision Medicine Inc., 2435 North 55th Street, Boulder, CO 80301, USA
e-mail: jcheroni@sourcepharma.com

J.A. Cidlowski
National Institute of Environmental Health Sciences, Laboratory of Signal Transduction, P.O. Box 12233, Research Triangle Park, NC 27709, USA
e-mail: cidlows1@niehs.nih.gov

K. De Bosscher
Unit of Eukaryotic Gene Expression and Signal Transduction,
Department of Molecular Biology, University of Gent – VIB,
K.L. Ledeganckstraat 35, 9000 Gent, Belgium
k.bosscher@icrf.icnet.ac.uk

List of Editors and Contributors

G. De Wilde
Unit of Eukaryotic Gene Expression and Signal Transduction,
Department of Molecular Biology, University of Gent – VIB,
K.L. Ledeganckstraat 35, 9000 Gent, Belgium
e-mail: gert.dewilde@devgen.com

D.D. DeFranco
University of Pittsburgh, Department of Biological Sciences,
Pittsburgh, PA 15260, USA
e-mail: dod1+@pitt.edu

A. Dostert
Chemotherapeutisches Forschungsinstitut, Georg-Speyer-Haus, Paul-Ehrlich-Str. 42-44, 60594 Frankfurt/Main, Germany
e-mail: dostert@em.uni-frankfurt.de

U. Egner
Research Laboratories, Schering AG, Müllerstr. 178, 13342 Berlin, Germany
e-mail: Ursula_2.Egner@schering.de

A. Erlandsson
Department of Medical Nutrition Karolinska Institute, Huddinge University Hospital, Novum F60, S-141 86 Huddinge, Sweden
e-mail: anna.erlandsson@csb.ki.se

D. Franchimont
Gastroenterology Department, Erasme University Hospital, ULB, 808,
LennikStreet, 1070 Brussels, Belgium
e-mail: dfranchimont@hotmail.com

J. Galon
INSERM U255, Centre de Recherches, Biomédicales des Cordeliers,
15 Rue de l'Ecole de Médecine, 75270 Paris Cedex 06, France
e-mail: jerome.galon@u255.bhdc.jussieu.fr

M.V. González
Instituto de Investigaciones Biomédicas "Alberto Sols" and Departamento de Bioquímica, Faculdad de Medicina, Consejo Superior de Investigaciones Científicas, University of Madrid, 28029 Madrid, Spain
e-mail: mvmeana@correo.uniovi.es

J.M. González-Sancho
Instituto de Investigaciones Biomédicas "Alberto Sols" and Departamento de
Bioquímica, Faculdad de Medicina, Consejo Superior de Investigaciones Científicas, University of Madrid, 28029 Madrid, Spain
e-mail: jog2008@med.cornell.edu

M. Göttlicher
Forschungszentrum Karlsruhe, Institute of Toxicology and Genetics,
P.O. Box 3640, 76021 Karlsruhe
e-mail: martin-goettlicher@itg.fzk.de

E. Greiner
Deutsches Krebsforschungszentrum Heidelberg, Abteilung Molekularbiologie
der Zelle I, Im Neuenheimer Feld 280, 69120 Heidelberg, Germany
e-mail: e.greiner@dkfz.de

B. Groner
Institute for Biomedical Research, Georg-Speyer-Haus,
Paul-Ehrlich-Str. 42-44, 60596 Frankfurt/Main, Germany
e-mail: groner@em.uni-frankfurt.de

G. Haegeman
Unit of Eukaryotic Gene Expression and Signal Transduction,
Department of Molecular Biology, University of Gent – VIB,
K.L. Ledeganckstraat 35, 9000 Gent, Belgium
e-mail: guy.Haegeman@dmb.rug.ac.be

G.L. Hager
National Cancer Institute, Laboratory of Receptor Biology and Gene Expression, Building 41, Room B602, Bethesda, MD 20892-5055, USA
e-mail: hagerg@exchange.nih.gov

P.L. Härkönen
Department of Anatomy, Institute of Biomedicine, University of Turku,
Kiinamyllynkatu 10, 20520 Turku, Finland
e-mail: pirkko.harkonen@utu.fi

E. Hedman
Department of Medical Nutrition Karolinska Institute, Huddinge University
Hospital, Novum F60, S-141 86 Huddinge, Sweden
e-mail: erik.hedman@mednut.ki.se

List of Editors and Contributors

T. Heinzel
Chemotherapeutisches Forschungsinstitut, Georg-Speyer-Haus,
Paul-Ehrlich-Str. 42-44, 60594 Frankfurt/Main, Germany
e-mail: heinzel@em.uni-frankfurt.de

H. Hennekes
TH Projectmanagement, Schering AG, Müllerstr. 178, 13342 Berlin, Germany
e-mail: hartwig.hennekes@schering.de

P. Herrlich
Forschungszentrum Karlsuhe, Institute of Toxicology and Genetics,
P.O. Box 3640, 76021 Karlsruhe, Germany
e-mail: peter.herrlich@itg.fzk.de

S. Jaroch
Medicinal Chemistry, Schering AG, Müllerstr. 178, 13342 Berlin, Germany
e-mail: stefan.jaroch@schering.de

B. Jiménez
Instituto de Investigaciones Biomédicas "Alberto Sols" and Departamento de
Bioquímica, Faculdad de Medicina, Consejo Superior de Investigaciones Científicas, University of Madrid, 28029 Madrid, Spain
e-mail: bjimenez@iib.uam.es

O. Kassel
Forschungszentrum Karlsruhe, Institute of Toxicology and Genetics,
P.O. Box 3640, 76021 Karlsruhe, Germany
e-mail: Olivier.Kassel@itg.fzk.de

C. Kellendonk
Howard Hughes Medical Institute, Research Laboratories,
Center For Neurobiology and Behavior, 722 West 168 Street,
6th Floor Res. Annex, New York, NY 10032, USA
Ck@fido.cpmc.columbia.edu

M. Lehmann
Medicinal Chemistry, Schering AG, Müllerstr. 178, 13342 Berlin, Germany
e-mail: manfred.lehmann@schering.de

P.R. Mittelstadt
Laboratory of Immune Cell Biology, NCI, National Institute of Health,
Bethesda, MD 20892, USA
e-mail: mittel@pop.nci.nih.gov

M. Morales
Departament de Bioquímica i Biologia Molecular, Div. IV, Faculatat de Farmàcia, Universitat de Barcelona, Avenida Diagonal, 643, 08028 Barcelona,
Spain
e-mail: mmorales@farmacia.far.ub.es

A. Muñoz
Instituto de Investigaciones Biomédicas "Alberto Sols" and Departamento de
Bioquímica, Faculdad de Medicina, Consejo Superior de Investigaciones Científicas, University of Madrid, 28029 Madrid, Spain
e-mail: amunoz@iib.uam.es

J.J. O'Shea
National Institute of Arthritis and Metabolic Diseases, NIH,
Bethesda, MD 20892, USA
e-mail: osheaj@arb.niams.nih.gov

U. Pleyer
Department of Ophthalmology, University Hospital Charité, Campus
Virchow-Klinik, Augustenburger Platz 1, 13353 Berlin, Germany
e-mail: uwe.pleyer@charite.de

H. Rehwinkel
Medicinal Chemistry, Schering AG, Müllerstr. 178, 13342 Berlin, Germany
e-mail: hartmut.rehwinkel@schering.de

H.M. Reichardt
Institut für Virologie und Immunbiologie, Universität Würzburg,
Versbacher Strasse 7, 97078 Würzburg, Germany
Holger.reichardt@mail.uni-wuerzburg.de

P. Reinke
Department of Nephrology and Internal Intensive Medicine, Charité, Campus
Virchow Clinic, Augustenburger Platz 1, 13353 Berlin, Germany
e-mail: petra.reinke@charite.de

List of Editors and Contributors

M.J.M. Schaaf
Laboratory of Signal Transduction, National Institute of Environmental Health Sciences, National Institutes of Health, MD F307, 111 Alexander Drive, P.O. Box 12233, Research Triangle Park, NC 27709, USA
e-mail: schaaf@niehs.nih.gov

N. Schmees
Medicinal Chemistry, Schering AG, Müllerstr. 178, 13342 Berlin, Germany
e-mail: norbert.schmees@schering.de

W. Schmid
Deutsches Krebsforschungszentrum Heidelberg, Abteilung Molekularbiologie der Zelle I, Im Neuenheimer Feld 280, 69120 Heidelberg, Germany
e-mail: w.schmid@dkfz.de

A. Schottelius
Research Business Area Dermatology, Schering AG, Müllerstr. 178, 13342 Berlin, Germany
e-mail: arndt.schottelius@schering.de

G. Schütz
Deutsches Krebsforschungszentrum Heidelberg, Abteilung Molekularbiologie der Zelle I, Im Neuenheimer Feld 280, 69120 Heidelberg, Germany
e-mail: g.schuetz@dkfz.de

C.C. Shemanko
Institute for Biomedical Research, Georg-Speyer-Haus, Paul-Ehrlich-Str. 42-44, 60596 Frankfurt/Main, Germany
e-mail: shemanko@em.uni-frankfurt.de

Z. Sherif
Department of Ophthalmology, University Hospital Charité, Campus Virchow-Klinik, Augustenburger Platz 1, 13353 Berlin, Germany

W. Sterry
Department of Dermatoloy, University Hospital Charité, Humboldt University, Schumannstr. 20-21, 10117 Berlin, Germany
e-mail: wolfram.sterry@charite.de

F. Tronche
CNRS FRE2401, Génetique Moléculaire, Neurophysiologie et Comportement, Institut de Biologie, Collège de France, 11 place Marcelin Berthelot, 75231 Paris Cedex 5, France
e-mail: Francois.troche@college-de-france.fr

V. Tyron
Source Precision Medicine Inc., 2425 North 55th Street, Boulder, CO 80301, USA
e-mail: vtryon@sourcepharma.com

K. Väänänen
Department of Anatomy, Institute of Biomedicine, University of Turku, Kiinamyllynkatu 10, 20520 Turku, Finland
e-mail: kalervo.vaananen@utu.fi

W. Vanden Berghe
Unit of Eukaryotic Gene Expression and Signal Transduction, Department of Molecular Biology, University of Gent – VIB, K.L. Ledeganckstraat 35, 9000 Gent, Belgium
e-mail: wimvdb@dmb001.rug.ac.be

L. Vermeulen
Unit of Eukaryotic Gene Expression and Signal Transduction, Department of Molecular Biology, University of Gent – VIB, K.L. Ledeganckstraat 35, 9000 Gent, Belgium
e-mail: linda@dmb.rug.ac.be

H.-D. Volk
Institute of Medical Immunology, Charité, Humboldt University Berlin, 10098 Berlin, Germany
e-mail: hans-dieter.volk@charite.de

U. Wahn
Klinik für Pädiatrie der Charité, Campus Virchow, Augustenburger Platz 1, 13353 Berlin, Germany
e-mail: ulrich.wahn@charite.de

List of Editors and Contributors

C. Widén
Department of Medical Nutrition Karolinska Institute, Huddinge University Hospital, Novum F60, S-141 86 Huddinge, Sweden
e-mail: christina.widen@csb.ki.se

A.-C. Wikström
Department of Medical Nutrition Karolinska Institute, Huddinge University Hospital, Novum F60, S-141 86 Huddinge, Sweden
e-mail: Lotta.Wikstrom@mednut.ki.se

J. Zilliacus
Department of Medical Nutrition Karolinska Institute, Huddinge University Hospital, Novum F60, S-141 86 Huddinge, Sweden
e-mail: Johanna.zilliacus@mednut.ki.se

1 Glucocorticoids and Asthma

P.J. Barnes

1.1	Molecular Mechanisms	1
1.2	Target Genes in Allergic Inflammation Control	6
1.3	Effects on Cell Function	10
1.4	Effects on Asthmatic Inflammation	14
1.5	Corticosteroid Resistance	15
1.6	Therapeutic Implications	17
References		18

1.1 Molecular Mechanisms

Corticosteroids are highly effective anti-inflammatory therapy in asthma, and the molecular mechanisms involved in suppression of allergic inflammation are now better understood (Barnes 1998, 2001). Corticosteroids are effective clinically because they block many of the inflammatory pathways that are abnormally activated in asthma and they have a very broad spectrum of anti-inflammatory actions.

1.1.1 Molecular Basis of Asthma

In asthma there is a persistent inflammation in the airways characterised by infiltration of eosinophils and activation of T-lymphocytes, particular T helper-2 cells and mast cells. There are characteristic structural changes including shedding of airway epithelial cells, fibrosis under the epithelial cells and proliferation of airway smooth muscle cells (Busse and Lemanske 2001). This is an increased expression of multiple inflammatory proteins in structural cells of the airways, including cytokines, enzymes that synthesise inflammatory mediators, adhesion molecules and inflammatory receptors. This is largely due to increased gene expression, which has suggested that transcription factors play a critical role in orchestrating the inflammatory process in asthma (Barnes and Adcock 1998). These transcription factors include nuclear factor (NF)-κB and activator protein (AP)-1, both of which are activated in asthmatic airways and result in the expression of many of the inflammatory genes abnormally expressed in the airways of asthmatic patients (Demoly et al. 1992; Hart et al. 1998). In addition, other transcription factors, including nuclear factor of activated T-cells (NF-AT), signal transduction activated transcription factors (STATs) and GATA transcription factors are involved in inflammatory gene expression in asthma (Caramori et al. 1998; Holtzman et al. 1998). Since corticosteroids are very effective in suppressing inflammation in asthmatic airways, it is likely that they are working at the level of inflammatory gene transcription.

1.1.2 Increased Gene Transcription

Corticosteroids produce their effect on responsive cells by activating the glucocorticoid receptor (GR) to directly or indirectly regulate the transcription of certain target genes (Reichardt et al. 1998). In the airways, GR is expressed in all cells, but in a particularly high concentration in airway epithelial and endothelial cells (Adcock et al. 1996). There is no evidence for reduced expression of GR in airways of asthmatic patients, even after treatment with inhaled corticosteroids.

The number of genes per cell *directly* regulated by corticosteroids is estimated to be between 10 and 100, but many genes are indirectly regulated through an interaction with other transcription factors. GR dimers bind to DNA at consensus sites termed glucocorticoid response elements (GREs) in the 5′-upstream promoter region of steroid-responsive genes. This interaction changes the rate of transcription, resulting in either induction or repression of the gene. Interaction of the activated GR homodimer with GRE usually increases transcription, resulting in increased protein synthesis. GR may increase transcription by interacting with large coactivator molecules, such as CREB-binding protein (CBP). CBP is bound at the start site of transcription and this leads via a series of linking proteins to the binding and activation of RNA polymerase II, resulting in formation of messenger RNA and then synthesis of protein. Binding of activated GR to CBP results in increased acetylation of core histones around which DNA is wound within the chromosomal structure (Ito et al. 2000), and this is critical for the subsequent activation of RNA polymerase II. For example, high concentrations of corticosteroids increase the secretion of the antiprotease secretory leukoprotease inhibitor (SLPI) from epithelial cells (Abbinante-Nissen et al. 1995). This is associated with a selective acetylation of lysine residues 5 and 16 on histone 4, resulting in increased gene transcription (Ito et al. 2000). This same mechanism is likely to also apply to other anti-inflammatory genes that are switched on by corticosteroids, but may also apply to the side-effects of corticosteroids which are largely mediated by this transactivation.

1.1.3 Decreased Gene Transcription

In controlling inflammation, the major effect of corticosteroids is to inhibit the synthesis of multiple inflammatory proteins. This was originally believed to be through interaction of GR with negative GREs, resulting in repression of transcription. However, negative GREs have only very rarely been demonstrated and are not a feature of the promoter region of inflammatory genes that are suppressed by steroids in the treatment of asthma.

1.1.4 Interaction with Transcription Factors

Activated GR may bind directly with several other activated transcription factors as a protein–protein interaction (Pfahl 1993). This could be an important determinant of corticosteroid responsiveness and is a key mechanism whereby corticosteroids switch off inflammatory genes. Most of the inflammatory genes that are activated in asthma do not appear to have GREs in their promoter regions yet are repressed by corticosteroids. There is persuasive evidence that corticosteroids inhibit the effects of transcription factors that regulate the expression of genes that code for inflammatory proteins, such as cytokines, inflammatory enzymes, adhesion molecules and inflammatory receptors, as discussed above (Barnes and Adcock 1998). It was once believed that activated GR interacted directly with activated transcription factors through a protein–protein interaction, but this may be a feature that is seen in transfected cells, rather than primary cells at therapeutic concentrations of corticosteroids. Thus, in a chronically transfected epithelial cell line with a NF-κB-driven reporter gene, there is relatively little effect of corticosteroids on transcription (Newton et al. 1998). Furthermore, treatment of asthmatic patients with high doses of inhaled corticosteroids that suppress airway inflammation is not associated with any reduction in NF-κB binding to DNA (Hart et al. 2000). This suggests that corticosteroids act downstream of the binding of inflammatory transcription factors to DNA, and attention has now focussed on their effects on chromatin structure and histone acetylation.

1.1.5 Effects on Chromatin Structure

There is increasing evidence that corticosteroids may have effects on the chromatin structure. DNA in chromosomes is wound around histone molecules in the form of nucleosomes (Grunstein 1997; Wolffe and Hayes 1999). Several transcription factors interact with large co-activator molecules, such as CBP and the related molecule p300, which bind to the basal transcription factor apparatus (Fig. 1). Several transcription factors bind directly to CBP, including AP-1, NF-κB, STATs and GR (Kamei et al. 1996; Ogryzko et al. 1996). At a microscopic level, chromatin may become dense or opaque due to the winding or unwinding of DNA around the histone core. Co-activator molecules, including CBP and p300, have histone acetylation activity which is stimulated by the binding of transcription factors, such as AP-1 and NF-κB. Acetylation of lysine residues in the N-terminal tails of core histones results in unwinding of DNA that is tightly coiled around the histone core of the resting gene, thus opening up the chromatin structure. This hyperacetylation allows transcription factors and RNA polymerase II to bind more readily, thereby switching on or increasing transcription.

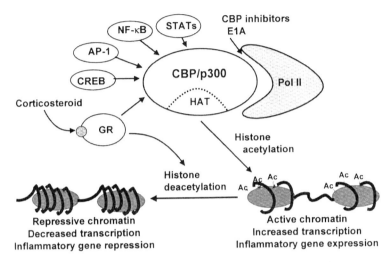

Fig. 1. Effect of corticosteroids on chromatin structure. Transcription factors, such as STATs, AP-1 and NF-κB bind to co-activator molecules, such as CREB binding protein (*CBP*) or p300, which have intrinsic histone acetyltransferase (*HAT*) activity, resulting in acetylation (*–Ac*) of histone residues. This leads to unwinding of DNA and this allows increased binding of transcription factors resulting in increased gene transcription. Glucocorticoid receptors (*GR*) after activation by corticosteroids bind to a glucocorticoid receptor co-activator which is bound to CBP. This results in deacetylation of histone, with increased coiling of DNA around histone, thus preventing transcription factor binding leading to gene repression

This inflammatory mechanism is reversed by deacetylation of the hyperacetylated histones induced by inflammatory stimuli, due to inhibition of hypoxanthine-aminopterin-thymidine (HAT) activity of co-activator molecules and also by increased deacetylation (Imhof and Wolffe 1998; Ito et al. 2001a). Deacetylation of histones increases the winding of DNA round histone residues, resulting in dense chromatin structure and reduced access of transcription factors and RNA polymerase to their binding sites, thereby leading to repressed transcription of inflammatory genes. Activated GR may bind to several transcription co-repressor molecules that associate with proteins that have histone deacetylase (HDAC) activity, resulting in deacetylation of hyperacetylated histones, increased winding of DNA round histone residues and thus reduced access of transcription factors to their binding sites and therefore repression of inflammatory genes. In addition, activated GR recruits HDACs to the transcription start site, resulting in deacetylation of histones, and a decrease in inflammatory gene transcription (Ito et al. 2000). Over ten distinct HDACs are now recognised and these are differentially expressed and regulated in different cell types (Fischle et al. 2001; Kuo and Allis 1998). This could contribute to the differences in responsiveness to corticosteroids between different genes and between different cell types.

1.1.6 Non-transcriptional Effects

It is increasingly recognised that GR may also affect the synthesis of some proteins by reducing the stability of mRNA, through effects on ribonucleases that break down mRNA. Some inflammatory genes, such as the gene encoding granulocyte-macrophage (GM)-CSF, produce mRNA that has a sequence rich in AU nucleotides at the 3'-untranslated end. It is this region that interacts with ribonucleases that break down mRNA, thus switching off protein synthesis (Bergmann et al. 2000; Newton et al. 2001).

1.2 Target Genes in Allergic Inflammation Control

Corticosteroids may control inflammation by inhibiting many aspects of the inflammatory process in allergy through increasing the transcription of anti-inflammatory genes and decreasing the transcription of many inflammatory genes (Table 1).

1.2.1 Anti-inflammatory Proteins

Corticosteroids may suppress inflammation by increasing the synthesis of anti-inflammatory proteins. For example, corticosteroids increase the synthesis of lipocortin-1, a 37-kDa protein that has an inhibitory effect on phospholipase A_2 (PLA_2), and therefore may inhibit the production of lipid mediators. Corticosteroids induce the formation of lipocortin-1

Table 1. Effect of corticosteroids on gene transcription

Increased transcription
Lipocortin-1 (phospholipase A_2 inhibitor)
β_2-Adrenoceptor
Secretory leukoprotease inhibitor
Clara cell protein (CC10, phospholipase A_2 inhibitor)
IL-1 receptor antagonist
IL-1R2 (decoy receptor)
IκB-α (inhibitor of NF-κB)

Decreased transcription
Cytokines
  (IL-1, IL-2, IL-3, IL-4, IL-5, IL-6, IL-9, IL-11, IL-12, IL-13, IL-16, IL-17, IL-18, TNF-α, GM-CSF, SCF)
Chemokines
  (IL-8, RANTES, MIP-1α, MCP-1, MCP-3, MCP-4, eotaxin)
Inducible nitric oxide synthase (iNOS)
Inducible cyclooxygenase (COX-2)
Cytoplasmic phospholipase A_2 (cPLA$_2$)
Endothelin-1
Neurokinin (NK)$_1$-receptors, NK$_2$-receptors
Bradykinin (B)$_1$-receptors, B$_2$-receptors
Adhesion molecules (ICAM-1, E-selectin)

in several cells and recombinant lipocortin-1 has acute anti-inflammatory properties. However, lipocortin-1 does not appear to be increased by inhaled corticosteroid treatment in asthma (Hall et al. 1999). Corticosteroids increase the expression of other potentially anti-inflammatory proteins, such as interleukin (IL)-1 receptor antagonist (which inhibits the binding of IL-1 to its receptor), SLPI (which inhibits proteases, such as tryptase), neutral endopeptidase (which degrades bronchoactive peptides such as kinins), CC-10 (an immunomodulatory protein), an inhibitor of NF-κB (IκB-α) and IL-10 (an anti-inflammatory cytokine). The expression of IL-10 in macrophages from asthmatic patients is decreased, and this may increase the expression of several inflammatory genes. Corticosteroids increase secretion of IL-10 and may therefore overcome this defect (John et al. 1998).

1.2.2 β₂-Adrenoceptors

Corticosteroids increase the expression of β_2-adrenoceptors by increasing the rate of transcription, and the human β_2-receptor gene has three potential GREs (Collins et al. 1988). Corticosteroids double the rate of β_2-receptor gene transcription in human lung in vitro, resulting in increased expression of β_2-receptors (Mak et al. 1995a). This also occurs in vivo in nasal mucosa with treatment with topical corticosteroids (Baraniuk et al. 1997). This may be relevant in asthma as corticosteroids may prevent down-regulation of β-receptors in response to prolonged treatment with β_2-agonists. In rats, corticosteroids prevent down-regulation and reduced transcription of β_2-receptors in response to chronic β-agonist exposure (Mak et al. 1995b).

1.2.3 Cytokines

The inhibitory effect of corticosteroids on cytokine synthesis is likely to be of particular importance in the control of inflammation in allergic inflammation, as cytokines play a critical role in the chronic inflammatory process of asthma (Chung and Barnes 1999). Corticosteroids inhibit the transcription of many cytokines and chemokines that are relevant in allergy (Table 1). These inhibitory effects are due, at least in part, to an inhibitory effect on the transcription factors that regulate induction of these cytokine genes, including AP-1 and NF-κB. For example, eotaxin, which is important in selective attraction of eosinophils from the circulation into the airways, is regulated in part by NF-κB, and its expression in airway epithelial cells is inhibited by corticosteroids (Lilly et al. 1997). Many transcription factors are likely to be involved in the regulation of inflammatory genes in asthma, in addition to AP-1 and NF-κB. IL-4 and IL-5 expression in T-lymphocytes plays a critical role in allergic inflammation, but NF-κB does not play a role, whereas NF-AT is important (Rao et al. 1997). AP-1 is a component of the NF-AT transcription complex, so that corticosteroids inhibit IL-5, at least in part, by inhibiting the AP-1 component of NF-AT.

There may be marked differences in the response of different cells and of different cytokines to the inhibitory action of corticosteroids and this may be dependent on the relative abundance of transcription factors within different cell types. Thus in alveolar macrophages and peripheral blood monocytes, GM-CSF secretion is more potently inhibited by corticosteroids than IL-1β or IL-6 secretion (Linden et al. 1998).

1.2.4 Inflammatory Enzymes

Nitric oxide (NO) synthase may be induced by proinflammatory cytokines, resulting in NO production. NO may amplify asthmatic inflammation and contribute to epithelial shedding and airway hyperresponsiveness through the formation of peroxynitrite. The induction of the inducible form of NOS (iNOS) is inhibited by corticosteroids. In cultured human pulmonary epithelial cells, pro-inflammatory cytokines result in increased expression of iNOS and increased NO formation, due to increased transcription of the iNOS gene, and this is inhibited by corticosteroids acting partly through inhibition of NF-κB (Robbins et al. 1994). Corticosteroids inhibit the synthesis of several other inflammatory mediators implicated in asthma through an inhibitory effect on the induction of enzymes, such as cyclo-oxygenase-2 and cytosolic PLA_2 (Newton et al. 1997).

1.2.5 Inflammatory Receptors

Corticosteroids also decrease the transcription of genes coding for certain receptors. Thus the gene for the natural killer $(NK)_1$-receptor, which mediates the inflammatory effects of tachykinins in the airways, has an increased expression in asthma and is inhibited by corticosteroids, probably via an inhibitory effect on AP-1 (Adcock et al. 1993). Corticosteroids also inhibit the transcription of the NK_2-receptor which mediates the bronchoconstrictor effects of tachykinins (Katsunuma et al. 1998). Corticosteroids also inhibit the expression of the inducible bradykinin B_1-receptor and bradykinin B_2-receptors (Haddad et al. 2000).

1.2.6 Adhesion Molecules

Adhesion molecules play a key role in the trafficking of inflammatory cells to sites of inflammation. The expression of many adhesion molecules on endothelial cells is induced by cytokines and corticosteroids may lead indirectly to a reduced expression via their inhibitory effects on cytokines, such as IL-1β and tumour necrosis factor (TNF)-α. Corticosteroids may also have a direct inhibitory effect on the expression of adhesion molecules, such as intercellular adhesion molecule (ICAM)-1 and E-selectin at the level of gene transcription. ICAM-1 and vascular cell adhesion molecule (VCAM)-1 expression in bronchial epithelial cell lines and monocytes is inhibited by corticosteroids (Atsuta et al. 1999).

1.2.7 Apoptosis

Corticosteroids markedly reduce the survival of certain inflammatory cells, such as eosinophils. Eosinophil survival is dependent on the presence of certain cytokines, such as IL-5 and GM-CSF. Exposure to corticosteroids blocks the effects of these cytokines and leads to programmed cell death or apoptosis, although the corticosteroid-sensitive molecular pathways have not yet been clearly defined (Walsh 1997). By contrast, corticosteroids decrease apoptosis in neutrophils and thus prolong their survival (Meagher et al. 1996). This may contribute to the lack of anti-inflammatory effects of corticosteroids in chronic obstructive pulmonary disease (COPD) where neutrophilic inflammation is predominant.

1.3 Effects on Cell Function

Corticosteroids may have direct inhibitory actions on several inflammatory cells and structural cells that are implicated in asthma (Fig. 2).

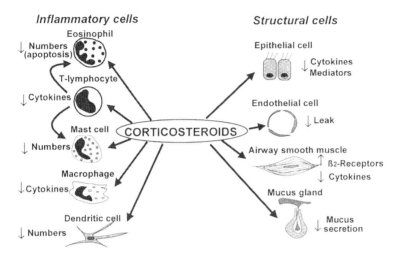

Fig. 2. Cellular effect of corticosteroids

1.3.1 Macrophages

Corticosteroids inhibit the release of inflammatory mediators and cytokines from alveolar macrophages in vitro (Linden et al. 1998). Inhaled corticosteroids reduce the secretion of chemokines and proinflammatory cytokines from alveolar macrophages from asthmatic patients, whereas the secretion of IL-10 is increased (John et al. 1998).

1.3.2 Eosinophils

Corticosteroids have a direct inhibitory effect on mediator release from eosinophils, although they are only weakly effective in inhibiting secretion of reactive oxygen species and eosinophil basic proteins. More importantly, corticosteroids induce apoptosis by inhibiting the prolonged survival due to IL-3, IL-5 and GM-CSF (Meagher et al. 1996; Walsh 1997), resulting in an increased number of apoptotic eosinophils in induced sputum of asthmatic patients (Woolley et al. 1996). There is a delay in the apoptosis of eosinophils in asthma, which is reversed by treatment with corticosteroids (Kankaanranta et al. 2000). One of the best-described actions of corticosteroids in asthma is a reduction in circulating eosinophils, which may reflect an action on eosinophil production in the bone marrow.

1.3.3 T-lymphocytes

T helper 2 lymphocytes (Th2) play an important orchestrating role in asthma through the release of the cytokines IL-4, IL-5, IL-9 and IL-13 and may be an important target for corticosteroids in asthma therapy. For example, corticosteroids inhibit IL-5 gene expression in Th2 cells by blocking binding of NF-AT (Quan et al. 2001). Corticosteroids also induce apoptosis in T-cells.

1.3.4 Mast Cells

While corticosteroids do not appear to have a direct inhibitory effect on mediator release from lung mast cells, chronic corticosteroid treatment is associated with a marked reduction in mucosal mast cell numbers. This may be linked to a reduction in IL-3 and stem cell factor (SCF) production, which are necessary for mast cell expression at mucosal surfaces (Finotto et al. 1997). Mast cells also secrete various cytokines (TNF-α, IL-4, IL-5, IL-6 and IL-8), and this may also inhibited by corticosteroids (Williams and Galli 2000).

1.3.5 Dendritic Cells

Dendritic cells in the epithelium of the respiratory tract appear to play a critical role in antigen presentation in the lung as they have the capacity to take up allergen, process it into peptides and present it via MHC molecules on the cell surface for presentation to uncommitted T-lymphocytes. In experimental animals the number of dendritic cells is markedly reduced by systemic and inhaled corticosteroids, thus dampening the immune response in the airways (Nelson et al. 1995).

1.3.6 Neutrophils

Neutrophils, which are not prominent in the biopsies of asthmatic patients, are not sensitive to the effects of corticosteroids. Indeed, systemic corticosteroids increase peripheral neutrophil counts, which may reflect an increased survival time due to an inhibitory action of neutrophil apoptosis (Meagher et al. 1996).

1.3.7 Endothelial Cells

GR gene expression in the airways is most prominent in endothelial cells of the bronchial circulation and airway epithelial cells. Corticosteroids do not appear to directly inhibit the expression of adhesion molecules, although they may inhibit cell adhesion indirectly by suppression of cytokines involved in the regulation of adhesion molecule expression. Corticosteroids may have an inhibitory action on airway microvascular leak induced by inflammatory mediators. This appears to be a direct effect on postcapillary venular epithelial cells. Although there have been no direct measurements of the effects of corticosteroids on airway microvascular leakage in asthmatic airways, regular treatment with inhaled corticosteroids decreases the elevated plasma proteins found in bronchoalveolar lavage fluid of patients with stable asthma.

1.3.8 Epithelial Cells

Epithelial cells may be an important source of many inflammatory mediators in asthmatic airways and may drive and amplify the inflammatory response in the airways through the secretion of proinflammatory cytokines, chemokines and inflammatory peptides. Airway epithelium may be one of the most important cellular targets for inhaled corticosteroids in asthma (Barnes 1996; Schweibert et al. 1996). Inhaled corticosteroids inhibit the increased expression of many inflammatory proteins in airway epithelial cells. An example is iNOS, which has an increased expression in airway epithelial and inflammatory cells in asthma and is reduced by inhaled corticosteroids (Saleh et al. 1998).

This is reflected by a reduction in the elevated levels of exhaled NO in asthma after inhaled corticosteroids (Kharitonov et al. 1996)

1.3.9 Mucus Secretion

Corticosteroids inhibit mucus secretion in airways and this may be a direct action of corticosteroids on submucosal gland cells. Corticosteroids may also inhibit the expression of mucin genes, such as MUC2 and MUC5AC (Kai et al. 1996). In addition, there are indirect inhibitory effects due to the reduction in inflammatory mediators that stimulate increased mucus secretion.

1.4 Effects on Asthmatic Inflammation

Corticosteroids are remarkably effective in controlling the inflammation in asthma and it is likely that they have multiple cellular effects. Biopsy studies in patients with asthma have now confirmed that inhaled corticosteroids reduce the number and activation of inflammatory cells in the airway mucosa and in bronchoalveolar lavage (Barnes 1996). These effects may be due to inhibition of cytokine synthesis in inflammatory and structural cells and suppression of adhesion molecules. The disrupted epithelium is restored and the ciliated to goblet cell ratio is normalised after 3 months of therapy with inhaled corticosteroids. There is also some evidence for a reduction in the thickness of the basement membrane, although in asthmatic patients taking inhaled corticosteroids for over 10 years the characteristic thickening of the basement membrane is still present.

1.4.1 Effects on Airway Hyperresponsiveness

By reducing airway inflammation inhaled corticosteroids consistently reduce airway hyperresponsiveness (AHR) in asthmatic adults and children (Barnes 1990). Chronic treatment with inhaled corticosteroids reduces responsiveness to histamine, cholinergic agonists, allergen (early and late responses), exercise, fog, cold air, bradykinin, adenosine and irritants (such as sulphur dioxide and metabisulphite). The reduction in AHR takes place over several weeks and may not be maximal until several months of therapy. The magnitude of reduction is variable between patients and is in the order of one to two doubling dilutions for most challenges and often fails to return to the normal range. This may reflect suppression of the inflammation but persistence of structural changes which cannot be reversed by corticosteroids. Inhaled corticosteroids not only make the airways less sensitive to spasmogens, but they also limit the maximal airway narrowing in response to spasmogens.

1.5 Corticosteroid Resistance

Although corticosteroids are highly effective in the control of asthma and other chronic inflammatory or immune diseases, a small proportion of patients with asthma fail to respond even to high doses of oral corticosteroids (Barnes et al. 1995; Szefler and Leung 1997; Barnes 2000). Resistance to the therapeutic effects of corticosteroids is also recognised in other inflammatory and immune diseases, including rheumatoid arthritis and inflammatory bowel disease. Corticosteroid-resistant patients, although uncommon, present considerable management problems. Recently, new insights into the mechanisms whereby corticosteroids suppress chronic inflammation have shed new light on the molecular basis of corticosteroids-resistant asthma.

1.5.1 Clinical Features

Corticosteroid-resistant asthma is defines as a failure to improve lung function by greater than 15% after treatment with high doses of prednisolone (30–40 mg daily) for 2 weeks, providing the oral steroid is taken (verified by plasma prednisolone level or a reduction in early morning cortisol level). These patients are not Addisonian and they do not suffer from the abnormalities in sex hormones described in the very rare familial glucocorticoid resistance. Plasma cortisol and adrenal suppression in response to exogenous cortisol is normal in these patients, so they suffer from side-effects of corticosteroids.

Complete corticosteroid resistance in asthma is very rare, with a prevalence of less than 1:1000 asthmatic patients. Much more common is a reduced responsiveness to corticosteroids, so that large inhaled or oral doses are needed to control asthma adequately (corticosteroid-dependent asthma). It is likely that there is a range of responsiveness to corticosteroids and that corticosteroid resistance is at one extreme of this range.

It is important to establish that the patient has asthma, rather than COPD, "pseudoasthma" (a hysterical conversion syndrome involving vocal cord dysfunction), left ventricular failure or cystic fibrosis that do not respond to corticosteroids. Asthmatic patients are characterised by a variability in peak expiratory flow (PEF) and, in particular, a diurnal variability of greater than 15% and episodic symptoms. It is also important to identify provoking factors (allergens, drugs, psychological problems) that may increase the severity of asthma and its resistance to therapy. Biopsy studies have demonstrated the typical eosinophilic inflammation of asthma in these patients (Szefler and Leung 1997).

1.5.2 Molecular Mechanisms of Corticosteroid Resistance

There may be several mechanisms for resistance to the effects of corticosteroids. Certain cytokines (particularly IL-2, IL-4 and IL-13) may induce a reduction in affinity of glucocorticoid receptors in inflammatory cells such as T-lymphocytes, resulting in local resistance to the anti-inflammatory actions of corticosteroids (Szefler and Leung 1997). Another mechanism is an increased activation of the transcription factor AP-1 by inflammatory cytokines, so that AP-1 may consume activated glucocorticoid receptors and thus reduce their availability for suppression of inflammation at inflamed sites (Adcock et al. 1994). There is an increased expression of c-Fos, one of the components of AP-1 (Lane et al. 1998). The reasons for this excessive activation of AP-1 by activating enzymes is currently unknown, but may be genetically determined. Another proposed mechanism is an increase in expression of GR-β, which then interferes with DNA binding of GR (Hamid et al. 1999), but any increase in GR-β is insufficient to account for reduced responsiveness to corticosteroids (Gagliardo et al. 2000).

1.5.3 Corticosteroid Resistance in COPD

Although inhaled corticosteroids are highly effective in asthma, they provide little benefit in COPD, despite the fact that airway and lung inflammation is present. This may reflect that the inflammation in COPD is not suppressed by corticosteroids, with no reduction in inflammatory cells, cytokines or proteases in induced sputum even with oral corticosteroids (Keatings et al. 1997; Culpitt et al. 1999). Corticosteroids do not suppress neutrophilic inflammation in the airways and corticosteroids may prolong the survival of neutrophils (Nightingale et al. 2000). There is some evidence that the airway inflammation in COPD is corticosteroid resistant, as corticosteroids have no inhibitory effect on inflammatory proteins, such as cytokines, that are normally suppressed by corticosteroids. This lack of response to corticosteroids may be explained in part by an inhibitory effect of cigarette smoking on HDACs, thus interfering with an important anti-inflammatory action of corticosteroids (Ito et al. 2001b).

1.6 Therapeutic Implications

Inhaled corticosteroids are now used as first-line therapy for the treatment of persistent asthma in adults and children in many countries, as they are the most effective treatments for asthma currently available (Barnes et al. 1998). They are also used widely for the treatment of perennial and seasonal rhinitis.

1.6.1 Dissociated Corticosteroids

All currently available inhaled corticosteroids are absorbed from the lungs into the systemic circulation and therefore inevitably have some systemic component. Understanding the molecular mechanisms of action of corticosteroids has led to the development of a new generation of corticosteroids. As discussed above, a major mechanism of the anti-inflammatory effect of corticosteroids appears to be inhibition of transcription factors, such as NF-κB and AP-1, that are activated by proinflammatory cytokines (trans-repression) via an inhibitory action on histone acetylation and stimulation of histone deacetylation. By contrast, the endocrine and metabolic effects of steroids that are responsible for the systemic side-effects of corticosteroids are likely to be mediated via DNA binding (trans-activation) (Reichardt and Schutz 1998). This has led to a search for novel corticosteroids that selectively trans-repress, thus reducing the potential risk of systemic side-effects. Since corticosteroids bind to the same GR, this seems at first to be an unlikely possibility, but while DNA binding involves a GR homodimer, interaction with transcription factors AP-1 and NF-κB involves only a single GR. A separation of trans-activation and trans-repression has been demonstrated using reporter gene constructs in transfected cells using selective mutations of the glucocorticoid receptor. Furthermore, some steroids, such as the antagonist mifepristone (RU486), have a greater trans-repression than trans-activation effect. Indeed, the topical steroids used in asthma therapy today, such as fluticasone propionate (FP) and budesonide, appear to have more potent trans-repression than trans-activation effects, which may account for their selection as potent anti-inflammatory agents (Adcock et al. 1999). Recently, a novel class of steroids has been described in which there is potent trans-repression with relatively little trans-activation. These "dissociated" steroids, including RU24858 and RU40066 have anti-inflammatory effects in vitro (Vayssiere et al. 1997), although there is little separation of anti-inflammatory effects and systemic side effects in vivo (Belvisi et al. 2001). This suggests that the development of steroids with a greater margin of safety is possible and may even lead to the development of oral steroids that do not have significant adverse effects.

References

Abbinante-Nissen JM, Simpson LG, Leikauf GD (1995) Corticosteroids increase secretory leukocyte protease inhibitor transcript levels in airway epithelial cells. Am J Physiol 12:L601–L606

Adcock IM, Peters M, Gelder C, Shirasaki H, Brown CR, Barnes PJ (1993) Increased tachykinin receptor gene expression in asthmatic lung and its modulation by steroids. J Mol Endocrinol 11:1–7

Adcock IM, Brown CR, Shirasaki H, Barnes PJ (1994) Effects of dexamethasone on cytokine and phorbol ester stimulated c-Fos and c-Jun DNA binding and gene expression in human lung. Eur Respir J 7:2117–2123

Adcock IM, Gilbey T, Gelder CM, Chung KF, Barnes PJ (1996) Glucocorticoid receptor localization in normal human lung and asthmatic lung. Am J Respir Crit Care Med 154:771–782

Adcock IM, Nasuhara Y, Stevens DA, Barnes PJ (1999) Ligand-induced differentiation of glucocorticoid receptor trans-repression and transactivation: preferential targetting of NF-κB and lack of I-κB involvement. Br J Pharmacol 127:1003–1011

Atsuta J, Plitt J, Bochner BS, Schleimer RP (1999) Inhibition of VCAM-1 expression in human bronchial epithelial cells by glucocorticoids. Am J Respir Cell Mol Biol 20:643–650

Baraniuk JN, Ali M, Brody D, Maniscalco J, Gaumond E, Fitzgerald T, Wonk G, Mak JCW, Bascom R, Barnes PJ, Troost T (1997) Glucocorticoids induce b_2-adrenergic receptor function in human nasal mucosa. Am J Respir Crit Care Med 155:704–710

Barnes PJ (1990) Effect of corticosteroids on airway hyperresponsiveness. Am Rev Respir Dis 141:S70-S76

Barnes PJ (1996) Molecular mechanisms of steroid action in asthma. J Allergy Clin Immunol 97:159–168

Barnes PJ (1998) Anti-inflammatory actions of glucocorticoids: molecular mechanisms. Clin Sci 94:557–572

Barnes PJ (2000) Steroid-resistant asthma. Eur Respir Rev 10:74–78

Barnes, PJ (2001) Molecular mechanisms of corticosteroids in allergic diseases. Allergy 56:1–9

Barnes PJ, Adcock IM (1998) Transcription factors and asthma. Eur Respir J 12:221–234

Barnes PJ, Greening AP, Crompton GK (1995) Glucocorticoid resistance in asthma. Am J Respir Crit Care Med 152:125S-140S

Barnes PJ, Pedersen S, Busse WW (1998) Efficacy and safety of inhaled corticosteroids: an update. Am J Respir Crit Care Med 157:S1-S53

Belvisi MG, Wicks SL, Battram CH, Bottoms SE, Redford JE, Woodman P, Brown TJ, Webber SE, Foster ML (2001) Therapeutic benefit of a dissociated glucocorticoid and the relevance of in vitro separation of transrepression from transactivation activity. J Immunol 166:1975–1982

Bergmann M, Barnes PJ, Newton R (2000) Molecular regulation of granulocyte macrophage colony-stimulating factor in human lung epithelial cells by interleukin (IL)-1β, IL-4, and IL-13 involves both transcriptional and post-transcriptional mechanisms. Am J Respir Cell Mol Biol 22:582–589

Busse WW, Lemanske RF (2001) Asthma. N Engl J Med 344:350–362

Caramori G, Lim S, Ito K, Tomita K, Oates T, Jazrawi E, Chung KF, Barnes PJ, Adcock IM (2001) Expression of GATA family of transcription factors in T-cells, monocytes and bronchial biopsies. Eur Respir J 18:466–473

Chung KF, Barnes PJ (1999) Cytokines in asthma. Thorax 54:825–857

Collins S, Caron MG, Lefkowitz RJ (1988) β-Adrenergic receptors in hamster smooth muscle cells are transcriptionally regulated by glucocorticoids. J Biol Chem 263:9067–9070

Culpitt SV, Nightingale JA, Barnes PJ (1999) Effect of high dose inhaled steroid on cells, cytokines and proteases in induced sputum in chronic obstructive pulmonary disease. Am J Respir Crit Care Med 160:1635–1639

Demoly P, Basset-Seguin N, Chanez P, Campbell AM, Gauthier-Rouviere C, Godard P, Michel FB, Bousquet J (1992) c-Fos proto-oncogene expression in bronchial biopsies of asthmatics. Am J Respir Cell Mol Biol 7:128–133

Finotto S, Mekori YA, Metcalfe DD (1997) Glucocorticoids decrease tissue mast cell number by reducing the production of the c-kit ligand, stem cell factor, by resident cells: in vitro and in vivo evidence in murine systems. J Clin Invest 99:1721–1728

Fischle W, Kiermer V, Dequiedt F, Verdin E (2001) The emerging role of class II histone deacetylases. Biochem Cell Biol 79:337–348

Gagliardo R, Chanez P, Vignola AM, Bousquet J, Vachier I, Godard P, Bonsignore G, Demoly P, Mathieu M (2000) Glucocorticoid receptor a and b in glucocorticoid dependent asthma. Am J Respir Crit Care Med 162:7–13

Grunstein M (1997) Histone acetylation in chromatin structure and transcription. Nature 389:349–352

Haddad EB, Fox AJ, Rousell J, Burgess G, McIntyre P, Barnes PJ, Chung KF (2000) Post-transcriptional regulation of bradykinin B_1 and B_2 receptor gene expression in human lung fibroblasts by tumor necrosis factor-a: modulation by dexamethasone. Mol Pharmacol 57:1123–1131

Hall SE, Lim S, Witherden IR, Tetley TD, Barnes PJ, Kamal AM, Smith SF (1999) Lung type II cell and macrophage annexin I release: differential effects of two glucocorticoids. Am J Physiol 276:L114-L121

Hamid QA, Wenzel SE, Hauk PJ, Tsicopoulos A, Wallaert B, Lafitte JJ, Chrousos GP, Szefler SJ, Leung DY (1999) Increased glucocorticoid receptor beta in airway cells of glucocorticoid-insensitive asthma. Am J Respir Crit Care Med 159:1600–1604

Hart LA, Krishnan VL, Adcock IM, Barnes PJ, Chung KF (1998) Activation and localization of transcription factor, nuclear factor-κB, in asthma. Am J Respir Crit Care Med 158:1585–1592

Hart L, Lim S, Adcock I, Barnes PJ, Chung KF (2000) Effects of inhaled corticosteroid therapy on expression and DNA-binding activity of nuclear factor-κB in asthma. Am J Respir Crit Care Med 161:224–231

Holtzman MJ, Look DC, Sampath D, Castro M, Koga T, Walter MJ (1998) Control of epithelial immune-response genes and implications for airway immunity and inflammation. Proc Assoc Am Physicians 110:1–11

Imhof A, Wolffe AP (1998) Transcription: gene control by targeted histone acetylation. Curr Biol 8:R422–R424

Ito K, Barnes PJ, Adcock IM (2000) Glucocorticoid receptor recruitment of histone deacetylase 2 inhibits IL-1b-induced histone H4 acetylation on lysines 8 and 12. Mol Cell Biol 20:6891–6903

Ito K, Jazwari E, Cosio B, Barnes PJ, Adcock IM (2001a) p65-activated histone acetyltransferase activity is repressed by glucocorticoids: Mifepristone fails to recruit HDAC2 to the p65/HAT complex. J Biol Chem 276:30208–30215

Ito K, Lim S, Caramori G, Chung KF, Barnes PJ, Adcock IM (2001b) Cigarette smoking reduces histone deacetylase 2 expression, enhances cytokine expression and inhibits glucocorticoid actions in alveolar macrophages. FASEB J 15:1100–1102

John M, Lim S, Seybold J, Robichaud A, O'Connor B, Barnes PJ, Chung KF (1998) Inhaled corticosteroids increase IL-10 but reduce MIP-1α, GM-CSF and IFN-g release from alveolar macrophages in asthma. Am J Respir Crit Care Med 157:256–262

Kai H, Yoshitake K, Hisatsune A, Kido T, Isohama Y, Takahama K, Miyata T (1996) Dexamethasone suppresses mucus production and MUC-2 and MUC-5AC gene expression by NCI-H292 cells. Am J Physiol 271:L484-L488

Kamei Y, Xu L, Heinzel T, Torchia J, Kurokawa R, Gloss B, Lin SC, Heyman RA, Rose DW, Glass CK, Rosenfeld MG (1996) A CBP integrator complex mediates transcriptional activation and AP-1 inhibition by nuclear receptors. Cell 85:403–414

Kankaanranta H, Lindsay MA, Giembycz MA, Zhang X, Moilanen E, Barnes PJ (2000) Delayed eosinophil apoptosis in asthma. J Allergy Clin Immunol 106:77–83

Katsunuma T, Mak JCW, Barnes PJ (1998) Glucocorticoids reduce tachykinin NK$_2$-receptor expression in bovine tracheal smooth muscle. Eur J Pharmacol 344:99–107

Keatings VM, Jatakanon A, Worsdell YM, Barnes PJ (1997) Effects of inhaled and oral glucocorticoids on inflammatory indices in asthma and COPD. Am J Respir Crit Care Med 155:542–548

Kharitonov SA, Yates DH, Barnes PJ (1996) Regular inhaled budesonide decreases nitric oxide concentration in the exhaled air of asthmatic patients. Am J Respir Crit Care Med 153:454–457

Kuo MH, Allis CD (1998) Roles of histone acetyltransferases and deacetylases in gene regulation. Bioessays 20:615–626

Lane SJ, Adcock IM, Richards D, Hawrylowicz C, Barnes PJ, Lee TH (1998) Corticosteroid-resistant bronchial asthma is associated with increased *c-Fos* expression in monocytes and T-lymphocytes. J Clin Invest 102:2156–2164

Lilly CM, Nakamura H, Kesselman H, Nagler Anderson C, Asano K, Garcia Zepeda EA, Rothenberg ME, Drazen JM, Luster AD (1997) Expression of eotaxin by human lung epithelial cells: induction by cytokines and inhibition by glucocorticoids. J Clin Invest 99:1767–1773

Linden J, Auchampach JA, Jin X, Figler RA (1998) The structure and function of A1 and A2B adenosine receptors. Life Sci 62:1519–1524

Mak JCW, Nishikawa M, Barnes PJ (1995a) Glucocorticosteroids increase β_2-adrenergic receptor transcription in human lung. Am J Physiol 12:L41-L46

Mak JCW, Nishikawa M, Shirasaki H, Miyayasu K, Barnes PJ (1995b) Protective effects of a glucocorticoid on down-regulation of pulmonary β_2-adrenergic receptors in vivo. J Clin Invest 96:99–106

Meagher LC, Cousin JM, Seckl JR, Haslett C (1996) Opposing effects of glucocorticoids on the rate of apoptosis in neutrophilic and eosinophilic granulocytes. J Immunol 156:4422–4428

Nelson DJ, McWilliam AS, Haining S, Holt PG (1995) Modulation of airway intraepithelial dendritic cells following exposure to steroids. Am J Respir Crit Care Med 151:475–481

Newton R, Kuitert LM, Slater DM, Adcock IM, Barnes PJ (1997) Cytokine induction of cytosolic phosholipase A_2 and cyclooxygenase-2 mRNA by proinflammatory cytokines is suppressed by dexamethasone in human epithelial cells. Life Sci 60:67–78

Newton R, Hart LA, Stevens DA, Bergmann M, Donnelly LE, Adcock IM, Barnes PJ (1998) Effect of dexamethasone on interleukin-1β-(IL-1β)-induced nuclear factor-κB (NF-κB) and κB-dependent transcription in epithelial cells. Eur J Biochem 254:81–89

Newton R, Staples KJ, Hart L, Barnes PJ, Bergmann MW (2001) GM-CSF expression in pulmonary epithelial cells is regulated negatively by posttranscriptional mechanisms. Biochem Biophys Res Commun 287:249–253

Nightingale JA, Rogers DF, Chung KF, Barnes PJ (2000) No effect of inhaled budesonide on the response to inhaled ozone in normal subjects. Am J Respir Crit Care Med 161:479–486

Ogryzko VV, Schiltz RL, Russanova V, Howard BH, Nakatani Y (1996) The transcriptional coactivators p300 and CBP are histone acetyltransferases. Cell 87:953–959

Pfahl M (1993) Nuclear receptor/AP-1 interaction. Endocrine Revs 14:651–658

Quan A, McCall MN, Sewell WA (2001) Dexamethasone inhibits the binding of nuclear factors to the IL-5 promoter in human CD4 T cells. J Allergy Clin Immunol 108:340–348

Rao A, Luo C, Hogan PG (1997) Transcription factors of the NFAT family: regulation and function. Annu Rev Immunol 15:707–47:707–747

Reichardt HM, Schutz G (1998) Glucocorticoid signalling–multiple variations of a common theme [In Process Citation]. Mol Cell Endocrinol 146:1–6

Reichardt HM, Kaestner KH, Tuckermann J, Kretz O, Wessely O, Bock R, Gass P, Schmid W, Herrlich P, Angel P, Schutz G (1998) DNA binding of the glucocorticoid receptor is not essential for survival [see comments]. Cell 93:531–541

Robbins RA, Barnes PJ, Springall DR, Warren JB, Kwon OJ, Buttery LDK, Wilson AJ, Geller DA, Polak JM (1994) Expression of inducible nitric oxide synthase in human bronchial epithelial cells. Biochem Biophys Res Commun 203:209–218

Saleh D, Ernst P, Lim S, Barnes PJ, Giaid A (1998) Increased formation of the potent oxidant peroxynitrite in the airways of asthmatic patients is associated with induction of nitric oxide synthase: effect of inhaled glucocorticoid. FASEB J 12:929–937

Schweibert LM, Stellato C, Schleimer RP (1996) The epithelium as a target for glucocorticoid action in the treatment of asthma. Am J Respir Crit Care Med 154:S16-S20

Szefler SJ, Leung DY (1997) Glucocorticoid-resistant asthma: pathogenesis and clinical implications for management. Eur Respir J 10:1640–1647

Vayssiere BM, Dupont S, Choquart A, Petit F, Garcia T, Marchandeau C, Gronemeyer H, Resche-Rigon M (1997) Synthetic glucocorticoids that dissociate transactivation and AP-1 transrepression exhibit antiinflammatory activity in vivo. Mol Endocrinol 11:1245–1255

Walsh GM (1997) Mechanisms of human eosinophil survival and apoptosis. Clin Exp Allergy 27:482–487

Williams CM, Galli SJ (2000) The diverse potential effector and immunoregulatory roles of mast cells in allergic disease. J Allergy Clin Immunol 105:847–859

Wolffe AP, Hayes JJ (1999) Chromatin disruption and modification. Nucleic Acids Res 27:711–720

Woolley KL, Gibson PG, Carty K, Wilson AJ, Twaddell SH, Woolley MJ (1996) Eosinophil apoptosis and the resolution of airway inflammation in asthma. Am J Respir Crit Care Med 154:237–243

2 Immune Monitoring of Glucocorticoid Therapy

P. Reinke, M. Bevilacqua, V. Tryon, J. Cheronis, H.-D. Volk

2.1	Introduction	25
2.2	Immune Monitoring of GC Therapy	27
2.3	Summary	37

2.1 Introduction

Glucocorticoids (GC) are broadly used to inhibit undesired immune reactions. They are applied both topically (inhalation, ointment) and systemically (*per os*, i.v.). Although they have been therapeutically used for decades now, the molecular mechanism of action has continued to be obscure. Recently we have learned more about the molecular targets of GC. Following binding to the intracellular GC receptor complex, the ligand/receptor/heat shock protein complex translocates into the nucleus and interacts with both transcription factors (like NFκB) and DNA (via GC responsive elements in various promoter regions) mediating transactivating or transrepressing mechanisms (Table 1). Both pathways mediate the anti-inflammatory properties of GC, whereas the side effects are mainly induced via the transactivating mechanisms. Recent results (this volume) suggest that it might be possible to separate the transactivating and transrepressing effects of GC in order to promote anti-inflammatory properties in absence of most of the side effects.

Targeting important signal transduction pathways, such as NFκB activation, makes the GC powerful in combating undesired immune

Table 1. Immunological targets of steroids

Molecular level:
GC interact with NFkB activation
GC inhibit several enzymes
GC block transcription/translation of proinflammatory cytokines
GC induce transcription/translation of immunoregulatory cytokines, etc.

Cellular level:
GC inhibit monocyte functions
GC inhibit T lymphocyte functions
GC inhibit granulocyte functions, etc.

Table 2. Immune monitoring of GC therapy

What do we need?
Well standardised assays (reproducibility, precision)
Not labor-intensive
Sample easily collectable

What do we expect?
Prevention of overimmunosuppression
Early detection of non-responders
Optimising individual therapy regarding dose, drug combination, duration

reactions. On the other hand, interfering with these pathways also impairs normal physiological immune responses to infections, etc. Interestingly, the individual response to GC is quite different – at the same dose the undesired immune reaction is not sufficiently inhibited in some patients while others show signs of severe immunosuppression. In addition, some patients respond well to GC therapy, others with the same diagnosis do not respond at all suggesting distinct entities or different courses of the same disease.

Obviously there is a medical need for an individual monitoring to define responders and non-responders to GC as early as possible and to optimise the individual dose to prevent overimmunosuppression (Table 2). This might be particularly important for systemic GC therapy in autoimmune diseases and transplantation.

Recent progress in immunology and molecular biology opens new opportunities to develop such a program. In this chapter, we summarise these new options.

2.2 Immune Monitoring of GC Therapy

2.2.1 Flow Cytometric Monitoring

GC target almost all immune cell populations. Systemic application results in changes of immune cell homing as well functional properties. Flow cytometric analysis allows the monitoring of GC effects on peripheral immune cells (Table 3). GC induces lymphopenia, particularly of $CD4^+$ T cells. The expression of functional, relevant markers on monocytes, like HLA-DR, seems to have high diagnostic value for monitoring the effects of GC. The monocytic HLA-DR expression correlates directly with the antigen-presenting capacity of these cells. In addition, this expression is related to the monocytic function in general and is tightly regulated by endogenous and exogenous mediators. Therefore, HLA-DR monitoring can provide an overall impression of the status of cellular immunity (Table 4).

Monocytes are an intermediate form between bone-marrow-derived precursor cells and tissue macrophages. As the half-life of monocytes in peripheral blood is approximately 24 h, this population of cells can be used for monitoring the effects of GC therapy. Almost all monocytes express HLA-DR on their surface (mean: >20,000 molecules/cell). The

Table 3. Flowcytometric immune monitoring of GC therapy

Lymphocyte differentiation markers
Lymphocyte activation markers
Functional monocyte markers
e.g., monocytic HLA-DR expression

Table 4. Monocytic HLA-DR expression – marker of immunocompetence I

Positively regulated by:
Interferon-γ
GM-CSF
Negatively regulated by:
IL-10
TGF-β
Endotoxins
Catecholamines
Corticosteroids

Table 5. Monocytic HLA-DR expression – marker of immunocompetence II

Half-life of monocytes in blood: about 24 h
Monocytic HLA-DR expression
Healthy probands:
  85%–100% positive monocytes
  >20,000 molecules/cell
  Mild immunodepression
  45%–85% positive monocytes
  10,000–20,000 molecules/cell
Borderline group
  30%–45% positive monocytes
  5,000–10,000 molecules/cell
"Immunoparalysis"
  <30% positive monocytes
  <5,000 molecules/cell

Table 6. Monocytic HLA-DR expression – marker of immunocompetence III

Risk to develop severe infections following high-dose steroid therapy:
Monocytic HLA-DR expression
>45% (mild immunodepression)
2% infection
30%–45% (borderline to immunoparalysis)
10% infection
<30% (immunoparalysis)
27% infection
>50% if immunoparalysis long-lasting (>4 days)

extent of downregulation of monocytic HLA-DR expression by GC and other immunosuppressive drugs and events directly reflects the level of immunosuppression (Table 5). The individual response to the same GC dose is quite variable. For example, Fig. 1 illustrates the distinct sensitivity of two patients to 3×3 mg/kg i.v. methylprednisolone. In one patient the HLA-DR expression reached critical levels of "immunoparalysis" that is associated with a high incidence of infectious complications (particularly fungal and bacterial infections) within 4 weeks follow-up (Table 6). Based on this assay, GC treatments have been monitored and individually adapted in patients from our centre in order to optimise the risk/benefit ratio of GC (and other immunosuppressive)

Immune Monitoring of Glucocorticoid Therapy

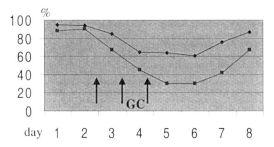

Fig. 1. GC (3×3 mg/kg methylprednisolone i.v.) inhibits monocytic HLA-DR expression. The inhibitory activity of GC differs between various patients. Some patients fall below the critical level of 30% HLA-DR expression (<5,000 molecules/cell)

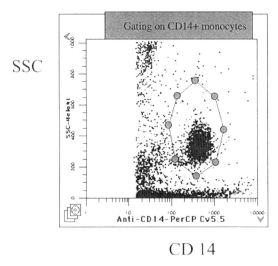

Fig. 2. Gating of monocytes from whole blood preparations by side scatter (*SSC*) and CD14 PerCP fluorescence signal properties

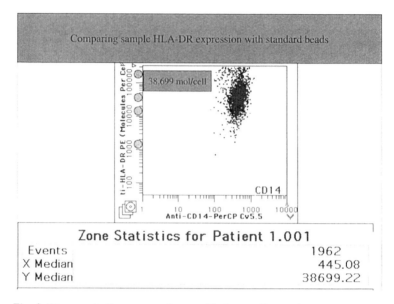

Fig. 3. The sample fluorescence is quantified according to the standard curve that is based on the fluorescence levels of four standard beads expressing defined molecules/bead (*grey points*). This figure shows the level of a patient before GC therapy

therapies for several years. The disadvantage of the conventional measurement of HLA-DR expression was the strong inter-laboratory variation. Results are strongly influenced by the sensitivity of individual flow cytometers and antibodies used. Very recently, a novel well-standardised assay for measuring HLA-DR expression was introduced into the market (Quantibrite HLA-DR, BD Bioscience, San Jose, Calif., USA). This assay is based on the use of standard fluorescence beads (with defined fluorescence molecules/bead) and a novel conjugation method of phycoerythrin to antibodies (1:1 ratio for all molecules). Thus the number of molecules per cell can be determined independent of the flow cytometric equipment used, making this assay better for comparisons between laboratories (Figs. 2–4). Monocytes from patients with "immunoparalysis" express less than 5,000 molecules per cell, whereas normal monocytes express more than 20,000 molecules/cell (Table 5). Figures 3 and 4 show a patient before and after GC therapy

Immune Monitoring of Glucocorticoid Therapy

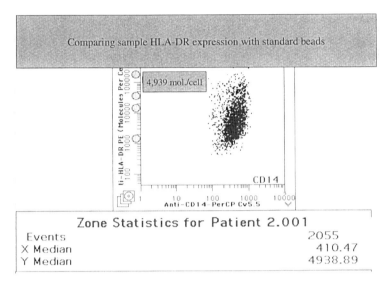

Fig. 4. This figure shows the level of the same patient (Fig. 3) after GC therapy (3×3 mg/kg i.v.). The patient falls into "immunoparalysis"

(3×3 mg/kg), respectively. The level of monocytic HLA-DR expression dropped by a factor 7.8 from greater than 38,000 to less than 5,000 molecules per cell. For this monitoring, a sample of less than 100 µl ethylenediaminetetraacetate (EDTA) blood is required. Both methodical reproducibility (about 8%) and biological reproducibility (about 20% in healthy probands) are excellent.

2.2.2 Ex Vivo Cytokine Secretion Assay

In addition to their antigen-presenting properties, monocytes/macrophages play an important role as effector cells in acute/chronic inflammation by secreting several cytokines. Tumour necrosis factor (TNF)-α is a key pro-inflammatory cytokine. Using a whole-blood secretion assay, it is possible to measure the capacity of monocytes to secrete TNF or other cytokines in response to powerful stimuli. Fig. 5 illustrates the principle of a well-standardised assay based on the ex vivo TNF release following lipopolysaccharide (LPS) stimulation. Tubes, medium and

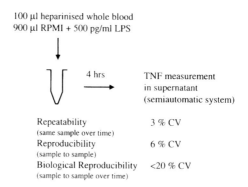

Fig. 5. Principle of the semi-automatic whole blood ex vivo cytokine secretion assay

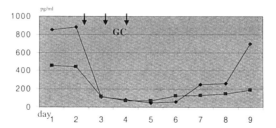

Fig. 6. GC (3×3 mg/kg methylprednisolone) inhibits 4 h-LPS induced TNF secretion capacity ex vivo relatively independent of the different pre-treatment levels (high/low responder). After tapering GC, ex vivo TNF secretion rises to pre-treatment levels of high/low response

endotoxin are well-characterised (Milenia Biotech, Bad Nauheim, Germany). In addition, TNF is measured by a semi-automatic assay (Immulite, DPC, Los Angeles, Calif., USA). For example, Fig. 6 shows the individually variable capacity of patient's monocytes to produce TNF before any therapy. Similarly, healthy probands show a broad variation (350–1800 pg/ml TNF secretion within 4 h). However, this phenotype of high or low response is very stable over time with less than 20% variation over one year follow-up in healthy probands. It has been reported that the TNF secretion is related to polymorphisms in coding and non-coding regions of the TNF gene.

GC therapy strongly inhibits ex vivo TNF secretion capacity and abolishes the genetically determined response variation at least partially (Fig. 6).

Monocytic HLA-DR expression and ex vivo TNF secretion appear to be related ($r=0.83$) in some, but not all cases.

For monitoring less than 100 µl heparinised blood is required.

2.2.3 Gene Expression Profiling

Applied genomics will be part of the new generation of health care. This approach is based on two powerful techniques – genetic mapping and gene expression profiling (Fig. 7). In the future it may be possible to predict genetically determined responders and non-responders to a particular therapy before starting the therapy using genetic mapping (DNA polymorphisms). This approach, however, relies on a map of highly relevant single nucleotide polymorphisms, which may not exist presently in the human genome database. Profiles of gene expression, on the other hand, illustrate the status of an individual's response to such a

Genetic Mapping

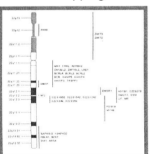

- "You vs. me"
- Static
- Marks predispositions

Gene Expression Profiling

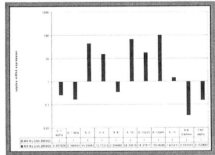

- "You today vs. you yesterday"
- Dynamic
- Marks disease progression and therapeutic response

Fig. 7. Genetic mapping (analysis of DNA polymorphisms) and gene expression profiling (analysis of mRNA expression) are two distinct approaches of applied genomics

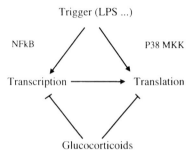

Fig. 8. GC inhibit cytokine production at both translational and transcriptional levels

therapy. Such profiles do not require prior detailed knowledge of the individual's genome.

An example of the power of gene expression profiling follows from examination of the effects of GC on the immune system. GC inhibit the production of agents such as cytokines and chemokines at both translational and transcriptional level (Fig. 8); thus it is possible to quantify their effects at both the protein and mRNA level. Because mRNA expression leads protein synthesis and release, mRNA is more rapid and sensitive to the condition of the patient than protein profiling. Gene expression profiling more closely reflects disease state, progression and therapeutic response at the individual level. The development of new technologies, such as real-time reverse transcriptase polymerase chain reaction (RT-PCR), allows a highly precise quantification of mRNA expression from small numbers of cells. Concerted efforts by Source Precision Medicine on methodology together with advances in preanalytic procedures (collecting cells, RNA stabilisers, RNA/cDNA preparation) substantially improved the repeatability and reproducibility during the last years. Over the last 3 years (1999 through 2001), CVs for repeatability and reproducibility have tightened tenfold (from 10%–20% to 1%–2% and 20%–50% to 2%–5%, respectively). Using these techniques, it is also possible to demonstrate that genes are tightly regulated within healthy individuals; for example, TNF mRNA exhibits a CV of less than 1%, while the CV for IL-1β was found to be 5% over a 4-week period. Owing to these advanced methods, the requisite sample size has been markedly reduced: For this test, less than 2.5 ml blood

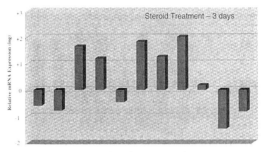

Fig. 9. GC treatment (3 days) induces both up- and downregulation of inflammation-related gene expression in peripheral blood immune cells

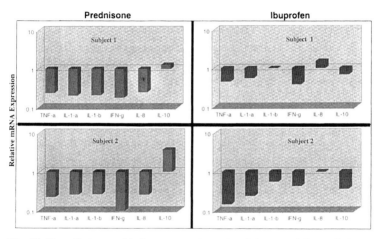

Fig. 10. Two distinct anti-inflammatory drugs (prednisone and ibuprofen) target different inflammation-related genes. Analysis was performed in two healthy probands before (*1*) and after a 3-day treatment with each drug (day 3). More than 4 week separated the two drug applications. Data shown as relative mRNA expression post- vs pretreatment level

is required. Clearly, the mRNA assay methods are available for highly precise gene expression profiling.

Several pilot trials demonstrate the power of precision profiling of gene expression. Figure 9 illustrates patient mRNA response to a 3-day course of GC for several inflammation-related genes. As expected, this

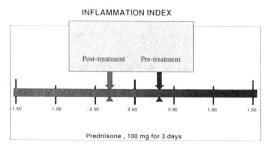

Fig. 11. Determination of the "inflammation score" according to an algorithm based on both preclinical and clinical studies using inflammatory stimuli with and without GC. The algorithm shows a change in the defined "inflammation index" before and after oral prednisone treatment for 3 days

regimen completely changed the gene expression profiling of this particular patient: Some genes were downregulated 5- to 50-fold while others were upregulated as much as 100-fold. Figure 10 demonstrates the differential effects of GC (prednisone) and a nonsteroidal anti-inflammatory drug (ibuprofen) on gene expression profile of peripheral immune cells from two healthy probands.

While a few carefully chosen genes may illustrate unequivocal changes in response to dramatic therapeutic agents, the complexity of expression profiles from 20 or more genes is difficult to understand and to manage. Source Precision Medicine has developed algorithms for analysing the data. The data illustrated on Fig. 11 were generated using an algorithm applied to data from Fig. 9. These algorithms allow the facile comparison of complex data sets, and exhibit great utility for monitoring GC therapy. One example of algorithm utility is the response of a normal subject to a 3-day course of oral GC (initial score 0.31 to –0.35 following treatment). In another case, a patient suffering from acute optic neuritis exhibited a "pro-inflammatory" score of 0.62–0.82 that dropped to 0.08 under GC bolus therapy (3×1 g/day). Algorithms, thus, are an extremely beneficial adjunct to gene expression profiling.

In summary, this novel technology reached now the methodical stability to be used in clinical trials.

2.3 Summary

There is a clear medical need to improve the GC therapy because of the individually variable responsiveness to GC.

Recently, several well-standardised cell-based assays focussing on monocyte functions (flowcytometric measurement of HLA-DR expression, semiautomatic measurement of ex vivo TNF release capacity) have been introduced into the clinical laboratory. These techniques allow for both the detection immunosuppression and the monitoring of therapeutic efficacy. The measurement of efficacy might be further improved by applying the novel technology of gene expression profiling that is ready for introducing into the diagnostics.

Acknowledgements. The authors would like to thank all the scientists, physicians and technicians who contributed to this review.

3 Topical Glucocorticoid Therapy in Dermatology

W. Sterry, K. Asadullah

3.1   Introduction 39
3.2   Molecular Aspects 41
3.3   Principles of Treatment 44
3.4   Side Effects 48
References 52

3.1 Introduction

In dermatology, glucocorticoids (GCs) are the most widely used drugs. Systemic, intralesional, and in particular topical applications are the daily practice of the dermatologist. Table 1 gives examples of frequent indications. Since cutaneous (topical) GC therapy is the domain of dermatologists, this will be within the focus of this review.

The introduction of topical hydrocortisone in the early 1950s represented a great advance on previously available therapies, but it was the first of the halogenated corticosteroids, triamcinolone acetonide, that began a revolution which cumulated in the appearance of the very potent agents available now. The enthusiasm for these highly effective agents was at its peak during the 1960s and 1970s, and perhaps inevitably, the more potent GCs were often used inappropriately and indiscriminately. Adverse effects became apparent and the subsequent backlash of opinion against topical GCs has created confusion and prejudice against all

Table 1. Examples of glucocorticoid therapy in dermatology

Application	Indication
Topical	Atopic dermatitis
	Contact eczema
	Psoriasis (specific forms)
	Prurigo
	Pityriasis rosea
	Lichen planus
	Alopecia areata
	Vitiligo
Intralesional	Keloid
	Granuloma annulare
	Alopecia areata
	Hypertrophic lichen planus
	Prurigo nodularis
Systemic	Bullous diseases
	Connective tissue diseases
	Anaphylactic reactions
	Vasculitis
	Systemic lupus erythematosus
	Severe lichen planus
	Acute, generalized eczema

steroid-containing preparations, in its extreme as "steroid phobia" is today still of considerable concern (Maibach and Suber 1992).

Despite this, topical GCs are successfully and commonly used in the treatment of several cutaneous diseases. On the one hand, however, they have great advantages in certain indications, localizations and types of disease, while they may have deleterious effects when used improperly. Their adequate use belongs to the "art of treatment" in dermatology, requiring a profound personal experience as well as solid knowledge concerning mode of action, indications as well as contraindications, different types of GCs available, possible combinations, and finally side effects. This review gives an overview on these aspects.

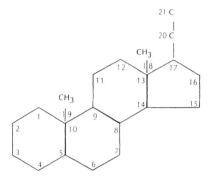

Fig. 1. Structure of the basic steroid molecule

3.2 Molecular Aspects

3.2.1 Chemical Structure

The steroid molecule consists of 17 carbon atoms in four rings, two methyl groups at positions 10 and 13, and an ethyl group at position 17 (Fig. 1) (Amin et al. 1998). To achieve therapeutic efficacy, the chemical structure has to be modified. The most commonly introduced chemical modifications are (Higgins and DuVivier 1994):

1. Reduction of the oxo group at position 11 to produce hydrocortisone
2. Insertion of a double bond at positions 1,2 to produce prednisolone
3. Fluorination at positions 6 and/or 9 to increase activity
4. Hydroxylation at position 16 and fluorination at position 9, which produces triamcinolone
5. Methylation at position 16 and fluorination at position 9, which produces betamethasone
6. Esterification at position 16, 17, 21, or 7, since both triamcinolone and betamethasone require, for full activity, the increase of their lipophilic properties
7. Addition of 16-hydroxy substituent, producing an acetonide
8. Substitution of the hydroxyl group at position 11 by a ketone group, which reduces the potential of systemic absorption in an otherwise powerful topical steroid, clobetasone butyrate

3.2.2 Potency: Assessment and Classification

The potency of GCs is estimated by several surrogate markers such as the vasoconstrictor assay. After the application of a topical GC to the normal forearm, a blanching occurs after 4–7 h, and may persist between 8 and 48 h. The effects seen in the vasoconstrictor assay correlate reliably with the clinical potency of GCs (MacKenzie and Stoughton 1962; Cornell and Stoughton 1985).

In clinical use, GCs are placed into four groups according to their clinical effects, both wanted and unwanted. Group I contains mild, group II moderate, group III strong, and group IV very strong topical GCs (Table 2). Since the brand names as well as the chemical entities differ in various countries, the reader is referred also to local reviews to get information of the brand names and their potency. One should always be familiar with several pharmaceutical products from each group in order to be able to treat different clinical situations adequately.

The local potency of a GC is determined by its concentration, penetration profile as well as its molecular structure; the latter will determine the interaction with the steroid receptor as well as the interaction of the steroid-receptor complex with the DNA of the target cell.

Table 2. Four categories of different strengths of topical glucocorticoids with typical examples

Category	Potency	Glucocorticoid
Class I	Mild	Dexamethasone 0.1%
		Hydrocortisone acetate 1.0%
		Hydrocortisone 1%–2.5%
Class II	Moderate	Fluocinolone acetonide 0.025%
		Hydrocortisone valerate 0.2%
Class III	Strong	Halcinonide 0.01%
		Betamethasone valerate 0.1%
		Fluocinonide 0.05%
		Amcinonide 0.1
Class IV	Very strong	Clobetasol propionate 0.1%

3.2.3 Mode of Action

Topical GCs are absorbed by lipophilic structures of the epidermis and penetrate in decreasing concentrations to the lower epidermis as well as the upper dermis. They will enter cells of epidermis and dermis by diffusion through the cell membrane. In the cytoplasm, they bind to GC receptors. These receptors belong to a group of DNA-binding molecules that include the receptors for sexual and mineralocorticoid hormones, thyroid hormones as well vitamin A and retinoids. Their effects are mainly exerted by the up- and/or downregulation of gene expression. Among the numerous genes having GC response elements in their promoter regions are structural proteins (collagen genes), enzymes (phospholipase A2), adhesion molecules, cytokines [interleukins, transforming growth factor alpha (Lee et al. 1991)] and many others. These complex interactions cause four effects that are relevant in clinical treatment (Higgins and DuVivier 1994):

- Anti-inflammatory
- Anti-mitotic
- Immunomodulatory
- Vasoconstrictive

The vasoconstrictive effect is clinically useful in acute inflammations, when the influx of further cells from circulation needs to be blocked. The anti-inflammatory and immunomodulatory effects are somewhat overlapping, but taking these two effects together, it is obvious that no other available drug has a similarly broad mode of action.

3.2.4 Percutaneous Penetration

Penetration is the sum of processes that allow a steroid to find its way through the horny layer into epidermis and dermis. Penetration depends on several factors inherent to normal or inflamed skin and the physicochemical properties of the steroid, and is further influenced by the vehicle. Methods to investigate the steroid penetration include the use of radiolabeled molecules, but also advanced tape-stripping techniques in combination with analytical techniques.

Resorption is defined as the systemic uptake of steroids following topical application. While treatment of small areas will not cause any problem related to steroid resorption, the use of high-potency steroids, in particular in conjunction with occlusion-using plastic foils, is associated with the risk of systemic side effects of GCs such as depression of endogenous cortisol production, adrenal cortex atrophy, bone demineralization, and growth retardation in children.

It is important to realize that the skin sites differ dramatically in penetration of topical steroids. As compared to palms and soles, a GC molecule will penetrate 40 times more effectively when applied to the forehead. Other areas with a high resorption are eyelids, scalp, axilla, and genital area.

Modern pharmacology has synthesized steroid molecules that are cleaved into relatively low potency steroids when they enter systemic circulation, thereby reducing or even abrogating the risks of resorption and systemic side effects.

3.3 Principles of Treatment

A wide range of skin diseases can be treated with topical GCs (Table 1). This, in particular, includes inflammatory disorders. One must keep in mind, however, that most of these diseases are chronic and need treatment over decades. Therefore, careful consideration is required to avoid long-term side effects that may add over the years. Several aspects need special consideration:

- Indication
- Age of the patient
- Extent and activity of the disease
- Sites and clinical types of psoriasis to be treated
- Control of compliance
- Combination with other treatment modalities
- Possible side effects

3.3.1 Indications

Topical corticosteroids are currently the most commonly prescribed treatment for eczema (including atopic dermatitis, allergic and non-allergic contact dermatitis) and, in North America (Lebwohl et al. 1995; Stern 1996), for psoriasis too (Table 1).

Actually, in all types of eczema the use of GCs is generally appropriate. In less severe cases monotherapy (once or twice daily) is sufficient, whereas combination with UV radiation can be useful in more severe cases. The class of the GCs to be chosen should depend on the disease activity. In acute eczema, a rapid improvement can frequently be achieved by treatment with a potent GC for just a few days. After clearing, a less potent GC can be applied or the application interval could be prolong (only every second day).

GC treatment of psoriasis is more complex. Indications are resistant plaques not following the regression of most lesions during conventional therapy (as UV radiation, anthralin, etc.), solitary lesions on elbows and knees that do not justify UV or anthralin therapy and are not responsive to vitamin D analogs, hyperkeratotic psoriasis of palms and soles, nail psoriasis, and intertriginous psoriasis (here in combination with clotrimazole). The short-term use of highly potent topical GCs under occlusion has been advocated for rapid control of erythrodermic psoriasis (Arbiser et al. 1994). Furthermore, GCs are the treatment of choice in psoriasis of the scalp, since atrophy has not been reported in this location. Besides these indications, GCs are even frequently used in chronic plaque psoriasis, although better therapeutic options exists.

Recently, different time schedules were compared regarding the antipsoriatic effect of augmented betamethasone: once daily and alternate day applications of augmented betamethasone were not significantly different in improving psoriasis over 15 days, while lower frequencies were less effective (Singh 1996). Intermittent corticosteroid maintenance treatment of psoriasis with a pulse dose treatment regimen was shown to effective (Katz et al. 1991). When used in three consecutive applications 12 h apart, this regimen could control psoriasis in 60% of patients over a period of 6 months, while in the control group 80% of the patients experienced exacerbation of the disease. No adverse local or systemic experiences were reported.

The combination of dithranol and topical GCs may be useful in patients with thick recalcitrant psoriatic plaques that respond only slowly to conventional treatment. In hyperkeratotic lesions, particularly on the scalp and in palmoplantar psoriasis, the combination of potent steroids with 5%–10% salicylic acid represents an effective modification. The keratolytic action of salicylic acid allows the steroid to penetrate to the inflamed skin, while sometimes in such situations the steroid does not come to function because of its inability to penetrate through the thick horny layer. Several studies have demonstrated that the combination of topical GCs and psoralen with UV-A (PUVA) gives more rapid clearing, and the UV dose can be reduced. In contrast, similar beneficial effects have not been observed, when UV-B and steroids were combined (Gould and Wilson 1977; Meola et al. 1991). Most remarkably, Schmoll and coworkers (Schmoll et al. 1978) showed that there is no increased relapse rate when topical steroids and PUVA are combined as compared to PUVA therapy; however, topical steroid treatment alone caused fast relapse (in 50% of the patients after 3 weeks). Finally, Calcipotriol and corticoids are often used in combination for treatment of psoriasis (Glade et al. 1996; Ruzicka and Lorenz 1998).

3.3.2 Localization

The sensitivity against GCs is different at different sites. As mentioned above, the skin sites differ dramatically in penetration of topical steroids. As compared to palms and soles, a GC molecule will penetrate much more effectively when applied to the forehead, eyelids, scalp, axilla, and genital area. Consequently, caution is necessary when treating such regions (shorter treatment periods, use of lower potent GCs). Topical unwanted side effects have not been reported associated with steroid treatment of the scalp. This is important since the scalp tends to be chronically involved by psoriasis and needs continuous treatment. Since other treatment modalities such as anthralin or UV treatment cannot be used in the scalp, the main antipsoriatic agent used in scalp psoriasis are topical steroids. Hyperproliferation is a prominent pathophysiological feature in this localization and requires the use of potent or very potent topical steroids. Patients will find steroid solutions without alcohol most agreeable, but in cases with extensive hyperkeratosis

Topical Glucocorticoid Therapy in Dermatology 47

and scale formation, the short-term use of steroid creams applied overnight will bring instant relief. The patients should be advised to treat the very early manifestations consequently rather than waiting until elevated plaques have developed.

3.3.3 Age of the Patients

GC penetration is enhanced in young children. Children are at special risk of developing systemic side effects such as growth retardation, Cushing's syndrome, and the development of intracranial hypertension (Freiwel 1969). Also, they develop local side effects, in particular steroid acne and striae cutis distensae. Since several inflammatory skin diseases such as atopic dermatitis and type 1 psoriasis have an early onset and carry a poor prognosis, and repeated treatment over prolonged periods are likely, one should carefully consider the use of steroids in children, and prefer other treatment modalities in psoriasis (Rasmussen 1986).

3.3.4 Control of Compliance

Use of topical steroids requires strict control of compliance. Such "golden rules" have been defined repeatedly (Sterry 1992; Higgins and DuVivier 1994) and include:

- Regular clinical review.
- No unsupervised repeat prescriptions.
- Limitation of the quantity prescribed to the affected area and the time period expected to be necessary for clearing or expected therapeutic effect.
- Periods of steroid-free treatment.
- High-potency steroids should be used under control of dermatologists.

Control of compliance is not only necessary to avoid unrestrictive, excessive use, which may cause severe side effects, but also in consideration of the steroid phobia. It is not uncommon that patients do not use the GCs prescribed because of concerns. It is the physician's task to explain to the patients the effect/potential side effect profile in order to achieve a realistic view.

3.3.5 Steroid Phobia

This phenomenon, which initially developed in the 1960s–1970s, is still of considerable concern (Maibach and Suber 1992). Recently, a questionnaire-based study of 200 dermatology outpatients with atopic eczema (age range 4 months–67.8 years) assessed the prevalence of topical corticosteroid phobia in Great Britain. Overall, 72.5% of people surveyed worried about using topical corticosteroids on their own or their child's skin, and 24% admitted to having been non-compliant with topical corticosteroid treatment because of these worries. The most frequent cause for concern was the perceived risk of skin thinning (34.5%). In addition, 9.5% of patients worried about systemic absorption leading to effects on growth and development. This indicates that a considerable number of patients do worry about using their prescribed GC (Charman et al. 2000).

3.4 Side Effects

3.4.1 Local Side Effects

3.4.1.1 Steroid Acne
Prolonged treatment with steroids may result in the development of steroid acne (Fig. 2). Most commonly, this will occur in adjunction of scalp treatment in the nuchal region as well as on the forehead. The localization and the lack of blackhead formation allow differentiation from normal acne. Treatment should consist in information for the patient and avoidance of contact with steroid ointment outside the areas intended for treatment, rather than formal topical acne treatment.

3.4.1.2 Perioral Dermatitis
Perioral dermatitis occurs frequently by self-abuse of topical steroid formulations of patients suffering from skin diseases in the face (Fig. 3).

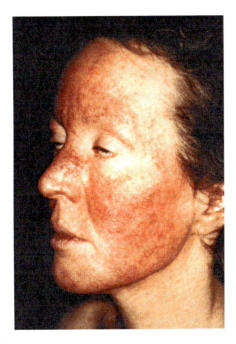

Fig. 2. Steroid acne

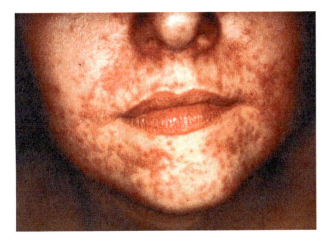

Fig. 3. Perioral dermatitis

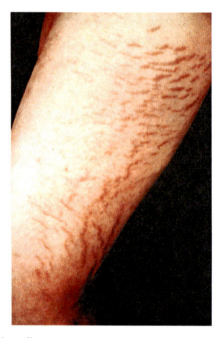

Fig. 4. Striae rubrae distensae

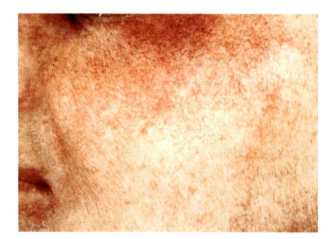

Fig. 5. Skin atrophy

Seborrheic dermatitis may be the most frequent skin disorder associated with steroid abuse in males, and comedogenic folliculitis in females.

3.4.1.3 Striae Distensae and Skin Atrophy
In adolescent patients with prolonged use of steroids in certain areas, particularly the axillary regions and the thighs, striae distensae will develop as a result of decreased collagen production along with increased connective tissue tension during body growth (Figs. 4, 5). Therefore, intertriginous psoriasis in adolescents should be treated, if at all, only with great care: short-term use and selection of middle or low potency steroids. Also, in such patients the "golden rules" should be followed strictly.

3.4.1.4 Glaucoma
Potent topical steroids may cause glaucoma when applied around the eye lids (Cubey 1976). Such preparations should not be used around the eyelids or on the face in patients.

3.4.1.5 Allergic Contact Dermatitis
Allergic contact dermatitis occurs with the use of many topical GCs (Alani and Alani 1972). The allergens are either the steroid molecules or resides in the vehicle. Careful observation of the patient and the therapeutic effects, including in particular those that are unexpected, will help to identify early the development of sensitization of the patient, and to identify the eliciting agent by patch testing.

3.4.1.6 Induction of Pustular Psoriasis
Many case reports published in the 1950s and 1960s have given evidence that not only systemic administration of corticosteroids but also their topical application may trigger generalized pustular psoriasis in patients having suffered from nonpustular forms of psoriasis (Baker 1976). Therefore, the topical use of psoriasis requires careful consideration of possible exacerbation and strict restriction to the indications and combinations recommended above.

3.4.2 Systemic Side Effects

With the advent of potent and highly potent topical GCs, the reports of systemic side effects become more frequent. Two clinical situations require the consideration that systemic side effects with suppression of plasma cortisol may occur: treatment of children and use of occlusive techniques in larger areas. For example, halcinonide cream applied for 5 days suppresses plasma cortisol in patients with extensive psoriasis with and without occlusion. This results from the increase absorption of steroids in the defective horny layer barrier situation in psoriatics. In contrast, no adrenal suppression was observed when betamethasonevalerate was applied to more than 50% of the body surface in normal healthy individuals (Gomez et al. 1977).

When highly potent steroids are used in hospitalized patients with psoriasis and eczema, clobetasol had a higher potency compared to betamethasone dipropionate and diflorasone diacetate in suppressing plasma cortisol (Amin et al. 1998). In our clinical impression, systemic side effects correlate with topical potency; every dermatologist treating large areas of the body for any skin disease should be aware of the fact that systemic GC effects will occur, and they may be, in part, relevant to the therapeutic effect as well as unwanted side effects.

References

Alani MD, Alani SD (1972) Allergic contact dermatitis to corticosteroids. Ann Allergol 30:181–185

Amin S, Maibach HI, Cornell RC, Stoughton RB (1998) Topical corticoids. In: Roenigk Jr HH, Maibach HI (eds) Psoriasis, 3rd edn. Marcel Dekker, New York, pp 453–467

Arbiser JL, Grossman K, Kaye E, Arndt KA (1994) Use of short-course class I topical glucocorticoid under occlusion for the rapid control of erythrodermic psoriasis. Arch Dermatol 130:704–706

Baker H (1976) Corticosteroids and pustular psoriasis. Br J Dermatol 94 [Suppl 12]:83–88

Charman CR, Morris AD, Williams HC (2000) Topical corticosteroid phobia in patients with atopic eczema. Br J Dermatol 142:931–936

Cornell RC, Stoughton RB (1985) Correlation of the vasoconstrictor assay and clinical activity in psoriasis. Arch Dermatol 121:63–67

Cubey RB (1976) Glaucoma following the application of corticosteroids to the skin of the eyelids. Br J Dermatol 95:207–208

Freiwel M (1969) Percutaneous absorption of topical steroids in children. Br J Dermatol 81 [Suppl 4]:113–116

Glade CP, van Erp PEJ, van de Kerkhof PCM (1996) Epidermal cell DNA content and intermediate filaments keratin 10 and vimentin after treatment of psoriasis with calcipotriol cream once daily, twice daily and in combination with clobetasone 17-butyrate cream or betamethasone 17-valerate cream: a comparative low cytometric study. Br J Dermatol 135:379–384

Gomez EC, Kaminester L, Frost P (1977) Topical halcinonide and betamethasone valerate effects on plasma cortisol. Arch Dermatol 113:1196–1202

Gould PW, Wilson L (1977) Psoriasis treated with clobetasol propionate and photochemotherapy. Br J Dermatol 98:133–136

Higgins EM, DuVivier A (1994) Glucocorticosteroids. In: Dubertret L (ed) Psoriasis. ISED, Brescia, pp 102–111

Katz HI, Prawer RS, Krueger GG, Mooney JJ, Jones ML, Samson CR (1991) Intermittent corticosteroid maintenance treatment of psoriasis: a double blind multicenter trial of augmented betamethasone dipropionate ointment in a pulse dose treatment regimen. Dermatologica 183:269–274

Lebwohl M, Abel E, Zanolli M, Koo J, Drake L (1995) Topical therapy for psoriasis. Int J Dermatol 44:673–684

Lee SW, Morhenn VB, Ilnicka M, Eugui EM, Allison A (1991) Autocrine stimulation of interleukin-1alpha and transforming growth factor alpha production in human keratinocytes and its antagonism by GCs. J Invest Dermatol 97:106–110

MacKenzie AW, Stoughton RB (1962) Method for comparing percutaneous absorption of steroids. Arch Dermatol 86:608–610

Maibach HI, Surber C (1992) Topical corticosteroids. Karger, Basel

Meola T, Soter NA, Lim HW (1991) Are topical corticosteroids useful adjunctive therapy for the treatment of psoriasis with ultraviolet radiation? Br J Dermatol 127:1708–1713

Rasmussen E (1986) Psoriasis in children. Dermatol Clin 4:99–106

Ruzicka T, Lorenz B (1998) Comparison of calcipotriol monotherapy and a combination of calcipotriol and betamethasone valerate after 2 weeks' treatment with calcipotriol in the topical therapy of psoriasis vulgaris: a multicentre, double-blind, randomized study. Br J Dermatol 138:254–258

Schmoll M, Henseler T, Christophers E (1978) Evaluation of PUVA, topical corticosteroids and the combination of both in the treatment of psoriasis. Br J Dermatol 99:693–702

Singh S (1996) Augmented betamethasone: efficacy in psoriasis with different dosing frequencies. Arch Dermatol 132:1525–1526

Stern RS (1996) Utilization of outpatient care for psoriasis. J Am Acad Dermatol 35:543–545
Sterry W (1992) Therapy with topical corticosteroids. Arch Dermatol Res 284 [Suppl 1]:27–29

4 Bone Effects of Glucocorticoid Therapy

K.H. Väänänen, P.L. Härkönen

4.1	Introduction	55
4.2	Glucocorticoid-Induced Osteoporosis	56
4.3	Glucocorticoid-Induced Aseptic Osteonecrosis	57
4.4	Indirect Effects of Glucocorticoids on Bone	58
4.5	Direct Effects of Glucocorticoids on Bone	59
References		62

4.1 Introduction

Since the beginning of pharmacological use of glucocorticoids it has been clear that they have serious adverse effects in bone. Chronic administration of glucocorticoids causes a loss of bone mass and pathological fractures, especially in the skeletal areas that are rich in trabecular bone. In addition to osteoporosis, large doses of glucocorticoids could lead to aseptic osteonecrosis affecting mainly proximal femur.

Although the effects of excess glucocorticoids and endogenous hypercortisolism on bone have been known for decades, cellular and molecular details of the mechanism of action have started to be clarified only recently when it has become clear that glucocorticoids have mainly direct effects on bone. In the following chapter, we will first describe clinical consequences of glucocorticoid excess in bone, and then we will address what is known about the cellular processes underlying the glucocorticoid-induced osteoporosis and osteonecrosis.

4.2 Glucocorticoid-Induced Osteoporosis

Glucocorticoid-induced osteoporosis is the third most common cause of osteoporosis after postmenopausal and senile osteoporosis (Lukert and Kream1996). The deleterious effects of glucocorticoid excess on human skeleton were recognized almost 70 years ago (Cushing 1932). Bone loss in glucocorticoid-induced osteoporosis is general but usually more pronounced in axial skeleton than in appendicular skeleton. Compression fractures of the spine are often the first sign of glucocorticoid-induced osteoporosis (Seeman et al. 1983). In addition to vertebral bodies, spontaneous fractures of ribs, wrist, proximal femur and other skeletal sites rich in trabecular bone are commonly seen as complications of chronic glucocorticoid administration. Bone density measurements with dual-energy X-ray absorptiometry have demonstrated that the bone loss is 0.6%–6.0% per year on average (Sambrook et al. 1990; Lukert et al. 1992). These studies also indicate that the speed of loss of bone is most rapid during the first months of the treatment, but it is obvious that it continues as long as the excess of glucocorticoids continues (Reid et al. 1992). It is also important to notice that the fracture threshold when judged on the basis of bone mineral density (BMD) is lower in glucocorticoid-treated patient than in other types of osteoporosis (Luengo et al. 1991). This may be due to a more severe local destruction of bone architecture in these patients compared to, for example, involution osteoporosis.

When blood and urine markers of bone metabolism have been measured in glucocorticoid-treated patients, it has become clear that a strong inhibition of bone formation dominates bone remodelling at the tissue level. Serum osteocalcin levels could decrease as much as 50% during the treatment, indicating decreased osteoblastic activity (Puolijoki et al. 1992). However, changes in bone resorption markers, for instance in urinary hydroxyproline and deoxypyridinoline, are less marked (Cosman et al. 1994). It should be noticed, however, that bone resorption markers could also be elevated, especially in the early phase of the glucocorticoid administration.

In addition to the predilection of fractures and changes in BMD and bone metabolic markers, histomorphometric analyses of bone biopsies from glucocorticoid-treated patients have also confirmed marked loss of trabecular bone. The most striking finding in bone biopsies from these patients is a marked inhibition in bone formation rate. This leads to a decrease in number and width of trabeculae and eventually to a marked decrease in the whole area of cancellous bone. Slightly increased cortical porosity is more likely due to enhanced bone resorption than disturbed bone formation.

In addition to human studies, the effect of glucocorticoids on bone has been studied in several animal species. All other studied species, except rat, seem to express rather similar skeletal responses to humans. Detailed analysis of bone has been done in glucocorticoid-treated mice. Again, the most marked finding is inhibition of bone formation and loss of cancellous bone. In contrast to mouse, glucocorticoid-treated rats sometimes reveal even enhanced bone formation and increased bone mass. Thus, the mouse seems to be a much better rodent model for glucocorticoid-induced osteoporosis than rat. Besides rodents, dog, rabbit and sheep, among other species, have been used to study skeletal effects of glucocorticoids. All these species express rather similar skeletal responses, the inhibition of bone formation being a dominant feature.

4.3 Glucocorticoid-Induced Aseptic Osteonecrosis

In addition to osteoporosis, glucocorticoids often cause another severe skeletal complication, namely, local aseptic osteonecrosis (avascular necrosis). The risk for osteonecrosis increases with both the dose of glucocorticoids and the duration of treatment (Zizic et al. 1985). However, it may also develop in the patients who receive large doses of glucocorticoids for a restricted period of time or even after some intra-articular injections. Most often this complication affects the femoral head or the head of the humerus. The frequency of aseptic osteonecrosis among patients on long-term glucocorticoid treatment could be very high (Zizic et al. 1985), and as many as 1/3 cases of idiopathic osteonecrosis have been connected to previous glucocorticoid treatment (Fisher and Bickel 1971).

Several clinical studies as well as experimental studies with different animal species have tried to clarify the pathogenesis of aseptic osteonecrosis associated with glucocorticoid use. These studies have indicated that the continuous glucocorticoid treatment causes increased fat accumulation in the bone marrow gradually displacing other marrow elements. Two different theories have arisen from this observation to explain the pathogenesis of aseptic necrosis. Jones (1985) suggested that microscopic fat emboli in bone marrow vessels are a reason for ischaemia and subsequent necrosis. Some other studies indicated that osteonecrosis is associated with an increased hydrostatic pressure in the bone marrow cavity and this could lead to occlusion in the blood vessels and thus to increased intraosseous pressure. The high hydrostatic intraosseous pressure then leads to disturbed circulation and finally to cell death and tissue necrosis.

In addition to the vascular theory and pressure theory, a mechanical theory for aseptic osteonecrosis has also been presented (Cruess et al. 1975). It is based on the fact that collapse of the epiphysis is a consequence of osteoporosis and accumulation of microfractures.

More recent studies have, however, challenged all above-mentioned theories and suggested that massive bone cell death, especially in osteocytes, is a direct consequence of glucocorticoid administration (Weinstein et al. 1998, 2000; Plotkin et al. 1999). Although only a limited number of patients have been analysed, data from experimental animal studies strongly support this new hypothesis that the primary mechanism is apoptosis of osteocytes. Further studies are obviously needed to make the final judgement, but if this turns out to be the case then it becomes important to screen all new glucocorticoid analogues, keeping this effect in mind.

4.4 Indirect Effects of Glucocorticoids on Bone

Clinical and animal studies have not been able to give a conclusive answer to the question of whether glucocorticoid effects in bone are mainly direct or indirect. Glucocorticoid excess has been linked to several endocrinological disturbances, which may provide at least a partial explanation to accompanying osteoporosis. These include, for instance, disturbances of growth hormone and insulin-like growth factor (IGF)-I secretion and production, respectively (Reid et al. 1989). IGF-I is a potent stimulant of osteoblasts that increases rate of bone formation in bone remodelling units (Binz et al. 1994). One possibility is that glucocorticoids regulate IGF-I activity by modulating the circulating levels of IGF binding protein (IGFBP).

Another possible route of indirect regulation of bone remodelling is inhibition of gonadal steroids. This could be obtained by affecting luteinizing hormone/ follicle-stimulating hormone (LH/FSH) secretion from the pituitary or by inhibiting gonadal function directly (Sakakura et al. 1975). Pituitary suppression could additionally lead to the suppression of the production of anabolic hormones by the adrenals.

Most likely, however, indirect effects of glucocorticoids on bone are mediated via modulation of the parathyroid hormone (PTH), Vitamin D and calcium/phosphate homeostasis (Cosman et al. 1994). Several studies have indicated that glucocorticoids inhibit calcium absorption from the intestine, which would in turn stimulate PTH secretion.

It is obvious that the above-mentioned and several other indirect mechanisms play a role in the pathogenesis of glucocorticoid-induced osteoporosis. We think, however, that direct effects of glucocorticoids on bone cells at the level of the bone remodelling unit are the most important mechanism in the development of osteoporosis.

4.5 Direct Effects of Glucocorticoids on Bone

Recent in vitro studies either using tissue culture or cell culture techniques have revealed that glucocorticoids have directs effects on various bone-cell populations. The expression of specific glucocorticoid receptors (GRs) in one or more main bone cell types is obviously a prerequisite for direct effects of glucocorticoids on bone remodelling. There is also evidence suggesting that mineralocorticoid receptors (MRs) could mediate some of the glucocorticoid effects in various cell types (Arriza et al. 1988). The structure, regulation, transcriptional effects and signalling pathways of GR in general are discussed in detail elsewhere in this book. Thus, we concentrate here on discussing the expression of GR in bone cells and the effects of glucocorticoids on bone cells in different in vitro models, such as isolated cells, primary cell cultures and tissue culture.

4.5.1 Expression of GRs in Bone Cells

Excellent studies on the expression of GR and MR in various bone cells have been recently published (Abu et al. 2000; Beavan et al. 2001). They show that GRalfa, GRbeta and MR are present in developing and adult human bone. They further demonstrate that different receptors show a distinct but overlapping pattern of expression. Both immunolocalization and RT-PCR show the expression of all three receptors in osteoblasts. However, only GRbeta expression was found in osteoclasts. MR protein has been detected at least in human neonatal osteoclasts.

Osteocytes are most abundant bone cells although their role in the bone metabolism and in the regulation of bone remodelling is still very incompletely understood. There are, however, reasons to believe that they are important mechanosensors in bone and may mediate various signals to osteoclasts and osteoblasts thus regulating bone resorption and bone formation, respectively. A low level of GRalfa has been detected in human osteocytes (Beavan et al. 2001). As mentioned earlier, recent in vivo studies strongly suggest that glucocorticoids may induce apoptosis in osteocytes. In addition to GRalfa, MR expression was also observed in osteocytes.

4.5.2 Glucocorticoids Decrease Osteoblast Number and Activity

A number of in vitro studies have demonstrated direct effects of glucocorticoids on osteoblast differentiation and function. This has been firmly established using bone marrow cultures, isolated primary osteoblasts, various transformed osteoblastic cell lines as well as organ cultures. The most striking and the most obvious conclusion to be drawn from these studies is that glucocorticoids effectively inhibit the synthesis of bone matrix proteins in osteoblast. This is in good agreement with the in vivo findings, which show a marked inhibition of bone formation. Transcription of several matrix proteins, including type I collagen, bone sialoprotein, osteocalcin, etc., are regulated by glucocorticoids. Their promoter areas contain GR responsive elements as well as other regulatory elements, such as activator protein (AP)-1, and also mediate transcriptional effects of GR.

The inhibition of matrix synthesis is probably not the only mechanism by which the rate of bone formation is decreased, since proliferation of osteoblast precursors and their differentiation from mesenchymal stem cells is also inhibited by pharmacological doses of various glucocorticoids.

4.5.3 Effects of Glucocorticoids on Osteoclasts

In contrast to the relatively easy and unequivocal demonstration of direct effects of glucocorticoids on osteoblasts, their direct effects on osteoclasts are either much less striking or are more difficult to demonstrate. Recently Huang et al. (2001) reported that glucocorticoids stimulate osteoclastogenesis from stromal cells of giant cell tumour. So far, no studies on normal, isolated human osteoclasts are available.

Gronowicz et al. (1990) and Conaway et al. (1997) reported that glucocorticoids stimulate resorption in fetal rat parietal bone cultures and mouse calvarial bone cultures, respectively. Stimulation of osteoclast differentiation in mouse bone marrow cultures by dexamethasone has been shown and it could be due to inhibition of granulocyte-macrophage colony-stimulating factor GMCF production or stimulation of the receptor activator of NFκB ligand (RANKL) production or both (Shuto et al. 1994; Hofbauer et al. 1999). Thus, glucocorticoids most likely stimulate osteoclastogenesis, which could well explain the observed increase of the activation frequency in human and animal bone biopsies.

The direct effect of glucocorticoids could easily explain all pathological findings seen in bone biopsies from glucocorticoid-treated patients. Especially the strong inhibitory effect of glucocorticoids on osteoblasts and bone formation is most likely mediated via osteoblastic GRs. It may well be that stimulation of bone resorption in the early phase of osteoporotic process is, at least partly, mediated by increased PTH secretion and following increase in activation frequency of remodelling units. On the basis of current knowledge, we suggests that the glucocorticoid-induced osteoporosis is prevented best by developing glucocorticoid analogue compounds that have no inhibitory effects on the osteoblastic cell lineage.

References

Abu EO, Horner A, Kusec V, Triffitt JT, Compston JE (2000) The localization of the functional glucocorticoid receptor alpha in human bone. J Clin Endocrinol Metab 85:883–889

Arriza JL, Simerly RB, Swanson LW, Evans RM (1988) The neuronal mineralocorticoid receptor as a mediator of glucocorticoid response. Neuron 1:887–900

Beavan S, Horner A, Bord S, Ireland D, Compston J (2001) Colocalization of glucocorticoid and mineralocorticoid receptors in human bone. J Bone Miner Res 16:1496–1504

Binz K, Schmid C, Bouillon R, Froesch ER, Jurgensen K, Hunziker EB (1994) Interactions of insulin-like growth factor I with dexamethasone on trabecular bone density and mineral metabolism in rats. Eur J Endocrinol. 130:387–393

Conaway HH, Grigorie D, Lerner UH (1997) Differential effects of glucocorticoids on bone resorption in neonatal mouse calvariae stimulated by peptide and steroid-like hormones. J Endocrinol 155:513–521

Cosman F, Nieves J, Herbert J, Shen V, Lindsay R (1994) High-dose glucocorticoids in multiple sclerosis patients exert direct effects on the kidney and skeleton. J Bone Miner Res 9:1097–1105

Cruess RL, Ross D, Crawshaw E (1975) The etiology of steroid-induced avascular necrosis of bone. A laboratory and clinical study. Clin Orthop 113:178–183

Cushing H (1932) The basophil adenomas of the pituitary body and their clinical manifestations (pituitary basophilism). Bull Johns Hopkins Hosp 50:137–195

Fisher DE, Bickel WH (1971) Corticosteroid-induced avascular necrosis. A clinical study of seventy seven patients. J Bone Joint Surg Am 53:859–873

Gronowicz G, McCarthy MB, Raisz L (1990) Glucocorticoids stimulate resorption in fetal rat parietal bones in vitro. J Bone Miner Res 5:1223–1230

Hofbauer LC, Gori F, Riggs BL, Lacey DL, Dunstan CR, Spelsberg TC, Khosla S (1999) Stimulation of osteoprotegerin ligand and inhibition of osteoprotegerin production by glucocorticoids in human osteoblastic lineage cells: potential paracrine mechanisms of glucocorticoid-induced osteoporosis. Endocrinology 140:4382–4389

Huang L, Xu J, Kumta S, Zheng M (2001) Gene expression of glucocorticoid receptor alpha and beta in giant cell tumour of bone: evidence of glucocorticoid-stimulated osteoclastogenesis by stromal-like tumour cells. Mol Cell Endocrinol 181:199–206

Jones JP Jr (1985) Fat embolism and osteonecrosis. Orthop Clin North Am 16:595–633

Luengo M, Picado C, Piera C, Guanabens N, Montserrat JM, Rivera J, Setoain J (1991) Intestinal calcium absorption and parathyroid hormone secretion in asthmatic patients on prolonged oral or inhaled steroid treatment. Eur Respir J 4:441-444

Lukert, BP, Kream BP (1996) Clinical and basic aspects of glucocorticoid action in bone. In: Bilezikian JP, Raisz LG, Rodan GA (eds) Principles of bone biology. Academic Press, pp 533-548

Lukert BP, Johnson BE, Robinson RG (1992) Estrogen and progesterone replacement therapy reduces glucocorticoid-induced bone loss. J Bone Miner Res 7:1063-1069

Plotkin LI, Weinstein RS, Parfitt AM, Roberson PK, Manolagas SC, Bellido T (1999) Prevention of osteocyte and osteoblast apoptosis by bisphosphonates and calcitonin. J Clin Invest 104:1363-1374

Puolijoki H, Liippo K, Herrala J, Salmi J, Tala E (1992) Inhaled beclomethasone decreases serum osteocalcin in postmenopausal asthmatic women. Bone 13:285-288

Reid IR, Gluckman PD, Ibbertson HK (1989) Insulin-like growth factor 1 and bone turnover in glucocorticoid-treated and control subjects. Clin Endocrinol 30:347-353

Reid IR, Evans MC, Stapleton (1992) Lateral spine densitometry is a more sensitive indicator of glucocorticoid-induced bone loss. J Bone Miner Res 7:1221-1225

Sakakura M, Takebe K, Nakagawa S (1975) Inhibition of luteinizing hormone secretion induced by synthetic LRH by long-term treatment with glucocorticoids in human subjects. J Clin Endocrinol Metab 40:774-779

Sambrook PN, Shawe D, Hesp R, Zanelli JM, Mitchell R, Katz D, Gumpel JM, Ansell BM, Reeve J (1990) Rapid periarticular bone loss in rheumatoid arthritis. Possible promotion by normal circulating concentrations of parathyroid hormone or calcitriol (1,25-dihydroxyvitamin D3). Arthritis Rheum 33:615-622

Seeman E, Melton LJ 3rd, O'Fallon WM, Riggs BL (1983) Risk factors for spinal osteoporosis in men. Am J Med 75:977-983

Shuto T, Kukita T, Hirata M, Jimi E, Koga T (1994) Dexamethasone stimulates osteoclast-like cell formation by inhibiting granulocyte-macrophage colony-stimulating factor production in mouse bone marrow cultures. Endocrinology 134:1121-1126

Weinstein RS, Nicholas RW, Manolagas SC (2000) Apoptosis of osteocytes in glucocorticoid-induced osteonecrosis of the hip. J Clin Endocrinol Metab 85:2907-2912

Weinstein RS, Jilka RL, Parfitt AM, Manolagas SC (1998) Inhibition of osteoblastogenesis and promotion of apoptosis of osteoblasts and osteocytes by glucocorticoids. Potential mechanisms of their deleterious effects on bone. Clin Invest 102:274–282

Zizic TM, Marcoux C, Hungerford DS, Dansereau JV, Stevens MB (1985) Corticosteroid therapy associated with ischemic necrosis of bone in systemic lupus erythematosus. Am J Med 79:596–604

Zizic TM (1991) Osteonecrosis. Curr Opin Rheumatol 3:481–489

5 Corticosteroids in Ophthalmology

U. Pleyer, Z. Sherif

5.1	Historical Background	66
5.2	Routes of Administration	67
5.3	Topical Application	67
5.4	Ocular Pharmacological Effects	69
5.5	Indications	69
5.6	Complications of Corticosteroid Therapy	73
5.7	Corticosteroid-Induced Glaucoma	74
5.8	Pathophysiology of Corticosteroid-Induced Glaucoma	74
5.9	Corticosteroid-Induced Cataract	75
5.10	Future Directions on the Use of Corticosteroids in Ophthalmology	77
5.11	Summary	78
References		78

Few drugs have gained comparable importance, in almost any field of medical therapy, to corticosteroids. For the ophthalmologist, corticosteroids remain one of the basic therapeutics in a broad spectrum of inflammatory conditions. This review is intended to summarize some relevant aspects on the use of corticosteroids, their main indications, as well as their limitations in ophthalmology.

5.1 Historical Background

An endogen release of corticosteroids following an injection of attenuated typhoid vaccine lead to early substantial progress in the therapy of inflammatory and autoimmune disease states. With the development of adrenocorticotropic hormone (ACTH) and cortisone, and with their successful application in the treatment of chronic polyarthritis, case reports on control of ocular inflammatory diseases appeared. Elkinton et al. (1949) first described the effect of ACTH therapy on inflammatory hemorrhagic retinopathy associated to "generalized collagen disease." In 1950, ACTH was reported to be effective in various inflammatory conditions of the eye. Mann and Markson (1950) demonstrated positive results in patients treated with ACTH for uveitis and episcleritis. In 1950, Olson et al. reported ACTH to be successful in various inflammatory diseases of the eye in seven patients. An early landmark on the use of corticosteroids in patients with inflammatory ocular disease was set by AC Woods, treating 14 patients with either ACTH or cortisone with favorable outcome (Woods 1950). He furthermore was first to report on the clinical use of topical corticosteroids. With the introduction of prednisone and prednisolone ophthalmologists were given therapeutic tools five to ten times stronger in their anti-inflammatory potency compared to cortisone, whereas these drugs were shown to have significantly less metabolic adverse effects. Although the mechanisms of action were not fully understood at that time, clinical experience led to a rapidly increasing spectrum of indications for the usage of corticosteroids in ophthalmology.

However, ocular side effects were also reported shortly after introduction of corticosteroids.

Rising intraocular pressure as one of the most relevant ocular adverse effects was already noticed in the early 1950s. Blake et al. (1950) observed significantly increased intraocular pressure in patients receiving ACTH as Stern did in patients with cortisone therapy. Additionally, Armaly (1966) first detected a heritable nature of corticosteroid-induced ocular hypertension and glaucoma. It was established early that there is a close correlation between the anti-inflammatory potency of topically applied corticosteroids and the risk of rising intraocular pressure. This observation has generated not only strong efforts to develop strategies to minimize the risk of secondary glaucoma, but also provided new aspects on the genetic background in glaucoma.

5.2 Routes of Administration

Soon after corticosteroids were introduced as ocular therapeutics, their topical use was found to be equal or even superior to systemic administration.

The route of corticosteroid application depends on the site of ocular involvement. Still today, topical use remains the most important and effective route of administration for the anterior segment of the eye. Significant drug levels can be achieved in the conjunctiva, cornea, anterior chamber, and uvea. Following a single drop application of dexamethasone (50 µl), it is measurable in human aqueous humor (Watson 1988).

Advantages include a patient-friendly type of application that is inexpensive in the absence of systemic side effects. However, topical application of corticosteroids will be often ineffective in ocular diseases involving the posterior segment of the eye such as posterior scleritis, chorioretinitis, or optic neuritis. In these patients, systemic administration, mainly orally, is preferred. Other routes of administration include collagen shields, periocular injections, intraocular injections, or intravitreal implanted drug devices (Friedberg et al. 1991; Milani et al. 1993). Although these routes are effective in treating inflammatory conditions of the posterior segment, their invasive nature and potential risks limit their clinical acceptance (Carnahan and Goldstein 2000).

5.3 Topical Application

An important role that determines the use of any topically applied drug is corneal permeability and intraocular penetration. Following local application of various steroidal preparations, including fluorometholone, prednisolone acetate, prednisolone phosphate, and dexamethasone phosphate, significant differences on corneal permeability are found (see Table 1; Hull et al. 1974; Leibowitz Kupferman 1976). In addition, considerable difference exists in the intraocular penetration of pred-

Table 1. Corneal permeabilities (mean±SEM) of tritiated labeled steroids (from Hull et al. 1974)

Steroid preparation	Epithelium intact	W/O epithelium
	(nM cm^{-2} h^{-1})	(nM cm^{-2} h^{-1})
Dexamethasone phosphate	7.1±0.6	32.4±1.8
Prednisolone phosphate	12.1±0.5	37.7±2.6
Prednisolone acetate	14.3±1.1	11.5±1.2
Fluorometholone	15.4±2.2	14.4±3.3

Table 2. Decrease in corneal inflammation following topical corticosteroid therapy with various corticosteroid derivatives

Corneal Epithelium Intact	Decrease (%)	Corneal Epithelium Absent	Decrease (%)
Prednisolone acetate 1.0%	51	Prednisolone acetate 1.0%	53
Dexamethasone alcohol 0.1%	40	Prednisolone sodium phosphate 1.0%	47
Fluorometholone alcohol 0.1%	31	Dexamethasone alcohol 0.1%	42
Prednisolone sodium phosphate 1.0%	28	Fluorometholone alcohol 0.1%	37
Dexamethasone sodium phosphate 0.1%	19	Dexamethasone sodium phosphate 0.1%	22
Dexamethasone sodium phosphate 0.05% (ointment)	13		

nisone and prednisolone, depending on whether an alcohol or acetate form is applied.

A number of modifications have been introduced to increase intraocular penetration of steroids, including preparations in a microsuspension, gel, or viscous substance. Interestingly, even minor changes in formulation such as addition of benzalkonium chloride significantly enhanced ocular (corneal) penetration.

Several studies investigated the metabolism of corticosteroids within the eye, in particular in the cornea. The presence of 11-β-hydroxylase activity in the cornea is considered an important pathway to convert cortisone to the active product 11-β-hydrocortisone (Sugar et al. 1972; Rauz et al. 2001). In addition, the presence of phosphatase activity in the cornea, which converts phosphate derivatives of corticosteroids to the more active alcohol forms, is important (Table 2).

5.4 Ocular Pharmacological Effects

As in other tissues, corticosteroids do not appear to have specific effects but exert a broad spectrum of anti-inflammatory activity. In general, they are more effective in acute than in chronic inflammatory conditions. The beneficial effects on ocular inflammation include reduced capillary permeability, vasoconstriction, suppression of leukocyte adhesion and migration, inhibition of fibroblast proliferation, and reduced release of hydrolytic enzymes from leukocytes and macrophages. Their main effects as anti-inflammatory agents in the acute phase may rely on the inhibition of cytokine transcription including interleukin (IL)-1, IL-2, IL-6, IL-8 and tumor necrosis factor (TNF)-α (Barnes and Adcock 1993). These properties may in part explain not only the anti-inflammatory effects of corticosteroids, but also the use in inflammatory conditions that are accompanied by corneal scarring and neovascularization.

5.5 Indications

Topically applied corticosteroids are well-established in the therapy of acute allergic reactions of the eyelids and conjunctiva. Although there have been other anti-inflammatory drugs introduced like antihistamines and mast cell stabilizers, corticosteroids still remain powerful therapeutics in severe (acute) cases of allergic conjunctivitis. The selection of topically administered corticosteroids in these indications is determined by considerations favoring high anti-inflammatory effects at the conjunctiva but minimal intraocular action. Fluorometholone seems to meet these requirements as it is effective at the ocular surface but does not penetrate the cornea.

With their capability to control immune-mediated inflammatory conditions, topically administered corticosteroids are essential in the therapy of episcleritis, scleritis, and uveitis (Fig. 1). According to the severity and course of the disease as well as the patient's ocular and general conditions, topical treatment might be combined with systemic admini-

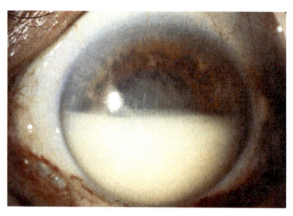

Fig. 1. Acute anterior uveitis presenting with fibrin exudation and hypopyon as consequence of blood–aqueous barrier breakdown

stration of corticosteroids or other immunomodulatory agents. As approximately 40%–50% of patients with scleritis have an associated systemic disorder like rheumatoid arthritis, vasculitis, inflammatory bowel disease, or systemic lupus erythematosus, the indication for long-term immunomodulatory treatment has to be closely monitored in an interdisciplinary collaboration (Table 3).

Multiple studies on uveitis patients have shown that, with a prevalence of 50%, idiopathic acute anterior uveitis is the most common form among acute intraocular inflammatory conditions (Weiner and BenEzra 1991). The underlying cause of idiopathic anterior uveitis still remains unknown, although both autoimmune and infectious mechanisms have been identified. Clinical observations such as increased circulating levels of IL-2 receptor, increased T-lymphocyte responsiveness to retinal S antigen and the presence of circulating immune complexes suggest underlying immune mechanisms (Dick 1999). Topically applied corticosteroids remain the mainstay to reduce inflammation in the acute episode. The rapid anti-inflammatory effect is particularly useful to minimize secondary damage to the eye, such as fibrin formation and posterior synechiae, or even secondary synechial angle closure glaucoma due to intraocular inflammation. Severe cases of idiopathic anterior uveitis may require periocular injections or systemic corticosteroid

Table 3. Indications for corticosteroids in ocular disorders

Allergic blepharitis and conjunctivitisMucocutaneous conjunctival lesions
Chemical burn of the cornea and conjunctiva
Phlyctenular conjunctivitis and keratitis
Contact dermatitis of the conjunctiva and eyelid
Vernal conjunctivitis
Ocular pemphigoid
Acne rosacea keratitis
Viral ocular diseases
Herpes simplex (disciform stage)
Herpes zoster
Adenovirus
Infiltrative corneal disease
Interstitial keratitis
Superficial punctate keratitis
Marginal corneal ulcers
Immune graft reaction
Scleritis and episcleritis
Iritis, iridocyclitis
Sympathetic ophthalmia
Posterior uveitis
Retinal vasculitis
Optic neuritis
Pseudotumor of the orbit
Progressive thyroid exophthalmopathy
Temporal arteritis

application. The long-term management of recurrent idiopathic anterior uveitis depends on frequency and severity of episodes and may include either short-term use of topically administered corticosteroids use or systemic immunosuppressive agents such as cyclosporin A or methotrexate (Pleyer et al. 1999).

Intermediate uveitis is an anatomic designation to inflammatory conditions of the peripheral uvea. As intermediate uveitis presents with chronic persistent inflammation, the treatment goal is not only to suppress inflammation but also to diminish long-term vision loss, which may develop due to cystoid macular edema (CME). A four-step treatment approach has been recommended using an initial trial of oral or

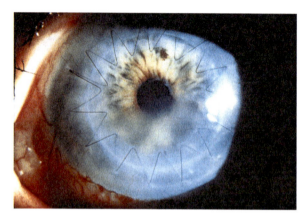

Fig. 2. Corneal graft rejection line. Destruction of corneal endothelial cells due to an allogeneic T-cell response

periocular injected corticosteroids (Kaplan 1984). Systemic corticosteroids may be given in doses up to 1.0 mg/kg daily for a period of 4–10 weeks and should be tapered depending on treatment response. Since all corticosteroid therapy carries the risk for ocular complications, especially rising intraocular pressure and cataract formation, patients and family should be well-informed concerning the therapeutic management. If corticosteroids fail to sufficiently suppress inflammation and CME does not resolve, surgical approaches, including cryotherapy and pars plana vitrectomy, might be considered. Long-term treatment may also depend on other systemic immunomodulatory agents like methotrexate, azathioprine, cyclophosphamide, cyclosporin A, or mycophenolate mofetil.

Furthermore, the anti-inflammatory effect of corticosteroid eye drops is regarded to be beneficial after surgical procedures like cataract extraction. Even when significant progress has been made in microsurgical technique, corticosteroids remain an important adjunct to reduce postoperative sequelae due to unspecific inflammatory conditions.

Postoperative management of patients undergoing penetrating keratoplasty includes topically applied corticosteroids (Pleyer et al. 1998). It has been shown that an additional systemic administration does not significantly improve the outcome of corneal allografts (Fig. 2) compared to a topical delivery of corticosteroids alone in eyes that do not meet high-risk criteria. However, the anti-inflammatory effect of topically applied corticosteroids in patients following corneal surgery is accompanied by its ability to inhibit corneal neovascularization. This effect is frequently shown to be beneficial in a vast spectrum of infectious or non-infectious corneal conditions that are at risk of developing corneal neovascularization and subsequent corneal scaring.

5.6 Complications of Corticosteroid Therapy

Ocular complications of corticosteroids have been documented secondary to any route of administration including oral, intravenous, inhaled, and topical application. They may occur in all structures of the eye including lids, cornea, trabecular meshwork, lens, and optic nerve (Carnahan and Goldstein 2000). Although a list of potential complications is provided at Table 4, the effect on the most frequent and sight-threatening side effects will be focused on.

Table 4. Ocular side effects of corticosteroids

Systemic administration
  Posterior capsular cataract
  Secondary glaucoma
  Delayed corneal wound healing
  Central serous retinopathy
  Microcysts (iris epithelium)
  Papilledema

Topical application
  Posterior capsular cataract
  Secondary glaucoma
  Delayed corneal/scleral wound healing
  Decreased resistance to infection
  Ptosis
  Mydriasis
  Paralysis to accommodation
  Eyelid depigmentation
  Corneal punctate keratitis

5.7 Corticosteroid-Induced Glaucoma

Glaucoma secondary to prolonged use of corticosteroids was suspected already in 1950 (McLean 1950). The concept of corticosteroid-induced glaucoma was further substantiated by a number of clinical observations. Taken together: An increase in intraocular pressure will be found in approximately 30% of individuals with an average rise of 10 mmHg. An additional 5% of patients will develop an increase of more than 16 mmHg.

This response is increased to 46% to 92% in patients with primary open-angle glaucoma (POAG). Patients over 40 years of age and with certain systemic diseases (e.g., diabetes mellitus, high myopia) as well as relatives of patients with POAG are more vulnerable to corticosteroid-induced glaucoma. The association of corticosteroid-induced ocular hypertension in other conditions which are considered risk factors for glaucoma (racial origins, hypertension, migraine, vasospasm) is likely but not fully established.

Secondary glaucoma due to corticosteroids can be induced by any route of administration (Mitchell 1999; Ozerdem et al. 2000).

The intensity of intraocular pressure elevation in a given individual depends on the potency of the corticosteroid, frequency and duration of application.

Pressure changes are completely reversible when treatment is discontinued. However, irreversible optic nerve damage may result from prolonged secondary hypertension.

5.8 Pathophysiology of Corticosteroid-Induced Glaucoma

The clinical and pharmacological links between the side effects of corticosteroids to increase intraocular pressure and primary open-angle glaucoma are of substantial interest given the fact that glaucoma remains a major cause of blindness in developed countries. Recent studies demonstrated that corticosteroid treatment of human trabecular meshwork cells produced delayed, progressive cellular and extracellular glycoprotein induction. From this system, a gene termed TIGR (trabecular meshwork inducible GC response) was cloned. The progressive induction of the TIGR gene combined with specific structural features of its cDNA, including features for interactions with glycosaminoglycans, glycoproteins, and signal sequences for secretion, suggests that it should be considered as a candidate gene for outflow obstruction in glaucoma (Stone et al. 1997).

Additional data suggest an activation of stress/apoptotic pathways in human trabecular meshwork cells. Pharmacological modulation showed that corticosteroid-induced TIGR gene expression is reduced approximately fourfold by basic fibroblast growth factor (bFGF) and transforming growth factor (TGF)-β, and certain nonsteroidal anti-inflammatory drugs protect against steroid-induced stress response (Johnson et al. 1997; Polansky et al. 2000).

These observations not only allow us to better understand the pathogenesis of steroid-induced secondary glaucoma, but they also will be helpful in investigating potential mechanisms in different types of glaucoma.

5.9 Corticosteroid-Induced Cataract

Posterior subcapsular cataract (PSC) as a complication of both topical and systemic corticosteroid therapy is well-documented (Derby and Maier 2000; Fig. 3). In 1960, Black et al. suggested a causative relation between systemic steroid administration and this type of cataract forma-

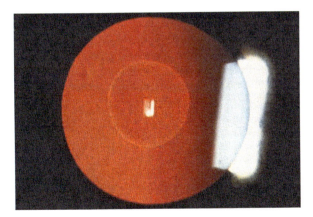

Fig. 3. Posterior subcapsular cataract following long-term (systemic) corticosteroid treatment

tion. Although the mechanism still is not understood in detail, this specific effect seems to be dependent on dosage and duration of therapy as well as an interindividual susceptibility (Hanania et al. 1995). The incidence of corticosteroid-induced cataracts varies between 6.4% and 38.7% after oral use (Urban and Cotlier 1986). As little as 5 mg of oral prednisolone per day has been reported to cause PSC within 2 months.

Other studies demonstrated the risk of significant PSC in patients already within 6 month after corticosteroids were given at a minimum dosage of 10 mg dexamethasone (Carnahan and Goldstein 2000). However, the pathophysiology of secondary cataracts remains unclear (Foreman et al. 1977), and several mechanisms may contribute.

The formation of covalent adducts of the steroid molecule with extracellular lysine residues may result in lens opacities (Chylack 1997).

Prolonged topical application of prednisolone caused marked changes in the aqueous humor, such as significant lowered ascorbic acid, an important antioxidant present in the aqueous humor.

Specific corticosteroid receptors have been identified in lens epithelial cells (Southern et al. 1976, 1977).

In addition, metabolic changes have been reported such as altered phospholipid metabolism in experimental animals (Tamada et al. 1980).

In contrast to steroid-induced glaucoma, there is no direct evidence that steroids directly cause PSC. Intraocular inflammation, the presence of posterior synechiae, aging processes, and other factors are also likely to play a role. Although in some animals cataract formation could be induced, cataractogenesis in experimental models is rather complex. Even high doses of systemically applied steroids did not cause cataract in rat, rabbit, or chicken (Bettmann et al. 1967). Topically applied steroids given daily for 6 months caused lens opacities in 50% of cortisone-treated, 35% of hydrocortisone-treated, and 20% of prednisolone-treated rabbits (Bettmann et al. 1967).

5.10 Future Directions on the Use of Corticosteroids in Ophthalmology

Because of their potential side effects, novel approaches for the use of corticosteroids are highly desirable. There are at least two main directions to achieve a high therapeutic index and to reduce the risk of undesired side effects:

- Development of "soft drugs"
- Application of long-term drug-release systems

A major approach to reduce side effects is based on the "retrometabolic design" technique. This concept has been already used for the development of safe antiglaucoma agents ("soft" beta blockers) or short-acting mydriatics ("soft" anticholinergics) (Bodor 1994). In principle, the design process is focusing on inactive drug metabolites which are metabolized at the target site or near the application site. The search for corticosteroids that suppress inflammation but do not cause secondary glaucoma goes on.

Loteprednol etabonate is a site-active corticosteroid synthesized through structural modifications of prednisolone-related compounds so that it will undergo a predictable transformation to an inactive metabolite (Howes 2000). In double-masked studies, loteprednol etabonate was effective in the treatment of giant papillary conjunctivitis, seasonal allergic conjunctivitis, postoperative inflammation, and uveitis. In a large double-masked study of corticosteroid responders, loteprednol etabonate demonstrated less propensity to cause clinically significant elevation in intraocular pressure when compared to prednisolone acetate (Novack et al. 1998).

Rimexolone is another candidate that has been launched onto the market as a "soft" corticosteroid. It appears that this agent induces a rise of intraocular pressure equivalent to that seen with fluorometholone, but less pressure rise than with dexamethasone phosphate or prednisolone acetate (Leibowitz et al. 1996).

At the same time, Rimexolone was as effective as an anti-inflammatory agents as prednisolone in a clinical uveitis trial.

Several approaches have been introduced to increase the bioavailability of ocular drugs including liposomes, microspheres, collagen shields or intraocular implants (Friedberg et al. 1991; Milani et al. 1993; Pleyer et al. 1993, 1994). Since it is particularly difficult to achieve effective levels of therapeutics in the vitreous/retina following topical, parabulbar, or even systemic administration, intraocular drug delivery devices have received considerable attention. Nonbiodegradable drug delivery devices containing 2 or 15 mg of fluocinolone that were implanted into the vitreous cavity of rabbit eyes demonstrated constant drug release over a 6 months test period. Based on these release rates the predicted life span of the 2- and 15-mg devices are 2.7 and 18.6 years, respectively (Jaffe et al. 2000). Since no drug toxicity was reported based on clinical, electrophysiological or histologic observation, these devices may show great promise in the treatment of chronic intraocular inflammation.

5.11 Summary

Corticosteroids are, even 50 years after introduction in ophthalmology, the best, and often only choice of treatment for acute inflammatory eye disorders. Their broad spectrum of actions may not only explain the greater potency compared with other anti-inflammatory agents but also be responsible for multiple serious side effects.

For future developments, several directions may be chosen including development of safer drugs with high therapeutic index and application of long-term drug-release systems.

Acknowledgements. The authors thank Christina Grigat for assistance in manuscript preparation.

References

Armaly MF (1966) The heritable nature of dexamethasone-induced ocular hypertension. Arch Ophthalmol 75:32–35

Barnes PJ, Adcock I (1993) Anti-inflammatory actions of steroids: molecular mechanisms. Trends Pharmacol Sci 172:436–441

Bettmann JW, Fund WE, Webster RG, Noyes PP, Vincent NJ (1967) Cataractogenic effect of corticosteroids on animals. Am J Ophthalmol 63:841–844

Black RL, Oglesby RB, von Sallman L, et al (1960) Posterior subcapsular cataracts induced by corticoids in patients with rheumatoid arthritis. JAMA 174:166–171

Blake EM, Fasanella RM, Wong AS (1950) The effect of ACTH in glaucoma. Am J Ophthalmol 33:1231–1235

Bodor B (1994) Designing softer ophthalmic drugs by soft drug approaches. J Ocul Pharmacol Ther 10:3–15

Carnahan MC, Goldstein DA (2000) Ocular complications of topical, peri-ocular and systemic corticosteroids. Curr Opin Ophthalmol 11:478–483

Chylack LT (1997) Cataracts and inhaled corticosteroids. N Engl J Med 337:46–48

Derby L, Maier WC (2000) Risk of cataract among users of intranasal corticosteroids. J Allergy Clin Immunol 105:912–916

Dick AD (1999) Immune regulation of uveoretinal inflammation. In: Pleyer U, Zierhut M, Behrens-Baumann W (eds) Immuno-ophthalmology. S. Karger AG, Basel, pp 187–203

Elkinton JR, Hunt AD Jr, Godfrey L, McCory WW, Stokes J Jr (1949) Effects of pituitary adrenocorticotropic hormone (ACTH) therapy. JAMA 141:1273–1279

Foreman AR, Loreto JA, Tina LV (1977) Reversibility of corticosteroid-associated cataracts in children with the nephrotic syndrome. Am J Ophthalmol 84:75–78

Friedberg M, Pleyer U, Mondino BJ (1991) Device drug delivery: collagen shields, iontophoresis, pumps. Ophthalmology 98:725–732

Garbe E, Suissa S, LeLorier J (1998) Association of inhaled corticosteroid use with cataract extraction in elderly patients. JAMA 280:539–543

Hanania NA, Chapman KR, Kesten S (1995) Adverse effects of inhaled corticosteroids. Am J Med 98:196–208

Howes JF (2000) Loteprednol etabonate: a review of ophthalmic clinical studies. Pharmazie 55:178–183

Hull DS, Hine JE, Edelhauser HF, Hyndius RA (1974) Permeability of the isolated rabbit cornea to corticosteroids. Invest Ophthalmol Vis Sci 13:457–459

Jaffe GJ, Yang CH, Guo H, Denny JP, Lima C, Ashton P (2000) Safety and pharmacokinetics of an intraocular fluocinolone acetonide sustained delivery device. Inv Ophthalmol Vis Sci 41:3569–3575

Johnson D, Gottanka J, Flugel C, Hoffmann F, Futa R, Lutjen Drecoll E (1997) Ultrastructural changes in the trabecular meshwork of human eyes treated with corticosteroids. Arch Ophthalmol 115:375–383

Kaplan HL (1984) Intermediate uveitis (pars planitis, chronic cyclitis): a four-step approach to treatment. In: Saari KM (ed) Uveitis update. Exerpta Medica 1984, Amsterdam, pp 169–172

Leibowitz HM, Bartlett JD, Rich R, McQuirter H, Stewart R, Assil K (1996) Intraocular pressure-raising potential of 1% rimexolone in patients responding to corticosteroids. Arch Ophthalmol 114:933–937

Leibowitz HM, Kupferman A (1976) Kinetics of topically administered prednisolone acetate. Arch Ophthalmol 94:1387–1389

Mann WA, Markson DE (1950) A case of recurrent iritis and episcleritis on a rheumatic basis treated with ACTH. Am J Ophthalmol 33:459–461

McLean JM (1950) Clinical and experimental observation on the use of ACTH and cortisone in ocular inflammatory disease. Trans Am Ophthalmol Soc 48:259

Milani JK, Verbukh I, Pleyer U, Sumner H, Adamu SA, Halabi HJ, Chou HJ, Lee DA, Mondino BJ (1993) Collagen shields impregnated with gentamycin-dexamethasone combination as a potential drug delivery device. Am J Ophthalmol 116:622–627

Mitchell P, Cumming RG, Mackey DA (1999) Inhaled corticosteroids, family history, and risk of glaucoma. Ophthalmology 106:2301–2306

Novack GD, Howes J, Crockett RS, Sherwood MB (1998) Change in intraocular pressure during long-term use of loteprednol etabonate. J Glaucoma 7:266–269

Olson JA, Steffensen EH, Margulis RR, Smith RW, Whitney EL (1950) Effect of ACTH on certain inflammatory diseases of the eye. JAMA 142:1276–1278

Ozerdem U, Levi L, Cheng L, Song MK, Scher C, Freeman WR (2000) Systemic toxicity of topical and periocular corticosteroid therapy in an 11-year-old male with posterior uveitis. Am J Ophthalmol 130:240–241

Pleyer U, Lutz S, Jusko WJ, Nguyen K, Narawane M, Rückert D, Mondino BJ, Lee VHL (1993) Ocular absorption of topically applied FK506 from liposomal and oil formulations in the rabbit eye. Invest Ophthalmol Vis Sci 34:2737–2742

Pleyer U, Elkins B, Rückert D, Lutz S, Grammer J, Chou J, Schmidt KH, Mondino BJ (1994) Ocular absorption of cyclosporine A from liposomes incorporated into collagen shields. Curr Eye Res 13:177–181

Pleyer U, Rieck P, Ritter T, Hartmann C (1998) Immunreaktion nach perforierender Keratoplastik. Ophthalmologe 95:444–459

Pleyer U, Liekfeld A, Baatz H, Hartmann C (1999) Pharmakologische Modulation immunmediierter Erkrankungen des Auges. Klin Monatsbl Augenheilkd 214:160–170

Polansky JR, Fauss DJ, Zimmerman CC (2000) Regulation of TIGR/MYOC gene expression in human trabecular meshwork cells. Eye 14:503–514

Rauz S, Walker EA, Shackleton CH, Hewison M, Murray PI, Stewart PM (2001) Expression and putative role of 11 beta-hydroxysteroid dehydrogenase isozymes within the human eye. Invest Ophthalmol Vis Sci 42:2037–2042

Southern AL, Altman K, Vittek J, Boniuk V, Gordon GG (1976) Steroid metabolism in ocular tissues of the rabbit. Invest Ophthalmol Vis Sci 15:222–228

Southern AL, Gordon GG, Yeh HS, Dunn MW, Weinstein BI (1977) Receptors for glucocorticoids in the lens epithelium of the calf. Science 200:1177–1178

Stern JJ (1953) Acute glaucoma during cortisone therapy. Am J Ophthalmol 36:389–390

Stone EM, Tingert JH, Wallace LM, Nguyen TD, Polansky JR, Sheffield VC (1997) Identification of a gene that causes primary open-angle glaucoma. Science 275:668–670

Sugar J, Burde RM, Sugar A, Waltman SR, Kripalani KJ, Weliky I, Becker B (1972) Tetrahydrotriamcinolone and triamcinolone I. Ocular penetration. Invest Ophthalmol Vis Sci 11:890–893

Tamada Y, Miyashita H, Ono S (1980) Studies on phospholipid metabolism of rabbit lens with special references to long-term topical administration of steroid. Jap J Ophthalmol 24:289–296

Urban RC Jr, Cotlier E (1986) Corticosteroid-induced cataracts. Surv Ophthalmol 31:102–110

Watson D, Noble MJ, Dutton GN, Midgley JM, Healey TM (1988) Penetration of topically applied dexamethasone alcohol into human aqueous humor. Arch Ophthalmol 106:686–687

Weiner A, BenEzra D (1991) Clinical patterns and associated conditions in chronic uveitis. Am J Ophthalmol 112:151–158

Woods AC (1950) Clinical and experimental observations on the use of ACTH and cortisone in ocular inflammatory disease. Am J Ophthalmol 33:1325–1349

6 Special Problems in Glucocorticoid Treatment in Children

U. Wahn

6.1	Topical Corticosteroids	84
6.2	Bioavailability of Topical Corticosteroids?	85
6.3	Hypothalamic–Pituitary–Adrenal Axis Function	86
6.4	Growth in Asthmatic Children	87
6.5	The Effects of Inhaled Corticosteroids on Growth	87
6.6	Conclusion	89
References		89

A variety of chronic diseases in infancy and childhood require long-term treatment with glucocorticoids including allergic diseases such as asthma, autoimmune diseases, or chronic inflammatory diseases of the gastrointestinal tract. While glucocorticoids are effective in controlling inflammatory processes, short- and long-term side effects are still a matter of concern for physicians as well as for the public. Side effects include the suppression of the hypothalamic–pituitary–adrenal (HPA) axis, skin atrophy, the incidence of cataracts and glaucoma, the reduction of bone mineral density as well as metabolic changes. A major concern in childhood is related to the possibility of reduced growth.

There is convincing evidence that daily maintenance treatment with systemic corticosteroids reduces growth in children. Alternate-day oral corticosteroid therapy is also associated with reduced growth but to a lesser degree than a daily schedule. The cumulative effect of systemic corticosteroids may amount to a growth reduction of more than 10 cm, although many children continue to grow normally even during long-term systemic steroid therapy.

6.1 Topical Corticosteroids

For the treatment of many patients with allergic rhinitis and asthma, topical glucocorticoids are now advocated as the preventive treatment of choice (International Consensus Report on Diagnosis and Treatment of Asthma 1992; National Institutes of Health 1995). Topical glucocorticoids like beclomethasone dipropionate (BDP), budesonide, or fluticason propionate (Fig. 1) have been demonstrated to be most effective for

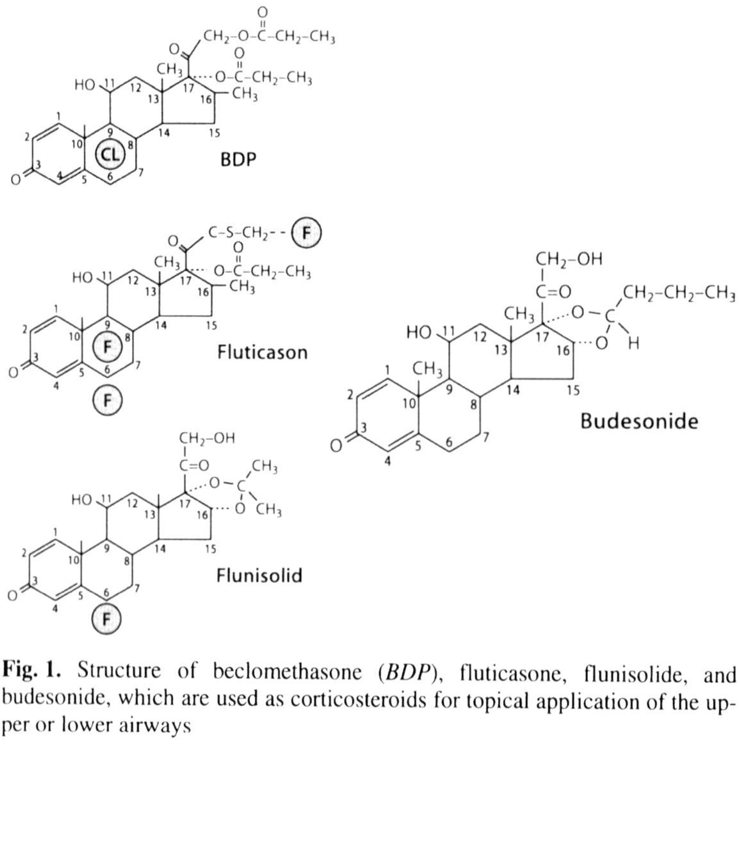

Fig. 1. Structure of beclomethasone (*BDP*), fluticasone, flunisolide, and budesonide, which are used as corticosteroids for topical application of the upper or lower airways

Special Problems in Glucocorticoid Treatment in Children 85

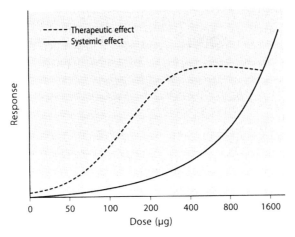

Fig. 2. Dose response curves for therapeutic and systemic effects of inhaled glucocorticoids

long-term control of persisting asthma in infancy and childhood (National Heart, Lung and Blood Institute 1997).

Both anti-inflammatory and systemic effects of corticosteroids are dose-dependent, but the dose response curves of pulmonary and systemic effects differ markedly (Fig. 2). The dose response curve for pulmonary effects appears steepest at relatively low daily doses, flattening off towards higher doses. For systemic effects the reverse pattern applies. Although this basic principle applies for all available inhaled corticosteroids, the exact dose response curves may differ between various compounds and even between different individuals using the same compound.

6.2 Bioavailability of Topical Corticosteroids?

Topical corticosteroids applied to the nose or the lower airways can only reduce growth after they become available systemically. Systemic side effects of inhaled corticosteroids are determined by absorption from the lung and the gut. The amounts of drug that are deposited in the lung and in the oral pharynx will depend largely on the type of inhaler device, the type of propellant and on the patient's inhalation technique. If a spacer device is being used, oropharyngeal deposition is low. With a metered dose inhaler without a spacer, a considerable proportion of the inhaled drug is deposited in the oral pharynx and swallowed and may thus contribute to systemic availability. Under these circumstances, systemic bioavailability depends on the degree of first pass inactivation.

Oropharyngeal deposition can be reduced by mouth rinsing. Regardless of the inhalation device and the specific drug inhaled, systemic bioavailability is predominantly determined by the amount of drug deposited in the lung. Improved lung deposition will, therefore, lead to increased systemic availability of inhaled corticosteroids (ICS).

The potential of systemically available ICSs to cause side effects is also determined by the pharmacokinetic properties. Fluticason propionate being the most lipophilic ICS not only has the largest corticosteroid receptor affinity, but also has a larger volume of distribution and a longer elimination half-life than other ICSs. This may help to explain while fluticasone propionate seams to cause more adrenal suppression than budesonide.

6.3 Hypothalamic–Pituitary–Adrenal Axis Function

Adrenal function can be assessed by sequential plasma cortisol levels as well as cortisol secretion stimulated by adrenocorticotrophic hormone (ACTH) or corticotropic releasing factor (Hofmann-Streb 1993). In general, basal and ACTH-stimulated cortisol levels are not adversely affected by treatment with topical corticosteroids within an acceptable dose range. Importantly, when comparing individuals treated with budesonide, fluticasone, or beclomethasone with placebo controls, there do not appear to be differences in the numbers showing a shift from normal to abnormal cortisol responses between baseline and several weeks after treatment.

It is not yet clear whether inhaled glucocorticoids accelerate bone loss, as is observed after systemic treatment. The results of prospective and cross sectional studies on the effects of inhaled glucocorticoids have been inconsistent, in that they have been associated with a decrease in bone density in some studies and no changes in others. Studies have usually been small or of short duration, have relied on retrospective data, or have not independently verified the doses of inhaled glucocorticoid taken by participants. In addition, the results may be confounded by a lack of compliance and differences between participants in the severity of asthma that may affect the patient's physical activity. A recent trial involving 109 pre-menopausal women from 18–45 years of age with asthma described a dose-related decline in bone density at both the total hip and the trochanter of 0.0044 g/m^2/PAF (100 µg) triamcinolone acetonide per year of treatment. The authors conclude that inhaled glucocorticoids do indeed lead to a dose-related loss of bone at the hip in pre-menopausal women (Israel et al. 2001).

6.4 Growth in Asthmatic Children

Like a variety of other factors, attenuation of growth is highly variable between individual patients. It appears to be related to the severity of the disease, being most pronounced in children with chronic, poorly controlled asthma. Records from 18-year-old Swedish military conscripts from 1883–1996 showed that subjects with asthma were significantly shorter than healthy subjects and that asthma severity was correlated with growth suppression. In order to evaluate the effect of treatment, it is important to assess height over a period of time (more than 12 months) as growth is inherently an erratic process.

6.5 The Effects of Inhaled Corticosteroids on Growth

6.5.1 Short-Term Studies

A number of well-designed controlled clinical trials have examined the short-term effect of inhaled corticosteroids on growth using knemometry. Since, by using this method, lower leg length can be measured with an accuracy of 0.1 mm, it is ideally suited to monitor growth over a period of days to weeks. Although such short-term growth rates do not predict longer term growth in any meaningful way the knemometer is considered to be the most valuable tool in determining short-term systemic side effects. There is a dose-dependent reduction of lower leg growth during short-term treatment and inhaled corticosteroids.

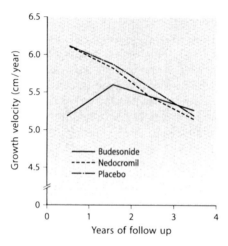

Fig. 3. Growth velocity in asthmatic children treated for several years with either inhaled glucocorticoids (budesonide) or non-steroidal anti-inflammatory drugs (nedocromil) (Agertoft and Pedersen, 2000)

Several controlled clinical trials studying the effect on growth over several months have been published to date (Agertoft and Pedersen 1994, 1998; Scott et al. 1999). Despite differences in study design, drugs used in control group, and inhaler device used, the results of all these studies were remarkably consistent. There is a small, but statistically significant reduction in growth rate during 1-year maintenance treatment making up 1.1 cm per annum (Fig. 3). Interestingly this small reduction in growth rate during the first year of treatment does not seem to persist during a follow-up of 4–6 years (Pederson et al. 2002). It appears therefore that the effect of ICSs on growth rate is temporary.

6.5.2 Long-Term Studies

There is a growing belief that the effect of inhaled corticosteroids on growth rate does not affect final height attainment (Fig. 4). However, the number of individuals being studied is still small. Most studies are retrospective. In addition, the use of a control group of asthmatics not using corticosteroids may not be valid, because it is likely that the

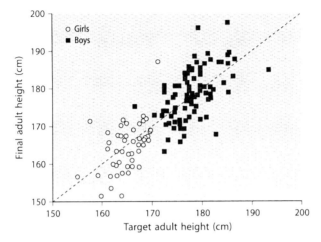

Fig. 4. Final adult height of asthmatic girls and boys after several years of treatment with inhaled glucocorticoids (Agertoft and Pedersen, 2000)

children in the treatment group had more severe asthma than the children in the control group, which may have effected their growth and final height.

6.6 Conclusion

Topical glucocorticoid treatment may suppress short- and medium-term growth in children with asthma; however, the effect is probably temporary. In any case, higher doses should be used with caution in any patient, as they only improve efficacy slightly, but may increase the risk of side effects (including reduced growth) quite substantially.

References

Agertoft L, Pedersen S (1994) Effects of long-term treatment with an inhaled corticosteroid on growth and pulmonary function in asthmatic children. Respir Med 88:373–381

Agertoft L, Pedersen S (1998) Bone mineral density in children with asthma receiving long-term treatment with inhaled budesonide. Am J Respir Crit Care Med 157:178–183

Agertoft L, Pedersen S (2000) Effect of long-term treatment with inhaled budesonide on adult height in children with asthma. N Engl J Med 343:1064–1069

Hoffmann-Streb A, L'Allemand D, Niggemann B, Büttner P, Wahn U (1993) Nebennierenrindenfunktion bei Kindern mit Asthma bronchiale unter Therapie mit Fluticason. Monatsschr Kinderheilkd 141:508–512

International Consensus Report on Diagnosis and Treatment of Asthma (1992) Eur Respir J 5:601–641

Israel E, Banerjee TR, Fitzmaurice GM, Kotlov TV, LaHive K, LeBoff MS (2001) Effects of inhaled glucocorticoids on bone density in premenopausal women. N Engl J Med 345:941–947

National Heart, Lung and Blood Institute (1997) National Asthma Education Program expert panel report. Guidelines for the diagnosis and management of asthma. Expert panel report No. 2 Bethesda, MD. NIH Publication No. 97-1051

National Institutes of Health (1995) National Heart, Lung and Blood Institute. Global Initiative for Asthma. NIH-Publication No. 95-3659

Pedersen S, Warner J, Wahn U, Staab D, Le Bourgeois M, van Essen-Zandvliet E, Arora S, Szefler SJ (2002) Growth systemic safety and efficacy during one year of asthma treatment with different beclomethasone dipropionate formulations: an open-label, randomized comparison of extrafine and conventional aerosols in children. Pediatrics (in press)

Scott MB, Skoner DP (1999) Short-term and long-term safety of budesonide inhalation suspension in infants and young children with persistent asthma. J Allergy Clin Immunol 104:200–209

7 Functional Implications of Glucocorticoid Receptor Trafficking

D.B. DeFranco

7.1	Introduction	91
7.2	Nucleocytoplasmic Shuttling of GR	92
7.3	Subnuclear Trafficking of Steroid Receptors	97
7.4	Regulation of Nuclear Receptor Degradation	100
References		103

7.1 Introduction

Glucocorticoid hormones affect a variety of cellular processes through their interactions with a soluble intracellular receptor protein, the glucocorticoid receptor (GR). The GR is a member of the nuclear receptor superfamily of ligand-activated transcription factors and regulates the transcriptional activity of a diverse subset of genes via its direct interaction with promoter-linked target sequences or interaction with unique transcription factors (McEwan et al. 1997). In the absence of ligand, GR is located primarily in the cytoplasm in close association with various chaperone proteins, including heat shock protein 90 (hsp90). Upon hormone binding, the hormone-receptor complex becomes activated, forms a homodimer, and moves into the nucleus. For transactivation, the hormone-receptor complex binds to glucocorticoid response elements (GRE) in the promoter region of target genes.

Two distinct mechanisms have been characterized for transrepression. GR can bind to promoter-linked negative GREs (nGREs) and adopt a distinct conformation that exposes a "repressive" domain whose interactions with unique components of the transcriptional machinery reduces the overall efficiency of transcription (Starr et al. 1996). Alternatively, GR can bring about transcriptional repression of specific genes without the necessity for DNA binding or homodimerization. In this case, direct interactions between the GR and other transcription factors or transcriptional coactivators may either sequester these gene regulatory proteins from their natural targets or limit their activity, even if they are promoter-bound (Glass and Rosenfeld 2000). The different phenotypes of transgenic mice with a targeted disruption of the GR gene compared to those with a point mutation in the DNA-binding domain of GR, which prevents formation of homodimers, emphasizes the importance of GR interactions with other transcription factors for physiologically relevant, GC-dependent transcriptional repression (Cole et al. 1995; Reichardt et al. 1998).

7.2 Nucleocytoplasmic Shuttling of GR

7.2.1 General Aspects of GR Trafficking

Steroid receptor proteins are not confined to a single subcellular compartment but constantly shuttle between the nucleus and cytoplasm (DeFranco et al. 1995). The rates of this bidirectional trafficking are not uniform and influenced by ligand occupancy (DeFranco et al. 1995). For example, the nuclear import of unliganded GR is limited predominantly by its association with chaperone proteins such as hsp90, a 23-kDa acidic protein, and an immunophilin (Pratt 1993; Smith and Toft 1993). When ligand bound, GR interactions with DNA, chromatin, and perhaps other subnuclear components limits its nuclear export, leading to its predominant localization within nuclei (DeFranco 1999). Once nuclear, GRs bind to high-affinity target sites on chromatin, resulting in altered transcriptional activity of target gene promoters (Yamamoto 1985). GR interactions with chromatin are not static (Mymryk and Archer 1995; McNally et al. 2000) as receptors are likely to disengage from high-affinity chromatin binding sites, even in the continuous presence of hormone. The reutilization of nuclear receptors can occur via a process that appears to require hsp90 (Liu and DeFranco 1999; and see below). Thus, appropriate trafficking of the receptor, even under conditions of chronic hormone exposure, may play a role in the maintenance of hormone-regulated transcription.

7.2.2 Role of Hsp90 in Cytoplasmic to Nuclear Transport of Steroid Receptors

Unliganded, cytoplasmic GR that is competent to bind hormone exists as a heteromeric complex that contains a dimer of hsp90, an immunophilin protein of the FK506-binding family (i.e., FKBP-52 or FKBP-54), and p23 (Bohen and Yamamoto 1994). A number of molecular chaperones participate in the highly ordered maturation process that is required for de novo translated steroid receptors to attain hormone-binding competence (Pratt and Toft 1997). Thus, other immunophilins (e.g., Cyp40) or heat shock proteins (e.g., hsp70) that are found associated with unliganded steroid receptors are likely to be involved as transient intermediates in the maturation of the receptor to its hormone-binding conformation (Pratt and Toft 1997). While the constitutive nuclear localization of ligand-binding domain (LBD)-deleted GRs suggested that nuclear import of these receptors are restricted by their association within heteromeric complexes (Picard and Yamamoto 1987; Pratt 1993), this view is now recognized as being overly simplistic. For example, some unliganded steroid receptors that appear to be localized predominantly within the nucleus are also assembled into heteromeric complexes (Smith and Toft 1994). This includes unliganded GR, which in some cells appears to accumulate within the nucleus (Sanchez et al. 1990).

How do we reconcile these results with the presumed role for steroid receptor heteromeric complexes in limiting nuclear import? As first shown for PR in vitro (Smith 1993), the assembly of steroid receptor heteromeric complexes is a dynamic process. Receptor association with chaperones such as hsp90 is transient even in the absence of hormone binding. As a result, the amino acid signals encoded within steroid receptors that are required for their nuclear import (Picard and Yamamoto 1987; Cadepond et al. 1992; Ylikomi at al. 1992; Tang et al. 1997) may be transiently exposed to appropriate nuclear transport proteins even in the absence of bound ligand. Thereafter, a productive interaction might ensue that would commit steroid receptor-nuclear transport protein complexes to associate with the nuclear pore complex proteins. It follows that the stability of steroid receptor-heteromeric complexes, which probably varies for individual receptors, and perhaps within different cell types, could have a direct impact on the cytoplasmic retention of unliganded receptors.

7.2.3 Role of Hsp90 in Nuclear Recycling of GR

In addition to their impact on hormone binding of unliganded cytoplasmic receptors, specific chaperones may target the receptors even when they are engaged with the transcriptional machinery. Overexpression of p23 exerts opposite effects, either stimulating or reducing transactivation, depending on the SR tested. For example, overexpression of p23 leads to stimulation of GR transactivation in both yeast and mammalian cells (Freeman et al. 2000). While such a stimulatory effect of p23 had previously been observed for estrogen receptor (ER) transactivation (Knoblauch and Garbedian 1999), p23 effects on GR transactivation are not apparent until a relatively long time after hormone addition (Freeman et al. 2000). Furthermore, p23 overexpression resulted in reduced transactivation activity of thyroid hormone receptor (T_3R) in yeast and mammalian cells, possibly because p23 facilitated the dissociation of the receptor from a specific DNA-binding site (Freeman et al. 2000).

What mechanism could account for the delayed effects of p23 on receptor transactivation in vivo? Since p23 is apparently capable of interacting with DNA-bound receptors, it may exert its effects on receptor function following their association with target sites within chromatin. For example, p23 could participate in nuclear receptor recycling and ensure the efficient rebinding of hormone to receptors that have released their bound hormone (Freeman et al. 2000). We and others have previously provided evidence for the existence of a nuclear recycling pathway for GR, which includes a hormone re-binding event and that does not require cytoplasmic transit of the recycled receptors (Orti et al. 1989; Liu and DeFranco 1999). If the acquisition and maintenance of hormone binding competence of nuclear GR were analogous to the process that occurs in the cytoplasm (Pratt and Toft 1997; Bohen and

Yamamoto 1997), a role for p23 and hsp90 in nuclear recycling would be essential.

In addition to aiding in the regeneration of hormone-bound, recycled nuclear receptors, p23 may play an active role in facilitating receptor release from DNA following engagement with the transcription machinery. As will be discussed below, steroid receptor interactions with target sites in native chromatin are dynamic and involve rapid binding and release of receptors (McNally et al. 2000). Such dynamic interactions with chromatin-embedded targets may be essential for receptors to confer a maximal transcriptional response. Alternatively, receptors may have reduced potency as transcriptional activators following multiple rounds of transcription due to their accrual of post-translational modifications or alteration in conformation. Thus, recycling of these "experienced" receptors might be required to reverse putative inhibitory modifications, or permit the nuclear chaperone machinery to restore the optimum conformation of the receptor for interactions with the transcription machinery. This hypothesis is actually consistent with the observed effects of p23 on ER transactivation, which were much more pronounced in cells that expressed low levels of receptor, or at low hormone doses. It seems likely that under conditions of limiting active ER, nuclear recycling would have more of an impact on the overall transactivation activity of the receptor than under conditions of receptor or hormone excess.

Additional results from our laboratory support the notion that molecular chaperones can act to regulate GR function in the nucleus. Treating cells with an inhibitor of hsp90 function [i.e., the benzoquinone ansamycin, geldanamycin (GA); Stebbins et al., 1997] disrupted GR nuclear recycling (Liu and DeFranco 1999). While the inability of recycled nuclear GR to rebind hormone in the presence of GA might have been expected given the role for hsp90 in promoting hormone binding of GR (Pratt and Toft 1997), GR interactions with chromatin in GA-treated cells were also affected (Liu and DeFranco 1999). In particular, although GA did not affect the release of bound hormone from GR upon hormone withdrawal, the release of unliganded receptors from chromatin was dramatically inhibited (Liu and DeFranco 1999).

Based upon these results, we previously hypothesized that hsp90, either alone or in combination with another partner (p23?) may be required to release unliganded receptors from high-affinity chromatin-binding sites (Liu and DeFranco 1999). Hsp90 avidly binds histones and modifies chromatin structure in vitro (Csermely et al. 1998). This raises the possibility that the interaction of chaperones with both steroid receptors and chromatin might be required to facilitate receptor recycling within nuclei. GA disrupts hsp90 interactions with p23 (Stebbins et al. 1997), suggesting that the release of unliganded GR from chromatin may be mediated primarily by p23, with hsp90 aiding in the recruitment of its partner to the experienced, chromatin-bound receptor. A direct role for hsp90 in GR release from chromatin cannot be excluded since purified hsp90 alone was able to dissociate GR from receptor/DNA complexes preformed in vitro (Kang et al. 1999).

Hsp90 and p23 are also known to function in the assembly of other nuclear macromolecular assemblies. For example, the formation of an active human telomerase enzyme containing a telomerase catalytic subunit (hTERT) and template mRNA (hTR) requires both hsp90 and p23 (Holt et al. 1999). In this case, it appears that hsp90 functions mainly to recruit p23 to hTERT complexes (Holt et al. 1999). Furthermore, since hsp90 and p23 remain associated with active telomerase complexes in vivo, they may function at steps subsequent to the assembly of active telomerase (Holt et al. 1999). Perhaps, analogous to their role in steroid receptor recycling, hsp90 and p23 may recognize "experienced" telomerase and facilitate its continued activity after repeated rounds of nucleotide addition.

Hsp90 and p23, along with other chaperones, regulate the activity of the heat shock transcription factor HSF-1 (Bharadwaj et al. 1999). Analogous to their effects on SRs, hsp90 and p23 may regulate HSF-1 activity at multiple levels. For example, in addition to their role in maintaining HSF-1 in an inactive state, hsp90 and p23 appear to participate in the attenuation of HSF-1 activity following stress (Bharadwaj et al. 1999). Since hsp90 and p23 are present in active HSF-1 trimers, they must somehow recognize some novel feature of "experienced" HSF-1 trimers and bring about their disassembly. The striking similarity noticed with hsp90 and p23 effects on SRs, telomerase, and HSF-1 is the recognition by the chaperones that their target has expressed its activity and requires their aid to either terminate further activity or regain full function. In all of these cases, it is unclear what exactly is meant by the chaperone partners being "experienced," but it is possible that subtle changes in conformation may be all that is required for hsp90 and p23 to bind.

What property of steroid receptors could account for the putative hsp90 requirements for chromatin release? Upon hormone binding, steroid receptor LBDs undergo a conformational change (Beekman et al. 1993) that is characterized by the movement of an exposed alpha helix (i.e., helix 12) towards the hormone binding pocket (Brzozowski et al. 1997; Tanenbaum et al. 1998). As a result, hydrophobic segments of the LBD, as well as the bound hormone itself, are no longer solvent-exposed. Furthermore, previously inaccessible LBD surfaces become exposed and available for interactions with appropriate co-activators and other components of the transcriptional machinery. It is unclear how the elaborately networked LBD structure responds to the release of bound hormone. Does the LBD simply "relax" to require the conformation it possessed when initially unliganded? In such a scenario, the exposure of hydrophobic segments of the unliganded receptor's hormone-binding pocket might increase the propensity for receptor aggregation unless molecular chaperones such as hsp90 are present to prevent such inappropriate interactions.

7.3 Subnuclear Trafficking of Steroid Receptors

7.3.1 Recognition of Specific Compartmentalization of Nuclear Receptors

Even though steroid hormone receptors can exert their effects on gene transcription quite rapidly (Yamamoto 1985), the mechanisms responsible for rapid location of target sites by the receptors within the crowded nuclear environment have proven difficult to discern. However, in recent years, a number of exciting cell biological, genetic, and biochemical experiments have yielded new insights regarding the mechanism of steroid receptor trafficking within the nucleus. Green fluorescence protein (GFP) chimeras have been invaluable tools for cell biologists and have allowed protein trafficking and protein–protein interactions to be visualized in live cells in real time (Lippincott-Schwartz et al. 2001). For some steroid receptor proteins, GFP chimeras provided definitive proof of their hormone-dependent cytoplasmic-nuclear transport (Htun et al. 1996, 1999; Fejes-Toth et al. 1998). Importantly, specific compartmentalization of steroid receptors within the nucleus could be discerned in live cells, confirming results obtained in earlier work with fixed cells, which implied that localization of bulk receptors within the nucleus was not random (Martins et al. 1991; Yang and DeFranco 1994).

The nature and importance of steroid receptor foci visible within the nucleus at the light microscope level has been controversial and subject to considerable debate (van Steensel et al. 1995). However, it does seem likely that, in addition to their concentration within regions of the nucleus associated with active transcription, receptors are able to concentrate into visible foci are transcriptionally inert and likely to represent storage sites, which transiently engage receptors that are destined for various alternative processing fates (DeFranco 1999).

7.3.2 Analysis of Steroid Receptor Movement Within Nuclei of Live Cells

Recently, both the Hager (McNally et al. 2000) and Mancini (Stenoien et al. 2001) laboratories have utilized steroid receptor GFP derivatives to reveal the dynamic nature of steroid receptor movement within the nucleus. Both groups used fluorescence recovery after photobleaching (FRAP), and other techniques, to provide real-time assessments of steroid receptor movement within the nucleus. The Hager laboratory took advantage of a cell line that contains a large array of integrated copies of a glucocorticoid responsive promoter [i.e., the mouse mammary tumor virus long terminal repeat (MMTV LTR)]. Thus, they could visualize in real time and in live cells the movement of a large amount of GR within a specific site where receptors were actively engaged in transcriptional regulation. These elegant studies confirmed previous work from traditional biochemical experiments that implied that GR interactions with specific target sites within chromatin templates are dynamic (Mymryk and Archer 1995).

The model put forth by Hager and co-workers (McNally et al. 2000) proposes that GR occupies its target sites only transiently relying on a "hit-and-run" mechanism to alter transcription. While bound, GR is likely to recruit essential coactivators and other cofactors to the target gene, but continued occupancy by the initial recruiter receptor may not be required for subsequent assembly of the active preinitiation complex. One expects that this elegant model system will be used in future

Fig. 1. Rapid intranuclear movement of GR-GFP in transfected Cos-1 cells. Cos-1 cells expressing transiently transfected GR-GFP were subjected to fluorescence recovery after photobleaching (FRAP) to assess intranuclear movement of the receptor. The movement of GR-GFP is so rapid that during the photobleach period (i.e., 15 s) of a small zone of the nucleus (*left panel*), the entire population of nuclear receptors is bleached (*right panel*)

experiments to directly visualize the kinetics of coactivator recruitment to active sites of transcription.

Remarkably, the Mancini team found that the kinetics of bulk ER movement within the nucleus was analogous to the rapid kinetics of GR exchange at a specific target site (Stenoien et al. 2001). As shown in Fig. 1, we have also found that ligand-bound, bulk GR is subjected to rapid internuclear movements (Fig. 1). This suggests that even those receptors that are not intimately involved in the transcriptional regulatory events at specific sites are nonetheless undergoing rapid movement. In fact unliganded ER was found to exhibit the most rapid subnuclear movement, suggesting that there may be some retardation of receptor trafficking within the nucleus as activated receptors are scanning the genome for specific target sites (Stenoien et al. 2001). ER bound to mixed antagonists are as mobile as agonist-bound ER while the movement of receptors bound to pure antagonists is even more retarded, implying that these receptors are limited by nonproductive interactions with some nuclear components (Stenoien et al. 2001). In this regard, it would be most informative when the dynamics of nuclear movement is followed for unliganded non-steroid nuclear receptors that are associated with corepressors.

7.4 Regulation of Nuclear Receptor Degradation

7.4.1 Homologous Downregulation of Nuclear Receptors

In most cell lines and tissues, long-term exposure to glucocorticoid hormones leads to a gradual reduction of GR levels (Oakley and Cidlowski 1993). This hormone-dependent, homologous downregulation of the receptor reflects both affects of glucocorticoids on GR gene transcription (Rosewicz et al. 1988; Burnstein et al. 1994) and protein turnover (McIntyre and Samuels 1985; Dong et al. 1988). Since glucocorticoid responsiveness in cell culture (Vanderbilt et al. 1987; Bellingham et al. 1992) and in vivo (Reichardt et al. 2000) is related to the relative abundance of GR, glucocorticoid regulated transcription is attenuated under conditions of homologous downregulation of the receptor. This form of receptor desensitization is not limited to GR and is characteristic of other steroid receptors as well (Saceda et al. 1988; Wei et al. 1988). Recent studies have established that GR are degraded under conditions of homologous downregulation by the ubiquitin–proteasome pathways (Wallace and Cidlowski 2001).

7.4.2 Degradation of Nuclear Receptors by the Ubiquitin–Proteasome Pathway

Proteasomes are a multisubunit complex that serve as one of the major protein degradative pathways within eukaryotic cells (Ciechanover et al. 2000). The targeting of proteins to the proteasome requires their covalent modification with multiple residues of the 76 amino acid ubiquitin protein (Ciechanover et al. 2000). While passage through the inner core of the proteasome leads to the degradation of target proteins into small peptides, intact ubiquitin moieties are liberated following target protein proteolysis and released for their subsequent reutilization. In addition to serving as the major degradative machinery to eliminate damaged and denatured proteins, proteasomes operate to degrade proteins with both short and long half-lives (Ciechanover et al. 2000). GR (Whitesell and Cook 1996; Connell et al. 2001; Wallace and Cidlowski 2001), like other nuclear receptors (Nawaz et al. 1999; Dace et al. 2000; Wijayaratne and McDonnell 2001), is degraded via the ubiquitin–proteasome pathway. Under conditions of homologous downregulation, the efficiency of GR degradation by the proteasome is enhanced (Wallace and Cidlowski 2001).

7.4.3 Role of Proteasomes in Nuclear Receptor Transactivation

Hormone-dependent downregulation of steroid hormone receptors, while limiting the duration of hormone responsiveness, also affects the efficiency of receptor transactivation. Thus, ER, progesterone receptor (PR), and thyroid hormone receptor (T_3R) receptor transactivation in transiently transfected cells is reduced when proteasome-mediated degradation of the receptors is inhibited (Lonard et al. 2000). In fact, model studies with chimeric transcriptional activators of differing potencies established a link between transcriptional activation and proteasome-mediated degradation. Specifically, the rate of activator degradation was found to directly correlate with transactivation potency (Molinari et al. 1999; Salghetti et al. 2000). However, GR appears to respond differently to MG132 than other nuclear receptors with enhanced transactivation accompanying proteasome inhibition (Wallace and Cidlowski 2001). Furthermore, nuclear receptor transactivation can be differentially responsive to receptor degradation. For example, an uncoupling of transactivation and degradation has recently been observed with specific mutants of the retinoid X receptor (Osburn et al. 2001). A PR mutant that does not undergo hormone-dependent degradation maintains some degree of hormone response in transfected HeLa cells, even though its ability to respond to the mitogen-activated protein kinase pathway is completely abrogated (Shen et al. 2001). Thus, the link between proteasome-mediated degradation and transactivation may be gene- and receptor-specific, and responsive to a unique subset of signal transduction pathways that affect nuclear receptor activity.

7.4.4 Nuclear Receptor Degradation and Nuclear Export

The efficiency of proteasome-mediated degradation of nucleocytoplasmic shuttling proteins has been linked in some cases with their rate of nuclear export (Freedman and Levine 1998; Rodriguez et al. 1999). For example, proteasome-mediated degradation of the cyclin-dependent kinase inhibitor protein p27[Kip1] is stimulated when its nuclear export is enhanced via its interaction with the Jab1 coactivator protein (Tomoda et al. 1999). The HDM2 RING-finger protein serves an analogous role to enhance the nuclear export and proteasome degradation of the p53 tumor-suppressor protein (Boyd et al. 2000; Geyer et al. 2000). When the rate of GR nuclear export is stimulated through linkage of a potent nuclear export signal sequence (NES) to its amino terminus, hormone-dependent downregulation of the chimeric NES-GR is enhanced (Liu and DeFranco 2000). This result implies that degradation of nuclear receptors may likewise be linked to their nuclear export.

7.4.5 Regulation of GR Degradation in Neurons

While most cells that have been examined exhibit homologous downregulation of GR in vitro or in vivo, the impact of chronic hormone treatment on GR expression and accumulation in neurons is not fully resolved (Sapolsky and McEwen 1985; Vedder et al. 1993; Herman and Spencer 1998). Furthermore, fetal exposure to glucocorticoids has been reported to have no effect on GR levels in whole embryos or embryonic liver (Ghosh et al. 2000). We have examined GR downregulation in a hippocampal cell line (i.e., HT22 cells) and primary hippocampal neurons for their response to chronic glucocorticoid treatment. Chronic glucocorticoid exposure did not lead to GR downregulation in primary

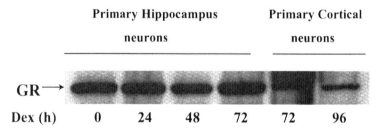

Fig. 2. GR is not downregulated in embryonic rat primary hippocampal or cortical neurons by dexamethasone treatment. Embryonic rat (E17) primary hippocampal or cortical neurons grown in serum-free conditions were treated with dexamethasone for the periods of time indicated. Western blot analysis shows GR levels in equivalent amounts of total cellular protein/lane

embryonic rat hippocampal and cortical neurons (Fig. 2) and in HT22 cells (not shown). Our results suggest that hormone-dependent downregulation of GR protein is not universal and may develop postnatally. Thus, GR processing pathways that contribute to hormone-dependent downregulation may not be fully operational in late term embryos.

The insensitivity of GR to downregulation in developing neurons may reflect an inability of the receptor to effectively engage the proteasome degradation machinery. The lack of GR downregulation in hippocampal neurons eliminates one widely used mechanism to limit GR action under conditions of chronic hormone exposure. Minimal (i.e., 20%–60%) forced overexpression of GR in transgenic mice leads to a number of effects on the HPA axis and stress responses (Reichardt et al. 2000) illustrating the physiological importance of maintaining homeostasis of GR expression.

Antenatal glucocorticoid treatment is widely used clinically in attempts to decrease mortality of premature infants (Banks et al. 1999). Despite the benefits to pulmonary function of premature infants from antenatal glucocorticoid treatment, some studies report detrimental effects on cognitive function in children or juvenile animals that had been exposed to chronic dexamethasone in utero (Barrington 2001). If GR downregulation is not fully developed, chronic activation of receptor may alter the transcription rate of genes whose precisely coordinated expression may be critical for the appropriate neuronal development. Using HT22 cells and primary embryonic hippocampal neurons as models, we may be able to delve more deeply into the functioning of GR in this critical phase of brain development and understand the mechanisms responsible for detrimental effects of fetal glucocorticoid exposure.

References

Banks BA, Cnaan A, Morgan MA, Parer JT, Merrill JD, Ballard PL, Ballard RA (1999) Multiple courses of antenatal corticosteroids and outcome of premature neonates. Am J Obstet Gynecol 181:709–717

Barrington KJ (2001) The adverse neuro-developmental effects of postnatal steroids in the preterm infant: a systematic review of RCTs. BMC Pediatrics 1:1

Beekman JM, Allan GF, Tsai SY, Tsai M-J, O'Malley BW (1993) Transcriptional activation by the estrogen receptor requires a conformational change in the ligand-binding domain. Mol Endocrinol 7:1266–1274

Bellingham DL, Sar M, Cidlowski JA (1992) Ligand-dependent down-regulation of stably transfected human glucocorticoid receptors is associated with the loss of functional glucocorticoid responsiveness. Mol Endocrinol 6:2090–2102

Bharadwaj S, Ali A, Ovsenek N (1999) Multiple components of the HSP90 chaperone complex function in regulation of heat shock factor 1 in vivo. Mol Cell Biol 19:8033–8041

Bohen SP, Yamamoto KR (1994) Modulation of steroid receptor signal transduction by heat shock proteins: In: Morimoto RI, Tissieres A, Georgopoulos C (eds) The biology of heat shock proteins and molecular chaperones. Cold Spring Harbor Laboratory Press, Cold Spring Harbor, pp 313–334

Boyd SD, Tsai KY, Jacks T (2000) An intact HDM 2 RING-finger domain is required for nuclear exclusion of p53. Nature Cell Biol 2:563–568

Brzozowski AM, Pike ACW, Z. D, Hubbard RE, Bonn T, Engstron O, Ohman L, Greene GL, Gustafsson J-A, Carlquist M (1997) Molecular basis of agonism and antagonism in the oestrogen receptor. Nature 389:753–758

Burnstein KL, Jewell CM, Sar M, Cidlowski JA (1994) Intragenic sequences of the human glucocorticoid receptor complementary DNA mediate hormone-inducible receptor messenger RNA down-regulation through multiple mechanisms. Mol Endocrinol 8:1764–1773

Cadepond F, Gasc JM, Delahaye F, Jibard N, Schweizer GG, Segard MI, Evans R, Baulieu EE (1992) Hormonal regulation of the nuclear localization signals of the human glucocorticosteroid receptor. Exp Cell Res 201:99–108

Ciechanover A, Orian A, Schwartz AL (2000). Ubiquitin-mediated proteolysis: biological regulation via destruction. BioEssays 22:442–451

Cole TJ, Blendy JA, Monaghan AP, Krieglstein K, Schmid W, Aguzzi A, Fantuzzi G, Hummler E, Unsicker K, Schutz G (1995) Targeted disruption of the glucocorticoid receptor gene blocks adrenergic chromaffin development and severely retards lung maturation. Genes Dev 9:1608–1625

Connell P, Ballinger CA, Jiang J, Wu Y, Thompson LJ, Hohfeld J, Patterson C (2001) The co-chaperone CHIP regulates protein triage decisions mediated by heat-shock proteins. Nat Cell Biol 3:93–96

Csermely P, Schnaider T, Soti C, Prohaszka Z, Nardai G (1998) The 90-kDa molecular chaperone family: structure, function and clinical applications. A comprehensive review. Pharmacol Ther 79:129–168

Dace A, Zhao L, Park KS, Furuno T, Takamura N, Nakanishi M, West BL, Hanover JA, Cheng S (2000) Hormone binding induces rapid proteasome-mediated degradation of thyroid hormone receptors. Proc Natl Acad Sci USA 97:8985–8990

DeFranco DB (1999) Regulation of steroid receptor subcellular trafficking. Cell Biochem Biophys 30:1–24

DeFranco DB, Madan, AP, Tang Y, Chandran UR, Xiao N, Yang J (1995) Nucleocytoplasmic shuttling of steroid receptors. In: Litwack G (ed) Vitamins and hormones, Vol. 51, Academic Press, New York, pp 315–338

Dong Y, Poellinger L, Gustafsson JA, Okret S (1988) Regulation of glucocorticoid receptor expression: evidence for transcriptional and posttranslational mechanisms. Mol Endocrinol 2:1256–1264

Fejes-Toth G, Pearce D, Naray-Fejes-Toth A (1998) Subcellular localization of mineralocorticoid receptors in living cells: effects of receptor agonists and antagonists. Proc Natl Acad Sci USA 95:2973–2978

Freedman DA, Levine AJ (1998) Nuclear export is required for degradation of endogenous p53 by MDM2 and human papillomavirus E6. Mol Cell Biol 18:7288–7293

Freeman BC, Felts SJ, Toft DO, Yamamoto KR (2000) The p23 molecular chaperones act at a late step in intracellular receptor action to differentially affect ligand efficacies. Genes Dev 14:422–434

Geyer RK, Yu ZK, Maki CG (2000) The MDM2 RING-finger domain is required to promote p53 nuclear export. Nature Cell Biol 2:569–573

Ghosh B, Wood CR, Held GA, Abbott BD, Lau C (2000) Glucocorticoid receptor regulation in the rat embryo: a potential site for developmental toxicity? Toxicol Appl Pharmacol 164:221–229

Glass CK, Rosenfeld MG (2000) The coregulator exchange in transcriptional functions of nuclear receptors. Genes Dev 14:121–141

Herman JP, Spencer R (1998) Regulation of hippocampal glucocorticoid receptor gene transcription and protein expression in vivo. J Neurosci 18:7462–7473

Holt SE, Aisner DL, Baur J, Tesmer VM, Dy M, Ouellette M, Trager JB, Morin GB, Toft DO, Shay JW, Wright WE, White MA (1999) Functional requirements of p23 and Hsp90 in telomerase complexes. Genes Dev 13:817–826

Htun H, Barsony J, Renyi I, Gould DL, Hager GL (1996) Visualization of glucocorticoid receptor translocation and intranuclear organization in living cells with a green fluorescent protein chimera. Proc Natl Acad Sci USA 93:4845–4850

Htun H, Holth LT, Walker D, Davie JR, Hager GL (1999) Direct visualization of the human estrogen receptor alpha reveals a role for ligand in the nuclear distribution of the receptor. Mol Biol Cell 10:471–486

Kang KI, Meng X, Devin-LeClerc J, Bouhouche I, Chadli A, Cadepond F, Baulieu E-E, Catelli M-G (1999) The molecular chaperone Hsp90 can negatively regulate the activity of a glucocorticosteroid-dependent promoter. Proc Natl Acad Sci USA 96:1439–1444

Knoblauch R, Garabedian MJ (1999) Role for Hsp90-associated cochaperone p23 in estrogen receptor signal transduction. Mol Cell Biol 19:3748–3759

Lippincott-Schwartz J, Snapp E, Kenworthy A (2001) Studying protein dynamics in living cells. Nat Rev Mol Cell Biol 2:444–456

Liu J, DeFranco DB (1999) Chromatin recycling of glucocorticoid receptors: implications for multiple roles of heat shock protein 90. Mol Endocrinol 13:355–365

Liu J, DeFranco DB (2000) Protracted nuclear export of glucocorticoid receptor limits its turnover and does not require the exportin 1/CRM1-directed nuclear export pathway. Mol Endocrinol 14:40–51

Lonard DM, Nawaz Z, Smith CL, O'Malley BW (2000) The 26S proteasome is required for estrogen receptor-alpha and coactivator turnover and for efficient estrogen receptor-alpha transactivation. Mol Cell 5:939–948

Martins VR, Pratt WB, Terracio L, Hirst MA, Ringold GM, Housley PR (1991) Demonstration by confocal microscopy that unliganded overexpressed glucocorticoid receptors are distributed in a nonrandom manner throughout all planes of the nucleus. Mol Endocrinol 5:217–225

McEwan IJ, Wright AP, Gustafsson JA (1997) Mechanism of gene expression by the glucocorticoid receptor: role of protein-protein interactions. Bioessays 19:153–160

McIntyre WR, Samuels HH (1985) Triamcinolone acetonide regulates glucocorticoid-receptor levels by decreasing the half-life of the activated nuclear-receptor form. J Biol Chem 260:418–427

McNally JG, Müller WG, Walker D, Wolford R, Hager GL (2000) The glucocorticoid receptor: rapid exchange with regulatory sites in living cells. Science 287:1262–1265

Molinari E, Gilman M, Natesan S (1999) Proteasome-mediated degradation of transcriptional activators correlates with activation domain potency in vivo. EMBO J 18:6439–6447

Mymryk JS, Archer TK (1995) Influence of hormone antagonists on chromatin remodeling and transcription factor binding to the mouse mammary tumor virus promoter in vivo. Mol Endocrinol 9:1825–1834

Nawaz Z, Lonard DM, Dennis AP, Smith CL, O'Malley BW (1999) Proteasome-dependent degradation of the human estrogen receptor. Proc Natl Acad Sci USA 96:1858–1862

Oakley RH, Cidlowski JA (1993) Homologous down regulation of the glucocorticoid receptor: the molecular machinery. Crit Rev Eukaryot Gene Expr 3:63–88

Orti E, Mendel DB, Smith LI, Bodwell JE, Munck A (1989) A dynamic model of glucocorticoid receptor phosphorylation and recycling in intact cells. J Steroid Biochem 34:85–96

Osburn DL, Shao G, Seidel HM, Schulman IG (2001) Ligand-dependent degradation of retinoid X receptors does not require transcriptional activity or coactivator interactions. Mol Cell Biol 21:4909–4918

Picard D, Yamamoto KR (1987) Two signals mediate hormone-dependent nuclear localization of the glucocorticoid receptor. EMBO J 6:3333–3340

Pratt WB (1993) The role of heat shock proteins in regulating the function, folding, and trafficking of the glucocorticoid receptor. J Biol Chem 268:21455–21458

Pratt WB, Toft DO (1997) Steroid receptor interactions with heat shock protein and immunophilin chaperones. Endo Rev 18:306–360

Reichardt HM, Kaestner KH, Tuckermann J, Kretz O, Wessely O, Bock R, Gass P, Schmid W, Herrlich P, Angel P, Schutz G (1998) DNA binding of the glucocorticoid receptor is not essential for survival. Cell 93:531–541

Reichardt HM, Umland T, Bauer A, Kretz O, Schutz G (2000) Mice with an increased glucocorticoid receptor gene dosage show enhanced resistance to stress and endotoxic shock. Mol Cell Biol 20:9009–9017

Rodriguez MS, Thompson J, Hay RT, Dargemont C (1999) Nuclear retention of IkBa protects it from signal-induced degradation and inhibits nuclear factor kB transcriptional activation. J Biol Chem 274:9108–9115

Rosewicz S, McDonald AR, Maddux BA, Goldfine ID, Miesfeld RL, Logsdon CD (1988) Mechanism of glucocorticoid receptor down-regulation by glucocorticoids. J Biol Chem 263:2581–2854

Saceda M, Lippman ME, Chambon P, Lindsey RL, Ponglikitmongkol M, Puente M, Martin MB (1988) Regulation of the estrogen receptor in MCF-7 cells by estradiol. Mol Endocrinol 2:1157–1162

Salghetti SE, Muratani M, Wijnen H, Futcher B, Tansey WP (2000) Functional overlap of sequences that activate transcription and signal ubiquitin-mediated proteolysis. Proc Natl Acad Sci USA 97:3118–3123

Sanchez ER, Hirst M, Scherrer LC, Tang HY, Welsh MJ, Harmon JM, Simons SSJ, Ringold GM, Pratt WB (1990) Hormone-free mouse glucocorticoid receptors overexpressed in Chinese hamster ovary cells are localized to the nucleus and are associated with both hsp70 and hsp90. J Biol Chem 265:20123–20130

Sapolsky RM, McEwen BS (1985) Down-regulation of neural corticosterone receptors by corticosterone and dexamethasone. Brain Res 339:161–165

Shen T, Horwitz KB, Lange CA (2001) Transcriptional hyperactivity of human progesterone receptors is coupled to their ligand-dependent down-regulation by mitogen-activated protein kinase-dependent phosphorylation of serine 294. Mol Cell Biol 15:6122–6131

Smith DF (1993) Dynamics of heat shock protein 90-progesterone receptor binding and the disactivation loop model for steroid receptor complexes. Mol Endocrinol 7:1418–1429

Smith DF, Toft DO (1993) Steroid receptors and their associated proteins. Mol Endocrinol 7:4–11

Starr DB, Matsui W, Thomas JR, Yamomoto KR (1996) Intracellular receptors use common mechanisms to interpret signaling information at response elements. Genes Dev 10:1271–1283

Stebbins CE, Russo AA, Schneider C, Rosen N, Hartl FU, Pavletich NP (1997) Crystal structure of an Hsp90-geldanamycin complex: targeting of a molecular chaperone by an antitumor agent. Cell 89:239–250

Stenoien DL, Patel K, Mancini MG, Dutertre M, Smith CL, O'Malley BW, Mancini MA (2001) FRAP reveals that mobility of oestrogen receptor-alpha is ligand- and proteasome-dependent. Nat Cell Biol 3:15–23

Tanenbaum DM, Wang Y, Williams SP, Sigler PB (1998) Crystallographic comparison of the estrogen and progesterone receptor's ligand binding domain. Proc Natl Acad Sci USA 95:5998–6003

Tang Y, Ramakrishnan C, Thomas J, DeFranco DB (1997) A role for HDJ-2/HSDJ in correcting subnuclear trafficking, transactivation and transrepression defects of a glucocorticoid receptor zinc finger mutant. Mol Biol Cell 8:795–809

Tomoda K, Kubota Y, Kato J-y (1999) Degradation of the cyclin-dependent-kinase inibibitor p27^{Kip1} is instigated by Jab1. Nature 398:160–165

van Steensel B, Brink M, van der Meulen K, van Binnendijk EP, Wansink DG, de Jong L, de Kloet ER, van Driel R (1995) Localization of the glucocorticoid receptor in discrete clusters in the cell nucleus. J Cell Sci 108:3003–3011

Vanderbilt JN, Miesfeld R, Maler BA, Yamamoto KR (1987) Intracellular receptor concentration limits glucocorticoid-dependent enhancer activity. Mol Endocrinol 1:68–74

Vedder H, Weiss I, Holsboer F, Reul JMHM (1993) Glucocorticoid and mineralocorticoid receptors in rat neocortical and hippocampal brain cells in culture: characterization and regulatory studies. Brain Res 605:18–24

Wallace AD, Cidlowski JA (2001) Proteasome mediated glucocorticoid receptor degradation restricts transcriptional signaling by glucocorticoids. J Biol Chem 276:42714–42721

Wei LL, Krett NL, Francis MD, Gordon DF, Wood WM, O'Malley BW, Horwitz KB (1988) Multiple human progesterone receptor messenger ribonucleic acids and their autoregulation by progestin agonists and antagonists in breast cancer cells. Mol Endocrinol 2:62–72

Whitesell L, Cook P (1996) Stable and specific binding of heat shock protein 90 by geldanamycin disrupts glucocorticoid receptor function in intact cells. Mol Endocrinol 10:705–712

Wijayaratne AL, McDonnell DP (2001) The human estrogen receptor-alpha is a ubiquitinated protein whose stability is affected differentially by agonists, antagonists, and selective estrogen receptor modulators. J Biol Chem 276:35684–35692

Yamamoto KR (1985) Steroid receptor regulated transcription of genes and gene networks. Annu Rev Genet 19:209–252

Yang J, DeFranco DB (1994) Differential roles of heat shock protein 70 in the in vitro nuclear import of glucocorticoid receptor and simian virus 40 large tumor antigen. Mol Cell Biol 14:5088–5098

Ylikomi T, Bocquel MT, Berry M, Gronemeyer H, Chambon P (1992) Cooperation of proto-signals for nuclear accumulation of estrogen and progesterone receptors. EMBO J 11:3681–3694

8 The Dynamics of Intranuclear Movement and Chromatin Remodeling by the Glucocorticoid Receptor

G.L. Hager

8.1 Introduction 111
8.2 Receptor-Directed Assembly of a Transcriptional Initiation
    Complex – Classic View 112
8.3 A System to Study Transcription in Living Cells 114
8.4 The Dynamic Interaction of Glucocorticoid Receptor
    with Hormone-Response Elements 118
8.5 Reconstitution of Glucocorticoid Receptor-Dependent Chromatin
    Remodeling In Vitro 121
8.6 Chromatin Remodeling In Vitro Is Accompanied by GR Loss
    from the Template 124
8.7 A Model for GR Mobility on the Template 124
8.8 Implications for Receptor Function 126
References 127

8.1 Introduction

Subcellular trafficking of nuclear receptors is an important component in the biological control of these regulatory molecules. The glucocorticoid receptor (GR) is found almost exclusively in the cytoplasm in the absence of ligand, and translocates rapidly to the nucleus after addition of hormone to the cells (Htun et al. 1996; Hager et al. 1998). Once located in the nucleus, the view from classic endocrinology has been that the receptor binds rapidly to glucocorticoid response elements (GREs), and remains bound to these elements in the continued presence of ligand. In this configuration, the chromatin-bound receptor is argued to serve a nucleating role, attracting a series of factors necessary for transcriptional activation through protein–protein interactions.

8.2 Receptor-Directed Assembly of a Transcriptional Initiation Complex – Classic View

A ligand-activated, template-bound GR is thought to recruit a variety of transcriptional coactivators and basal factors to promoters under control of specific GREs (Fig. 1). These factors include coactivators specific to the nuclear receptor superfamily [such as GRIP-1 and other members of the p160 family (Xu et al. 1999)], as well as more general coregulators, including p300, CBP, etc. Stabilization of these factors on the regulatory site brings into play a variety of activities that alter the chromatin template; these modifications include histone acetylation (Sterner and Berger 2000), methylation (Chen et al. 1999), and phosphorylation

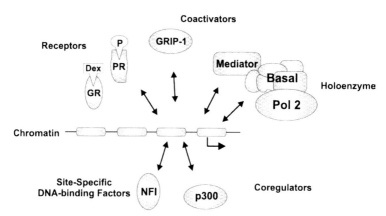

Fig. 1. Steroid receptors regulate transcription by the recruitment of transcription factor complexes to target promoters Steroid receptors interact with many cofactors to activate transcription. These factors include coactivators (members of the p160 family), coregulators (such as p300 and CBP), site-specific binding proteins (such as NF1), and members of the general (or basal) transcription factors

The Dynamics of Intranuclear Movement

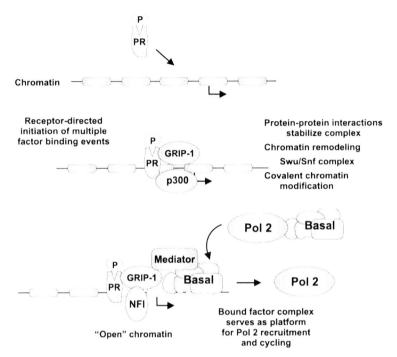

Fig. 2. "Classic" view of initiation complex assembly. In the current paradigm, receptors are thought to bind to specific sites in chromatin and attract many secondary regulators through protein–protein interactions. These cofactors in turn lead to template modification and promoter activation

(Cano et al. 1992). Modification of the template would lead (1) to a more open chromatin configuration (Fig. 2), increasing access for site-specific DNA-binding proteins and general transcription factors, and (2) provide site-specific protein tags for the recruitment of additional factors (Sterner and Berger 2000).

Under this model, the receptor serves essentially as an initiating factor to originate the formation of a large macromolecular complex, and this complex serves in turn as a platform for the eventual binding and initiation of transcription by RNA polymerase II. GR would remain bound to the regulatory site as long as ligand remains at sufficiently high concentration in the nucleoplasmic space (Fig. 2). Withdrawal of hor-

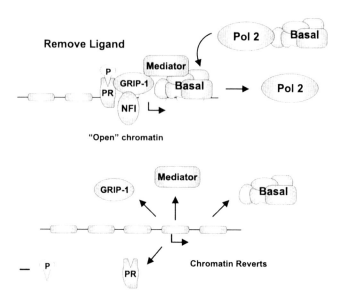

Fig. 3. Loss of ligand leads to disassembly of the initiation complex. A stable, template-bound receptor has been thought to form the critical nucleating event to establish a productive complex. Removal of ligand is believed to destabilize the complex and lead to disassembly of the initiation complex

mone from the cells would destabilize the receptor on the template, causing loss of receptor from chromatin and disruption of the complexes that require receptor for stable binding. In the absence of cofactors, chromatin would return to its pre-activation modification state, and the factors that require chromatin opening would no longer be capable of binding the template (Fig. 3). Transcription would then return to its ground state, and hormone-dependent activation would be reversed.

8.3 A System to Study Transcription in Living Cells

The model for GR function articulated above derives almost exclusively from in vivo experiments that employ transient transfection as the major methodology, and in vitro approaches involving the reconstruction of transcriptional activation with partially purified receptors and transcrip-

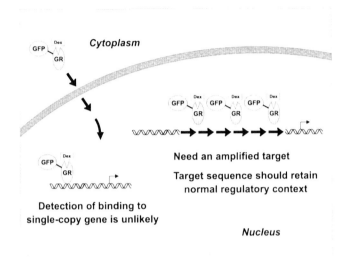

Fig. 4. Potential approach to detect gene targeting. Observation of a direct receptor-binding event to single copy genes in living cells cannot be achieved with current technology. To accomplish this goal, it is necessary to generate a tandemly amplified array of gene targets

tion extracts. We succeeded in 1996 in showing that subcellular location and movement of the GR could be effectively studied by tagging GR with the green fluorescent protein (GFP) (Htun et al. 1996). Given the success of this approach, we realized at the time that these fluorescent techniques opened the possibility of monitoring receptor binding to regulatory elements in living cells in real time. It seemed unlikely, however, that binding of a receptor to a single GRE would be detectable against the very high background of 30,000–40,000 receptors in the nucleoplasmic space (Fig. 4). If, however, an array of elements containing a local high density of GREs could be generated, it might be possible to visualize local binding of GR to these elements above the background of general nuclear receptor (Fig. 4).

Such a system became available with the development of a cell line containing a tandem array of mouse mammary tumor virus (MMTV) reporter elements integrated in chromosome 4 of a murine mammary carcinoma cell line (Kramer et al. 1999). In this cell line (3134) (Walker et al. 1999), a complete reporter cassette containing the MMTV pro-

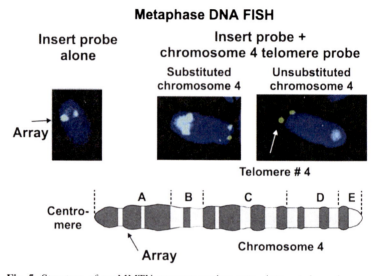

Fig. 5. Structure of an MMTV–reporter tandem array integrated on chromosome 4. A cell line was characterized that contains tandem array of MMTV LTR-reporter cassettes near the centromere of chromosome 4. The *left panel* shows metaphase FISH analysis with a probe to the MMTV insert; the *central panel* presents a FISH analysis with probes both to the insert and to the telomere of chromosome 4; and the *right panel* shows hybridization to the telomere probe alone

moter driving the ras oncogene is amplified as a perfect head-to-tail array of approximately 200 copies (Fig. 5). Fluorescence in situ hybridization analysis [DNA fluorescent in situ hybridization (FISH)] on metaphase chromosomes with the ras probe revealed the presence of a large ras-specific structure near the centromere of one chromosome. Hybridization analysis with telomere-specific probes showed that the array exists as a unique amplified element near the centromere in the A region of chromosome 4 (Fig. 5).

To detect interactions between a variety of transcription factors and regulatory elements in the MMTV promoter, we developed the general strategy shown in Fig. 6. Factors of interest are tagged with variants of the green fluorescent protein. These chimeras are then introduced into the 3134 cell line, generally under control of the tetracycline regulated

The Dynamics of Intranuclear Movement

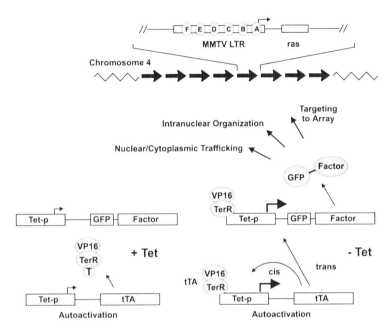

Fig. 6. A system to detect receptor binding to regulatory elements in living cells. Factors of interest are fused to GPF or other members of the fluorescence protein family. These chimeras are expressed in the cell line containing the MMTV tandem array, usually under conditional control of the tetracycline inducible promoter. Intracellular trafficking and gene targeting for the factor can then be observed

expression system (tet off). This approach overcomes two difficulties. First, it has been shown that receptors expressed by transient transfection can be defective with respect to activation of genes embedded in normal chromatin structures (Smith et al. 1997, 2000; Smith and Hager 1997). Stable expression of the GR overcomes this problem.

Secondly, expression of the GFP-tagged receptor from the tet-regulated promoter allows a controlled expression of the fluorescent protein. Cells can be grown for extended periods in the absence of the chimeric GR; it is only necessary to induce expression of the receptor for a few hours before observations are made. This approach assures a relatively uniform expression level for GR in the cell population and avoids problems associated with long-term high-level expression of the receptor. With this general approach, we can then examine several aspects of receptor behavior in living cells, including targeting to the array, intranuclear organization of labeled factors, and nuclear/cytoplasmic trafficking of the proteins (Fig. 6).

Using this system, we reported the direct, real-time recruitment of GR to regulatory elements in living cells (McNally et al. 2000). Receptor was observed to interact with one large structure in the nucleus of each hormone-stimulated cell. RNA FISH indicated that the large structures correspond to the MMTV tandem array (McNally et al. 2000). These findings demonstrated that receptor interactions with gene targets could be studied in real time, and opened the way to a direct analysis of the dynamic behavior of receptor on regulatory elements.

8.4 The Dynamic Interaction of Glucocorticoid Receptor with Hormone-Response Elements

With the ability to monitor GR interactions with gene targets in real time, we have been able to study the dynamic nature of GR binding events through the application of a variety of photobleaching technologies, including fluorescence recovery after photobleaching (FRAP) and fluorescence loss in photobleaching (FLIP) (Hager et al. 2000, 2002; McNally et al. 2000; Mueller et al. 2001). In the FRAP protocol (Fig. 7A), laser light is focused directly on the array, and bound GFP-GR is rapidly bleached. After the laser bleach is discontinued, potential exchange of bound GR with free GR in the nucleoplasm can be monitored by recovery of fluorescence. In the FLIP approach (Fig. 7B), the laser beam is focused not on the array, but at a separate position in the nuclear compartment. Photobleaching is then carried out in a pulsatile mode, gradually reducing the concentration of fluorescent GFP-GR in the nucleoplasm. If molecules bound to the array are mobile, they will be replaced by molecules from the nucleoplasm whose specific fluorescence is progressively decreased. Thus, the specific fluorescence on the array will decrease, and the rate of decrease will be a measure of the rate of exchange (Fig. 7B). If GFP-GR is bound to the array with a long half-life, then fluorescence at the array will not decrease during the bleaching period. This protocol provides two important advantages.

The Dynamics of Intranuclear Movement 119

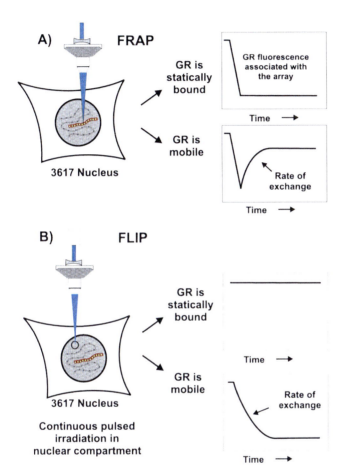

Fig. 7A, B. Genomic interactions monitored by photobleaching. Using the MMTV tandem array cell, the dynamics of GR binding to the hormone-response elements (HREs) can be observed in real time in living cells by photobleaching. In the FRAP protocol (**A**), GFP-labeled receptors on the array are bleached; after the bleach pulse is discontinued, the rate at which fluorescence receptors return to the array is a measure of the exchange rate. In the FLIP protocol (**B**), fluorescence receptors in the nucleoplasm are continuously bleached in a pulsatile mode. As the concentration of bleached receptors in the nucleoplasm decreases, the rate of depletion of fluorescence on the array is again a measure of exchange

First, fluorescent GFP-GR molecules bound to the array are not subject to bleaching. Thus, if molecules on the array are replaced by bleached GFP-GR from the nucleoplasm, it would be inferred that bleaching has little effect on the activity of the molecules, since they would be actively replacing GR on the array. Secondly, the chromatin structure of the gene target is also not subject to bleaching, again diminishing the likelihood of artifacts associated with the intense photobleaching laser pulse.

These two photobleaching approaches were used to study the real-time interaction of receptor with its regulatory sites in chromatin (McNally et al. 2000). When a bleaching beam is focused on the tandem array structure (FRAP, Fig. 7A), almost all GFP-GR molecules associated with the array are bleached after a laser pulse of 250 ms. When irradiation of the structure is discontinued, GFP-GR fluorescence is again detected on the array structure within 2 s. FLIP experiments (Fig. 7B) produce similar results; bleaching of GFP-GR in the nucleoplasm leads to a rapid loss of fluorescence from the array as receptor molecules on the MMTV hormone-response elements (HREs) are replaced with bleached molecules from the nucleoplasm. Thus, GFP-GR exchanges at a high rate between the array-bound state and the free-nucleoplasmic state.

These results are incompatible with standard models of initiation complex assembly as depicted in Figs. 1–3, wherein receptor occupancy of binding sites in chromatin is viewed as a long-lived event in the presence of ligand. It is clear, rather, that the receptor is cycling rapidly between a chromatin-bound and free-nucleoplasmic state (Fig. 8). Constant cycling of the receptor is observed in the continuous presence of hormone. The living cell experiments suggest, therefore, two general classes of models for the behavior of receptor in real-time. (1) Receptor may cycle continuously on the template, but initial recruitment of chromatin modifying activities would lead to a reorganized nucleoprotein state with a long half-life (Hager et al. 1993; Smith and Hager 1997; Fragoso et al. 1998). Secondary transcription factors could then bind to the modified structure. Under this model, GR could induce a long-lived "open" state through the recruitment of remodeling activities and secondary transcription factors, but continuous presence of the receptor may be unnecessary to maintain this state. (2) Alternatively, the complete process of nucleoprotein template modification and factor loading may be cycled each time the receptor binds to the template, and chromatin

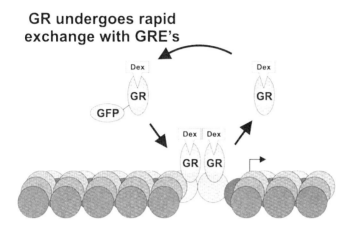

Fig. 8. Direct interaction of GR with target sites in living cells. The photobleaching experiments indicate that the GR is exchanging constantly with gene targets, even in the continuous presence of ligand

may cycle between the "open," enzyme accessible, and "closed" states very rapidly.

8.5 Reconstitution of Glucocorticoid Receptor-Dependent Chromatin Remodeling In Vitro

A detailed examination of the mechanism of GR-dependent chromatin remodeling and gene activation would be greatly facilitated by the availability of cell-free systems that support biologically accurate receptor-dependent nucleoprotein template modifications. The chromatin transition at the MMTV B/C nucleosome region has been characterized in considerable detail. Secondary transcription factors, including NF I and Oct 1, bind to the B/A region after receptor induction, and are continuously detected on the promoter by exonuclease footprinting analysis (Cordingley et al. 1987; Archer et al. 1992). In parallel, a specific nucleoprotein transition is induced in MMTV chromatin. This modified chromatin state is required for binding of secondary factors.

We have reconstituted this promoter in a cell-free system, and demonstrated GR-dependent modification of the nucleoprotein template

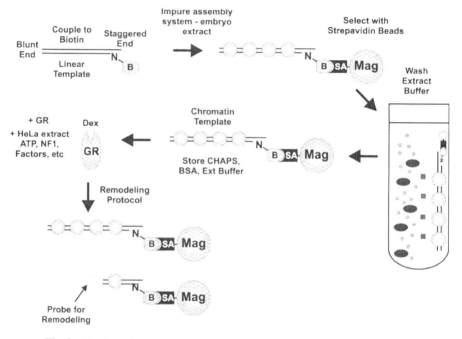

Fig. 9. GR-dependent chromatin remodeling – assembly system. DNA templates are first attached to paramagnetic beads, and then assembled into accurately positioned nucleosome arrays with an extract from drosophila embryos. After assembly, the template can be purified and manipulated through the use of intense field magnets. Chromatin remodeling experiments can be carried out with the template still attached to the beads

(Fig. 9) (Fletcher et al. 2000). The system utilizes the well-described chromatin assembly extracts isolated from rapidly dividing drosophila embryos (Becker and Wu 1992; Paranjape et al. 1994). To facilitate manipulation of the reconstituted chromatin, the MMTV template is first coupled to paramagnetic beads through a streptavidin-biotin linkage (Fig. 9). After assembly in the impure embryo extract, chromatin can be washed extensively to remove most of the soluble proteins from the extract. Modification reactions can then be performed with the template still bound to the magnetic beads.

Using this system, we showed (Fletcher et al. 2000) that nucleosomes deposited on the MMTV template adopt the same positions in vitro as

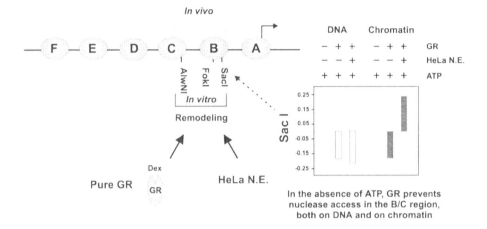

Fig. 10. GR is lost from the template during chromatin remodeling. Purified GR will support chromatin remodeling of the assembled template in the presence of a nuclear extract containing remodeling proteins. The position and extent of the remodeled region is accurate with respect to the GR-dependent transition that has been characterized in vivo. In the absence of energy, GR sterically blocks access of the diagnostic restriction enzyme to the template

observed for in vivo replicated chromatin (Fig. 10) (Richard-Foy and Hager 1987). Furthermore, addition of highly purified GR to the template results in the induction of a nucleoprotein transition that closely reproduces the well-described MMTV in vivo transition (Richard-Foy and Hager 1987; Truss et al. 1992; Fragoso et al. 1995, 1998; Smith and Hager 1997; Fragoso and Hager 1997). In particular, the region of chromatin remodeling corresponds to the complete nucleosome B family, but only the 3´ side of the C nucleosome family (see Fragoso et al. 1998 for a details of the MMTV chromatin). Since this reorganization event does not correspond to one or two nucleosomes, the MMTV transition cannot be easily resolved with a simple nucleosome "sliding" model (Hamiche et al. 2001; Langst and Becker 2001; Narlikar et al. 2001). Further details of the nucleoprotein remodeling event are needed to understand the molecular basis of the GR-dependent hypersensitive transition.

8.6 Chromatin Remodeling In Vitro Is Accompanied by GR Loss from the Template

The chromatin remodeling event induced in vitro at the MMTV promoter is remarkably analogous to the transition that has been characterized extensively in vivo. However, a key observation obtained in the in vitro system raised a paradox with respect to occupancy of the GR binding sites (Fletcher et al. 2000; Fletcher et al. 2002). When the remodeling system is completely supplemented with GR, nuclear extract, and ATP, a strong increase in nuclease sensitivity is observed throughout the remodeling region (Fig. 10). However, when either the source of remodeling factors or ATP is left out, a decrease is observed in restriction enzyme sensitivity. This GR-induced resistance to nuclease attack is observed on both DNA and chromatin templates, whereas the dramatic increase in sensitivity is found only with assembled chromatin. The enzyme sites whose access is monitored in the hypersensitivity experiments are, in fact, located in close proximity to the GR binding sites, suggesting that GR could interfere sterically with enzyme access. This model was confirmed in a series of experiments that compared access at sites throughout the template, both for deproteinized DNA and for chromatin. Enzyme access at sites distant from the GR is unaffected by GR binding, while access at sites overlapping the GREs is blocked by occupancy of the receptor binding site. The clear interpretation of these findings is that the residency time for GR on DNA, and on chromatin in the absence of remodeling, is sufficiently long, and the interaction is sufficiently stable, to block access for restriction enzymes whose sites are sterically hindered by the bound GR.

8.7 A Model for GR Mobility on the Template

The prototypic model for GR function suggests that receptor is bound to the template in the presence of ligand, and serves a nucleating role in the formation of large template-bound protein complexes (Figs. 1–3). In contrast to this model, we find that the receptor is surprisingly mobile on templates, both in vivo and in vitro. Importantly, rapid exchange in vitro is observed (1) only on nucleoprotein templates that accurately reflect the chromatin architecture of a natural promoter, and (2) only during the

The Dynamics of Intranuclear Movement

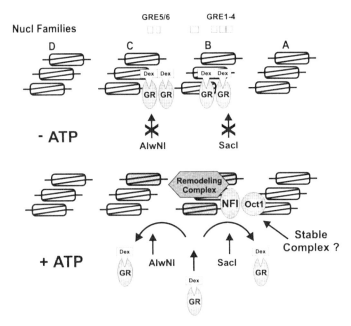

Fig. 11. The "hit-and-run" model. The GR binds to specific sites in chromatin in the absence of remodeling activities. When statically bound, the receptor blocks access of restriction enzymes by steric occlusion. During chromatin remodeling, the underlying nucleosome structures are reorganized, and the receptor is lost from the template

active process of chromatin remodeling. Based on these findings, we propose that receptor acts via a "hit-and-run" mechanism (Fig. 11). In the absence of a remodeling activity, or under ATP-free conditions, receptor binds to specific GREs in chromatin. Under these conditions, the receptor remains on the template with a relatively long half-life, and the access of restriction enzymes is inhibited. However, when energy and a remodeling system are provided, the receptor is no longer bound statically to the template. Specifically, we propose that the receptor is actively ejected from the template during the remodeling reaction. Thus, two separate processes are responsible for the increased enzyme access generally referred to as hypersensitivity. One process involves swi-/snf-directed nucleosome remodeling; during this reorganization, DNA sequences in the remodeled domain become intrinsically more accessible to nucleases. The second process involves actual loss of receptor from the template. Since the receptor will effectively block enzyme access on naked DNA templates, it seems highly unlikely that the marked increase in enzyme access could occur in chromatin, even if the DNA were completely released from the nucleosome structure. We have recently confirmed in a direct chromatin pull-down assay that the receptor is actively displaced from the template during chromatin remodeling (Fletcher et al. 2002).

8.8 Implications for Receptor Function

It now appears that the mechanism of GR action at regulatory elements involves rapid exchange with the template. Two general classes of mechanism can be envisioned that account for this mobility. Receptors could be viewed essentially as "triggering" factors. That is, the initial receptor binding event, even though transient, would set in motion a cascade of chromatin modifying and factor binding events. These secondary reactions would proceed without the necessary continued presence of template-bound receptor. Alternatively, the receptor may function in a "return-to-template" mode. Under this model, the receptor would continually return to the regulatory site, and successive binding events could have alternate consequences. For example, an initial event could lead to remodeling complex recruitment and local chromatin opening. Subsequent binding events could recruit coactivators responsible for histone modification and interactions with general transcription factors. In this class of models, the receptor would actually serve different functions at different times in the development of the transcription program.

This view of receptor function articulated here represents a significant departure from the more classic view of statically engaged receptors (Hager et al. 2002). A rigorous evaluation of these proposals will require an evolution in the methodology by which mechanisms of receptor action are studied. Most current approaches would not be sensitive to the very rapid exchange reactions that we have described. In vitro transcription experiments are typically carried out in an end-point mode. That is, a reaction is run for some period of time, and the amount of transcript accumulated is evaluated. This approach would not detect exchange events during the course of the reaction. In vivo localization experiments are also not designed to monitor rapid exchange, but in general monitor equilibrium states. One general approach that would seem to be useful in this regard is high-speed laser protein/DNA crosslinking (Hockensmith et al. 1991). This technique offers the unique ability to track template interactions as they occur. It is anticipated that the application of real-time methods to the study of receptor template interactions will considerably expand our knowledge of receptor function, and will confirm a dynamic behavior of these molecules that we are only beginning to understand.

Acknowledgments. I would like to thank the many colleagues who have contributed both to the experimental findings and the intellectual concepts advanced in this manuscript. Particular recognition goes to: James McNally and Hillary Mueller, who first identified the MMTV array; Dawn A. Walker, who created the array cell lines that made the photobleaching experiments possible; Barbour Warren, whose purification of GR was central to the in vitro system; and Terace Fletcher, who discovered "hit-and-run" in vitro.

References

Archer TK, Lefebvre P, Wolford RG, Hager GL (1992) Transcription factor loading on the MMTV promoter: A bimodal mechanism for promoter activation. Science 255:1573–1576

Becker PB, Wu C (1992) Cell-free system for assembly of transcriptionally repressed chromatin from Drosophila embryos. Mol Cell Biol 12:2241–2249

Cano E, Barratt MJ, Mahadevan LC (1992) Which histone kinase? Nature 360:116–116

Chen D, Ma H, Hong H, Koh SS, Huang SM, Schurter BT, Aswad DW, Stallcup MR (1999) Regulation of transcription by a protein methyltransferase. Science 284:2174–2177

Cordingley MG, Riegel AT, Hager GL (1987) Steroid-dependent interaction of transcription factors with the inducible promoter of mouse mammary tumor virus in vivo. Cell 48:261–270

Fletcher TM, Ryu B-W, Baumann CT, Warren BS, Fragoso G, John S, Hager GL (2000) Structure and dynamic properties of the glucocorticoid receptor-induced chromatin transition at the MMTV promoter. Mol Cell Biol 20:6466–6475

Fletcher TM, Xiao N, Mautino G, Baumann CT, Warren BS, Hager GL (2002) ATP-dependent mobilization of the glucocorticoid receptor during chromatin remodeling. Mol Cell Biol (in press)

Fragoso G, Hager GL (1997) Analysis of in vivo nucleosome positions by determination of nucleosome-linker boundaries in crosslinked chromatin. Methods 11:246–252

Fragoso G, John S, Roberts MS, Hager GL (1995) Nucleosome positioning on the MMTV LTR results from the frequency-biased occupancy of multiple frames. Genes Dev 9:1933–1947

Fragoso G, Pennie WD, John S, Hager GL (1998) The position and length of the steroid-dependent hypersensitive region in the mouse mammary tumor virus long terminal repeat are invariant despite multiple nucleosome B frames. Mol Cell Biol 18:3633–3644

Hager GL, Archer TK, Fragoso G, Bresnick EH, Tsukagoshi Y, John S, Smith CL (1993) Influence of chromatin structure on the binding of transcription factors to DNA. Cold Spring Harbor Symp Quant Biol 58:63–71

Hager GL, Smith CL, Fragoso G, Wolford RG, Walker D, Barsony J, Htun H (1998) Intranuclear trafficking and gene targeting by members of the steroid/nuclear receptor superfamily. J Steroid Biochem Mol Biol 65:125–132

Hager GL, Fletcher TM, Xiao N, Baumann CT, Muller WG, McNally JG (2000) Dynamics of gene targeting and chromatin remodeling by nuclear receptors. In: Steroid receptor coactivators and the remodeling of chromatin. Biochem Soc Trans 28:405–410

Hager GL, Elbi C, Becker M (2002) Protein dynamics in the nuclear compartment. Curr Opin Genet Dev 12:137–141

Hamiche A, Kang JG, Dennis C, Xiao H, Wu C (2001) Histone tails modulate nucleosome mobility and regulate ATP-dependent nucleosome sliding by NURF. Proc Natl Acad Sci USA 98:14316–14321

Hockensmith JW, Kubasek WL, Vorachek WR, Evertsz EM, von Hippel PH (1991) Laser cross-linking of protein-nucleic acid complexes. Methods Enzymol 208:211–236

Htun H, Barsony J, Renyi I, Gould DJ, Hager GL (1996) Visualization of glucocorticoid receptor translocation and intranuclear organization in living cells with a green fluorescent protein chimera. Proc Natl Acad Sci USA 93:4845–4850

Kramer P, Fragoso G, Pennie WD, Htun H, Hager GL, Sinden RR (1999) Transcriptional state of the mouse mammary tumor virus promoter can effect topological domain size in vivo. J Biol Chem 274:28590–28597

Langst G, Becker PB (2001) ISWI induces nucleosome sliding on nicked DNA. Mol Cell 8:1085–1092

McNally JG, Mueller WG, Walker D, Wolford RG, Hager GL (2000) The glucocorticoid receptor: Rapid exchange with regulatory sites in living cells. Science 287:1262–1265

Mueller WG, Walker D, Hager GL, McNally JG (2001) Large scale chromatin decondensation and recondensation in living cells and the role of transcription. J Cell Biol 154:33–48

Narlikar GJ, Phelan ML, Kingston RE (2001) Generation and interconversion of multiple distinct nucleosomal states as a mechanism for catalyzing chromatin fluidity. Mol Cell 8:1219–1230

Paranjape SM, Kamakaka RT, Kadonaga JT (1994) Role of chromatin structure in the regulation of transcription by RNA polymerase II. Annu Rev Biochem 63:265–297

Richard-Foy H, Hager GL (1987) Sequence specific positioning of nucleosomes over the steroid-inducible MMTV promoter. EMBO J 6:2321–2328

Smith CL, Hager GL (1997) Transcriptional regulation of mammalian genes in vivo: a tale of two templates. J Biol Chem 272:27493–27496

Smith CL, Htun H, Wolford RG, Hager GL (1997) Differential activity of progesterone and glucocorticoid receptors on mouse mammary tumor virus templates differing in chromatin structure. J Biol Chem 272:14227–14235

Smith CL, Wolford RG, O'Neill TB, Hager GL (2000) Characterization of transiently- and constitutively-expressed progesterone receptors: Evidence for two functional states. Mol Endocrinol 14:956–971

Sterner DE, Berger SL (2000) Acetylation of histones and transcription-related factors. Microbiol Mol Biol Rev 64:435–459

Truss M, Chalepakis G, Beato M (1992) Interplay of steroid hormone receptors and transcription factors on the mouse mammary tumor virus promoter. J Steroid Biochem Mol Biol 43:365–378

Walker D, Htun H, Hager GL (1999) Using inducible vectors to study intracellular trafficking of GFP-tagged steroid/nuclear receptors in living cells. Methods 19:386–393

Xu L, Glass CK, Rosenfeld MG (1999) Coactivator and corepressor complexes in nuclear receptor function. Curr Opin Genet Dev 9:140–147

9 Glucocorticoid Receptor Antagonism of AP-1 Activity by Inhibition of MAPK Family

C. Caelles, A. Bruna, M. Morales, J.M. González-Sancho,
M.V. González, B. Jiménez, A. Muñoz

9.1	Introduction	131
9.2	Results	134
9.3	Discussion	144
9.4	Materials and Methods	147
References		150

9.1 Introduction

Glucocorticoid hormones have multiple important physiological and therapeutic activities. They function by controlling the expression of target genes through the binding to specific glucocorticoid receptors (GR) which belong to the nuclear receptor superfamily of ligand-modulated transcription factors (Beato et al. 1995; Mangelsdorf et al. 1995). Glucocorticoids regulate their target genes mainly at the transcriptional level, though examples of post-transcriptional regulation have also been reported. Transcriptionally regulated genes can be directly modulated through the binding of ligand-bound GR homodimers to regulatory sequences termed glucocorticoid response elements (GREs). This mechanism is known as trans-activation. In addition, the transcription of target genes can be regulated via interactions ("cross-talk") of hormone-activated GR with other transcription factors. These interactions are often mutually antagonistic, but in other cases are unidirectionally antagonistic or even synergistic, depending on the factors involved and also the specific gene and cell type (Göttlicher et al. 1998). The inhibition of the gene regulatory effects of other factors by GR is commonly termed as trans-repression (Vayssière et al. 1997; Resche-Rigon and Gronemeyer 1998).

In recent years, several groups have demonstrated that the trans-repression activity of GR is at least as important as its trans-activation activity, being responsible for crucial developmental and physiological actions of glucocorticoids (Karin 1998; Reichardt et al. 1998). In addition, the pharmacological activities of glucocorticoids as anti-inflammatory, immunosuppressive and anti-proliferative agents also rely to a great extent on the repression of transcription factors (Vayssière et al. 1997; Gottlicher et al. 1998; Karin 1998; Resche-Rigon and Gronemeyer 1998). Consistently, many genes involved in the immune response, inflammation and proliferation coding for cytokines, proteases, oncogenes, adhesion molecules and others contain binding sites for activating protein (AP)-1 and/or nuclear factor (NF)kB transcription factors in their regulatory regions, both of which are repressed by activated GR.

AP-1 is a group of transcription factors composed of homodimers of members of the c-Jun protein family (c-Jun, JunB, JunD) or of heterodimers between them and members of the c-Fos family (c-Fos, FosB, Fra-1, Fra-2). AP-1 mediates the tumour-promoting activity of 12-*O*-tetradecanoyl phorbol 13-acetate (TPA) through binding to the sequences known as "TPA-response elements" (TREs), also called AP-1 sites. A variable basal level of AP-1 activity is found in all cell types as a result of the expression of specific dimers and other interacting proteins (Chinenov and Kerppola 2001; Mechta-Grigoriou et al. 2001). AP-1 activity is increased by growth factors and oncogenic stimulation, and by stress signals such as pro-inflammatory cytokines [tumour necrosis factor (TNF)-α and interleukin-1] and ultraviolet (UV) radiation (Karin et al. 1997). AP-1 is involved in the control of cell proliferation, transformation and apoptosis.

Activation of AP-1 may result from either the post-transcriptional modification (phosphorylation) of the c-Jun and c-Fos proteins or the increased expression of their corresponding genes (Karin 1995). Phosphorylation of the c-Jun protein in two serine residues (63 and 73) by the c-Jun N-terminal kinase (JNK) plays a major role in the enhancement of

AP-1 activity by enabling it to bind to the transcriptional co-activator CBP [cAMP response element-binding protein (CREB)-binding protein] (Arias et al. 1994). Transcription of *c-jun* is activated by c-Jun-ATF2 heterodimers or ATF2 homodimers through binding to a TRE located close to the promoter region. Similarly, *c-fos* is induced by the binding of the transcription factor Elk-1, together with the serum response factor (SRF) protein to the serum response element (SRE). Like c-Jun, ATF-2 and Elk1 are also phosphorylated and concomitantly activated by JNK. Extracellular-regulated kinase (ERK)1/2, another member of the mitogen-activated protein kinase (MAPK) family, may also phosphorylate c-Jun on serine 63 and 73 in some cell types (Leppa et al. 1998) and plays a major role in *c-fos* stimulation via the SRE by phosphorylation/activation of Elk-1 (Karin 1995; Minden and Karin 1997).

Activated GR and AP-1 usually show a mutual functional antagonism (Herrlich 2001). This cross-talk is also observed for other nuclear receptors such as those for retinoic acid (RAR) and thyroid hormone (TR) (Gottlicher et al. 1998). Several findings have emphasized the importance of this mechanism. For instance, defective AP-1 repression may explain resistance to the anti-inflammatory effect of glucocorticoids in asthma patients (Adcock et al. 1995). Moreover, AP-1 trans-repression is also the basis of the anti-tumour action of glucocorticoids in mouse skin (Tuckermann et al. 1999). Since its initial description in the early 1990s (Jonat et al. 1990; Schüle et al. 1990; Yang-Yen et al. 1990), several mechanisms have been proposed to explain the antagonism between activated GR and AP-1, including protein–protein interactions, either direct or indirect, and competition for binding to overlapping DNA-binding elements (Gottlicher et al. 1998; Herrlich 2001). However, neither of them has been confirmed in intact cells. In addition, Kamei et al. (1996) reported the competition between hormone-bound GR and activated AP-1 to recruit CBP which is present in the nucleus in limiting amounts and is required for both GR and AP-1 to induce transcription of their respective target genes. Again, this mechanism does not fully explain the situation and effects observed in vivo (Herrlich 2001).

9.2 Results

9.2.1 Dexamethasone Antagonizes AP-1 by Inhibiting Activation of the JNK and ERK Signalling Pathways

We have studied the mechanism of AP-1 repression by glucocorticoids in HeLa cells which express endogenous GR. To activate AP-1 we used two different inducers: UV radiation and TNF-α. AP-1 activity was measured in transient transfection assays using a luciferase reporter gene under the control of a fragment of the human collagenase I gene containing an active AP-1 site (–73Col-Luc). Treatment of HeLa cells with the synthetic glucocorticoid dexamethasone (Dex) for 45 min or longer before UV irradiation or TNF-α treatment caused a 50% inhibition of AP-1 activation (Fig. 1A). The inhibition was complete upon elevation of cellular GR content by ectopic expression of an exogenous GR gene (not shown; Caelles et al. 1997). In contrast, Dex did not significantly reduce the basal level of AP-1 activity in unstimulated cells.

Since c-Jun is the major component of AP-1 in HeLa cells and phosphorylation of its N-terminal trans-activation domain is critical for AP-1 activation (Devary et al. 1992), we examined whether Dex could affect this step. As seen in Fig. 1B (upper), TNF-α induced an increase in the level of c-Jun phosphorylated on serine 63 which peaked 25 min after cytokine stimulation and persisted for at least 75 min. Dex pretreatment caused a clear reduction in c-Jun phosphorylation at all the time points studied. A progressive accumulation of c-Jun protein followed the peak of c-Jun phosphorylation in stimulated cells, likely as a consequence of the regulation of the *c-jun* transcription by c-Jun caused by the presence of an AP-1 site in its 5'-upstream region (Fig. 1B, lower). Remarkably, this AP-1-dependent transcriptional response was also drastically inhibited in Dex-treated cells (Fig. 1B, lower).

In view of the inhibition of c-Jun phosphorylation, we explored whether Dex could be modulating the JNK activity. To this end, extracts from HeLa cells treated or not with TNF-α that had been subjected to pretreatment with either Dex or vehicle were immunoprecipitated with an antibody against JNK and the enzymatic activity in the immune complexes was assayed by incubation with glutathione S-transferase (GST)-c-Jun as substrate. It has been previously reported that glucocor-

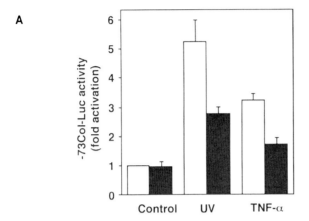

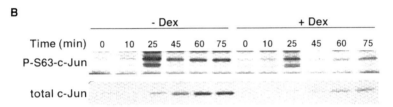

Fig. 1A, B. Dex blocks transcriptionally independent and transcriptionally dependent activation of AP-1 in HeLa cells in response to inducers of the JNK pathway. **A** Dex-inhibited AP-1-dependent transcriptional activation induced by either UV radiation or TNF-α. HeLa cells were transiently transfected with the –73Col-Luc reporter construct (3 µg). After serum starvation, cells were treated with vehicle (*white columns*) or Dex (*black columns*) and stimulated 45 min later. **B** Dex blocks c-Jun N-terminal phosphorylation and c-Jun accumulation in response to TNF-α. Western blot analysis of nuclear extracts (20 µg per lane) prepared form HeLa cells treated with vehicle (*–Dex*) or Dex (*+Dex*) and collected at the indicated time points after TNF-α stimulation. Specific antibodies to detect c-Jun phosphorylated on Ser63 (*P-S63-c-Jun*) (*upper*) or total c-Jun (*lower*) were subsequently used on the same membrane after stripping

A

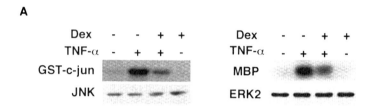

B

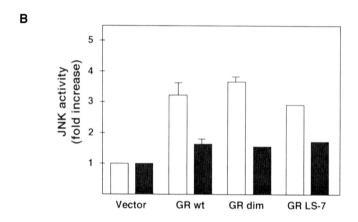

C

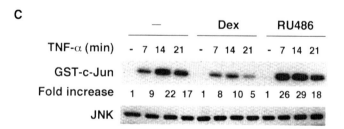

Fig. 2A–C. Legend see p. 137

ticoids inhibit AP-1 activation by TPA (Jonat et al. 1990; Schüle et al. 1990; Yang-Yen et al. 1990) as well as by a constitutively activated Ras (M. Morales et al., in preparation). Since both stimuli are poor inducers of the JNK pathway, though very good activators of the ERK pathway in HeLa cells (Minden et al. 1994), the same analysis was performed for ERK using a specific antibody and myelin basic protein (MBP) as substrate. Dex consistently inhibited the activation of JNK and ERK by TNF-α without affecting the cellular content of these enzymes (Fig. 2A). In agreement with the transcriptional activation assays shown in Fig. 1, Dex did not significantly inhibit the basal level of activity of any of these kinases. We also studied whether this synthetic glucocorticoid could modulate the p38MAPK. However, under the experimental conditions leading to JNK and ERK inhibition, the activation of this kinase by different various inducers (TNF-α, anisomycin, osmotic shock) was not affected by Dex in HeLa cells (not shown). The same specific down-regulation of JNK and ERK activities, but not of p38MAPK, by Dex has been reported in endothelial cells (González et al. 1999).

◄

Fig. 2A–C. Dex inhibits activation of MAPK signalling pathways independently of the trans-activation function of the GR. **A** Hormone-activated GR inhibits the activation of JNK and ERK signalling pathways by TNF-α. HeLa cells were serum-starved for 16 h, treated with dex (or vehicle) for 45 min and stimulated with TNF-α, as indicated. Cell extracts were prepared 20 min after TNF-α treatment. *Upper panels* show JNK (*left*) or ERK (*right*) activity as measured by the immune complex assay as described in "Materials and Methods", Sect. 9.4. *Bottom panels* show the total JNK (*left*) or ERK (*right*) amount present in each cell extract by Western blot analysis (20 µg per lane). **B** Inhibition of JNK activation by Dex is independent of GR dimerization, DNA binding, and transcriptional activation. Cos-7 cells were transiently co-transfected with expression plasmids for HA-JNK and GR-wt, GR-dim, GR-LS7 or with an empty vector (see "Materials and Methods", Sect. 9.4, for details). After serum starvation, cells were treated with vehicle (*white columns*) or Dex (*black columns*) for 45 min and then with TNF-α. After 20 min TNF-α stimulation, JNK activity in cell extracts was determined as above. **C** Glucocorticoid antagonist RU486 does not inhibit JNK activation. HeLa cells were pre-treated with vehicle (–), 10^{-6} M Dex or 10^{-5} M RU486 for 45 min before addition of TNF-α. JNK activity was measured at the indicated time points by the immune complex kinase assay. Quantitation of JNK enzymatic activity and control for equal loading by Western blotting using an anti-JNK antibody are shown below

9.2.2 Inhibition of JNK by Dexamethasone Is Independent of the Trans-activation Function of GR

Next we examined whether the inhibition of JNK required or not the trans-activating activity of GR. This was analysed by comparing the effect of Dex in Cos-7 cells that were transiently transfected with expression vectors for wild-type GR (GR-wt) or for two mutant receptors that retain their AP-1 trans-repressing activity upon hormone binding, while being defective in trans-activation. The first, GR-dim, that is unable to homodimerise and therefore so to bind DNA (Heck et al. 1994), and the second, GR-LS7, which shows a very reduced trans-activating activity due to a mutation in the DNA binding region (Schena et al. 1989; Helmberg et al. 1995). Remarkably, GR-dim and GR-LS7 mediated JNK inhibition by Dex to the same extent as GR-wt, indicating that the action of this hormone is independent of the GR trans-activation function and significantly, it correlates with its trans-repression function (Fig. 2B). This result was further confirmed by analysing hormone action on the JNK pathway in the presence of actinomycin D. Even in presence of this transcription inhibitor, Dex was equally efficient inhibiting JNK activation by UV irradiation (Caelles et al. 1997). Since the anti-hormone compound RU486 has been reported to act as a GR agonist or antagonist depending on the cell type and assay analysed (Wehle et al. 1995), we studied whether it could modulate JNK activity. RU486 did not inhibit the activation of JNK by TNF-α in HeLa cells (Fig. 2C), but rather the contrary, RU486 behaved as an antagonist causing a potentiation of JNK activity soon after TNF-α addition.

The trans-activation and trans-repression activities of GR are differentially induced by ligand concentration (Jonat et al. 1990). To examine whether this occurred in our system and to further confirm that the inhibition of JNK activity by hormone-bound GR can be separated from its capacity to trigger transcription through binding to a GRE, we carried out dose-response studies. Doses of Dex ranging between 10^{-7} and 10^{-11} M were tested in both assays. A dose-dependent inhibition of JNK activity was found, with a 30% inhibition already detectable at 10^{-9} M Dex (Fig. 3A). By contrast, at the same concentration Dex was unable to activate a construct that had the chloramphenicol acetyltransferase (CAT) reporter gene under the control of two copies of a consensus GRE (Fig. 3B). Our data show that the inhibition of JNK activation by GR

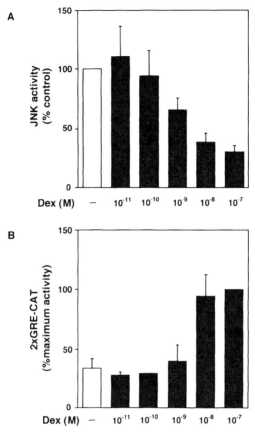

Fig. 3A, B. Inhibition of JNK activation and transactivation by GR are induced by different Dex doses. **A** Serum-starved HeLa cells were treated with the indicated doses of Dex (*black columns*) or vehicle (*white columns*) for 45 min, stimulated with TNF-α, and JNK activity was determined after 20 min by the immune complex kinase assay. **B** HeLa cells were transiently transfected with 2xGRE-TK-CAT reporter construct. After serum starvation, the indicated dosis of Dex (*black columns*) or vehicle (*white columns*) were added and CAT activity in cell extracts was determined after 6 h.

occurs at physiological concentrations of hormone (Felig et al. 1987) and requires a dose of Dex one order of magnitude lower than that required to induce the trans-activation function.

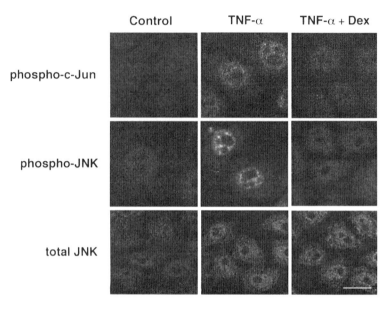

Fig. 4. Inhibition by Dex of c-Jun and JNK phosphorylation in TNF-α-treated HeLa cells. Dex has no effect on the TNF-α-induced nuclear accumulation of JNK in HeLa cells. Confocal laser microscopy analysis of HeLa cells incubated with specific antibodies was performed as described in "Materials and Methods", Sect. 9.4. Cells were treated with vehicle (*Control*), preincubated with vehicle for 45 min and then treated with TNF-α for 30 min (*TNF-α*), or preincubated with Dex for 45 min and then treated also with TNF-α for additional 30 min (*TNF-α+Dex*) as indicated. Scale bar, 15 µm

9.2.3 Dexamethasone Induces Accumulation of Inactive JNK in the Cell Nucleus

To examine whether the observed inhibition of JNK activity in extracts from Dex-treated cells takes place in intact cells, we employed immunofluorescence techniques using antibodies against total and phosphorylated c-Jun and JNK followed by confocal laser microscopy analysis. As expected, in line with the in vitro studies TNF-α treatment caused an increase in the level of phosphorylated c-Jun in the nucleus of HeLa cells (Fig. 4, upper). This increase was clearly blunted by Dex (Fig. 4, upper). TNF-α also increased nuclear and cytoplasmic content

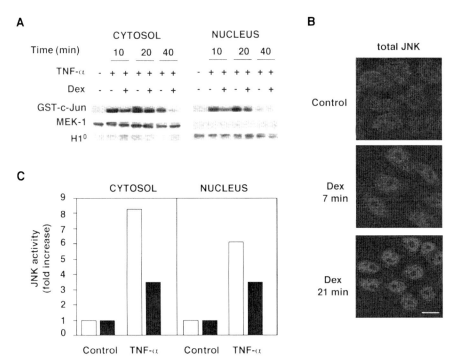

Fig. 5A–C. Dex inhibits TNF-α-induced JNK activity in nucleus and cytoplasm, and induces nuclear entry of inactive JNK. **A** Inhibition by Dex of JNK activity in nuclear and cytosolic fractions of TNF-α-treated HeLa cells. Cells were treated with vehicle or Dex 45 min before addition TNF-α. At the indicated times following TNF-α treatment subcellular fractionation was carried out. JNK activity in each fraction was estimated by the immune complex kinase assay. Purity of subcellular fractions was ensured by incubation with polyclonal anti-MEK-1 and monoclonal anti-histone H1^0 antibodies. **B** Dex alone induces JNK nuclear accumulation. Confocal microscopy analysis of HeLa cells incubated with vehicle (*Control*), or treated with Dex for 7 or 21 min. An antibody recognizing total JNK was used. Scale bar, 15 μm. **C** Lack of effect of Dex on JNK activity in unstimulated cells. Results obtained in cytosolic and nuclear fractions are shown. JNK activity in each subcellular fraction of HeLa cells incubated as indicated with vehicle (*Control*) or 10 ng/ml TNF-α, in the presence of vehicle (*white columns*) or 1 μM Dex (*black columns*) was measured by the immune complex kinase assay

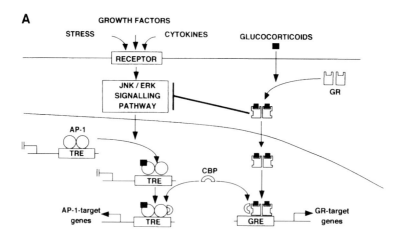

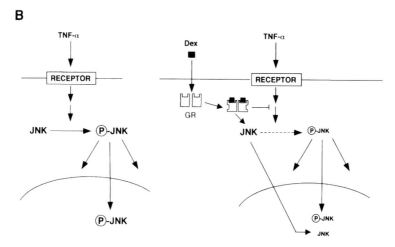

Fig. 6. Legend see p. 143

of phospho-JNK (Fig. 4, middle). Again, this effect was similarly inhibited by pre-treatment with Dex (Fig. 4, middle).

To study whether Dex inhibits the increase in phosphorylated c-Jun and JNK in the nucleus by affecting the translocation of JNK that follows cytokine stimulation, we used an antibody against total JNK. Dex did not affect JNK translocation. The accumulation of JNK in the cell nucleus induced by TNF-α occurred irrespective of Dex pre-treatment (Fig. 4, lower). The confocal images, corresponding to slices of the cells displaying highest nucleus:cytosol ratio could give the impression that TNF-α induced an increase in the cellular content of JNK. However, this possibility was discarded by using Western blotting analysis, which takes into account total cytoplasmic and nuclear JNK pools (not shown).

To estimate how Dex-induced changes affected the activation of JNK in the nucleus and cytosol, we measured JNK activity in both sub-cellular fractions. As shown in Fig. 5A, Dex caused a similar JNK inhibition in both nucleus and cytosol fractions. Having established that Dex did not affect the nuclear translocation of JNK induced by TNF-α, we examined whether it could exert an effect by itself. Unexpectedly, Dex

◄

Fig. 6. A Model for the antagonistic action of hormone-activated nuclear receptors on AP-1 activity through the inhibition of the JNK and ERK signal transduction pathways. Activation of gene transcription by either nuclear hormone receptors or AP-1 relies on an activation step: hormone binding for nuclear receptors and c-Jun phosphorylation on Ser63/73 for AP-1. According to our data, upon hormone activation, nuclear receptors inhibit the induction of the JNK signalling pathway and, consequently, c-Jun N-terminal phosphorylation. By this mechanism, glucocorticoid and possibly other steroid and thyroid hormones and retinoic acid may antagonize AP-1 activity and inhibit expression of AP-1 target genes. **B** Model for inhibition of JNK activation by glucocorticoid hormones. TNF-α signalling through its membrane-bound receptor leads to the activation by phosphorylation of JNK in the cytoplasm. A fraction of activated JNK is then able to translocate into the nucleus, where it phosphorylates c-Jun, causing the activation of AP-1 and stimulation of the transcription of AP-1 target genes. In the presence of Dex, the formation of hormone-receptor (*GR-Dex*) complexes in the cytoplasm inhibits the full activation of JNK. However, nuclear accumulation of JNK is not affected but, as a result of the inhibitory action of Dex-bound GR, a proportion (roughly 50%) of JNK within the nucleus is non-phosphorylated and therefore inactive. The result is a reduction in the level of c-Jun phosphorylation and AP-1 activation which causes a partial induction only of AP-1 target genes

alone induced a slight but consistent accumulation (around twofold) of JNK into the cell nucleus (Fig. 5B). Although we had previously shown that Dex did not induce JNK activity by itself, by assaying its activity in whole-cell extracts (Caelles et al. 1997), we considered that it might have cell compartment-specific effects. This possibility was excluded by performing JNK assays using sub-cellular fractions. Dex did not induce JNK activity in either compartment but, as explained above, inhibited the activation by TNF-α in both fractions (Fig. 5C). Analogous results were obtained in the case of ERK (M. Morales et al., in preparation).

In summary, our results indicated that hormone-activated GR inhibits the activation of JNK and ERK signalling pathways, which mediate AP-1 activation in response to a great variety of extracellular stimuli (Fig. 6A). This inhibition is independent of the transactivation function of GR, and correlates with its trans-repression function. The inhibition of JNK affects both the cytosolic and nuclear pools. Dex by itself causes the accumulation of inactive JNK in the cell nucleus. Upon TNF-α stimulation, the intracellular redistribution of total JNK is not affected by Dex but the ratio of active versus inactive JNK inside the nucleus is significantly decreased (Fig. 6B).

9.3 Discussion

Glucocorticoid regulation of gene expression is mediated by GR, which upon hormone binding modulate gene transcription by distinct mechanisms. These relay either in direct binding of GR to positive or negative GREs, giving rise to transcriptional activation or repression respectively, or in interference with other transcription factors, which is mostly independent of GR binding to DNA and results in the down-regulation of gene expression (Beato et al. 1995). Studies in GR knock-out mice have shown that the GR activities which are independent of its binding to DNA, such as trans-repression, are crucial for survival (Cole et al. 1995; Reichardt et al. 1998). Moreover, compelling evidence indicates that GR transcriptional interference mediates some of the most important pharmacological activities of glucocorticoids, such as anti-inflammatory, immunosuppressive and antineoplastic agents since AP-1 and NF-κB are among the transcriptional activators which are effectively antagonized by GR (Barnes and Karin 1998; Herrlich 2001).

The cross-talk between hormone-activated GR and AP-1 was first described in the early 1990s, and since then several different mechanisms have been proposed to explain it (Herrlich 2001). Among them, direct GR-AP-1 interaction (Jonat et al. 1990; Schüle et al. 1990; Yang-Yen et al. 1990) and competition for the transcriptional coactivator CBP (Kamei et al. 1996) are among the most well-documented. Neither of them, however, fully explain the whole set of effects observed in vivo (Herrlich 2001). Our results show that an additional mechanism underlying the inhibition of AP-1 by glucocorticoids is the blockade of the activation of JNK and ERK signalling pathways which leads to a reduction of c-Jun N-terminal phosphorylation. Phosphorylation of c-Jun on its N-terminal trans-activation domain is critical for AP-1 activation (Karin et al. 1997). Supporting this view, transgenic mice harbouring a mutant allele of c-Jun with Ser$^{63/73}$ mutated to alanines show defects in stress-induced apoptosis and cellular proliferation in vivo (Behrens et al. 1999). At the molecular level, c-Jun N-terminal phosphorylation is required to activated gene transcription since it enables binding to the transcriptional coactivator CBP (Arias et al. 1994). Therefore, its inhibition by glucocorticoids prevents AP-1 interaction with CBP and, consequently, promotes the binding of CBP to the hormone-activated GR instead. According to this view, and in line with to the model proposed by Kamei et al. (1996), inhibition of JNK by glucocorticoids may constitute a previous step in the competition for binding to CBP.

Moreover, we show that inhibition of JNK and ERK activation by glucocorticoids affects not only c-Jun but also other JNK- (ATF-2, Elk-1; Caelles et al. 1997) and ERK-targeted (Elk-1; M. Morales et al., in preparation) transcriptional activators involved in *c-jun* and *c-fos* induction. Consequently, activated GR causes a reduction in both the transcriptionally independent and transcriptionally dependent activation of AP-1. Significantly, neither the basal AP-1 nor JNK/ERK activities in unstimulated cells are inhibited by Dex, which may avoid undesirable cytotoxicity.

Our data do not clarify the exact step in the JNK or ERK pathways targeted by GR. However, additional experiments have shown that Dex inhibits the activation of either JNK in cells transfected with a constitutively active mitogen-activated ERK-activating kinase kinase (MEKK)1 (Caelles et al. 1997) or ERK in cells transfected with a constitutively active MEK1 (MEK-E) (M. Morales et al., in preparation). These results, together with the observation that the level of phosphorylated JNK or ERK (M. Morales et al., in preparation) is reduced in Dex-treated cells, indicate that either MAPKKs or the MAPKs themselves are the direct target of glucocorticoid action. Additionally, the inhibition of JNK and ERK by GR suggests the possibility of direct binding between GR and the MAPKKs or the MAPKs, which, however, has not been substantiated. One explanation for the negative results obtained in the studies of GR-MAPK(K)s interaction can be the use of inappropriate antibodies, the existence of a weak and/or transient interaction, or that this takes place in the frame of large complexes. Further work with mutant versions of these enzymes is in progress to elucidate among these possibilities. However, under our experimental conditions, no significant effect on p38MAPK induction was found, although this does not exclude that in other conditions the p38MAPK pathway may also be down-regulated by these hormones. Actually, Dex has been described as having an inhibitory effect on p38MAPK induction in HeLa cells which, in contrast to our results on JNK and ERK, it requires for de novo gene transcription (Krstic et al. 2001).

GR-dim and GR-LS7 mutants are efficient as GR-wt mediating the inhibition of the JNK pathway by Dex (Caelles et al. 1997; González et al. 2000). This observation demonstrates that monomeric GR which is unable to bind DNA and transactivate can inhibit JNK activity. In addition, it clearly shows that JNK inhibition correlates with the transrepression function of GR.

Our data reveal that the inhibition of c-Jun phosphorylation by JNK is not due to the blockade of the nuclear translocation of JNK induced by TNF-α. Rather unexpectedly, Dex induces by itself a certain level of nuclear accumulation of JNK and ERK.

Initial work showed that GR-AP-1 antagonism is mutual (Jonat et al. 1990; Schüle et al. 1990; Yang-Yen et al. 1990). Interestingly, studies done by other groups have shown that GR is subjected to phosphorylation by both JNK (Rogatsky et al. 1998) and ERK (Marija et al. 1997). In both cases, GR phosphorylation by these MAPKs result in inhibition of its activity as a transcription factor. Therefore, the balance between JNK and/or ERK action on GR and c-Jun, which perhaps depends on the amount of hormone-bound GR and c-Jun levels and on the activating signals, may define the antagonism between Dex and AP-1. We propose that the inhibition of JNK and ERK signalling pathways which lead to a decrease in the induction of AP-1 activity by stress signals may contribute to the pharmacological activities of glucocorticoid hormones.

9.4 Materials and Methods

9.4.1 Cell Culture, Plasmids and Transfections

HeLa and Cos-7 cells were grown in Dulbecco's modified Eagle's medium (DMEM) supplemented with 10% fetal calf serum (FCS). Cells were serum starved by changing the culture medium to DMEM supplemented with 0.5% FCS 16 h before treatment. Thereafter, vehicle (ethanol) or Dex (10^{-6} M if unspecified) were added 45 min before UV irradiation (40 Jm^{-2}, UV-C) or TNF-α addition (10 ng/ml). All the plasmids used in this study have been previously described (Caelles et al. 1997; González et al. 2000). Transfections were performed by the standard calcium phosphate precipitation method (Ausubel et al. 1994). Luciferase or CAT activity was determined 6 h after stimulation. In the HA-JNK activity assays, Cos-7 were transiently transfected with 3 µg of plasmid encoding HA-JNK (pCDNA3-JNK1) and 0.4 µg of those encoding either GR-wt (pSB-hGR), GR-dim (pSB-hGR(A458 T), GR LS-7 (pRC/βact-GR-LS7), or with the empty vector (pRSh$^-$R$^-$). After 24 h, transfection cells were pre-treated with Dex or vehicle (ethanol) for 45 min. After Dex pretreatment, TNF-α or its vehicle (ethanol) was added to the cells. Control cells were incubated with hormone during the pretreatment and throughout for the period of stimulation (Dex-alone cells). Average results from three independent experiments run in duplicates are shown.

9.4.2 Western Blotting

Total and phosphorylated c-Jun, JNK and ERK were detected in cytoplasmic, nuclear or total cell extracts by immunoblotting using specific antibodies (sc-045, sc-822, sc-474 and sc-154 from Santa Cruz, respectively). HA-JNK was detected using the 12CA5 antibody from BabCo. H1^0 was detected using a monoclonal antibody donated by Prof. A. Alonso, Deutches Krebsforschung Zentrum, Heidelberg. Western blots were performed and developed using the ECL detection system (Amersham).

9.4.3 Protein Kinase Assays

JNK and ERK kinase activity was measured by the immune complex kinase assay. For JNK activity assays, cells were lysed in 20 mM HEPES-NaOH pH 7.6, 10 mM ethyleneglycoltetraacetic acid (EGTA), 2.5 mM $MgCl_2$, 1% NP40, 1 mM dithiothreitol (DTT), 40 mM β-glycerophosphate, 100 μM sodium orthovanadate, 0.5 mM phenylmethylsulphonyl fluoride, 1 μg/ml aprotinin and 1 μg/ml leupeptin and lysates cleared by 10 min centrifugation at 12,000 rpm at 4°C. Extracts were immunoprecipitated with either an anti-JNK1 (sc-474) or an anti-HA (12CA5 from BabCo) and immune complexes were recovered with protein A-sepharose or protein G-sepharose, respectively. Beads were sequentially washed three times with 1% NP40 and 2 mM sodium orthovanadate in phosphate-buffered saline (PBS), once with 100 mM Tris-HCl pH 7.5 and 500 mM LiCl, and once with JNK-KB [20 mM hydroxyethylpiperazine ethanesulfonic acid (HEPES)-NaOH pH 7.6, 2.5 mM $MgCl_2$, 2 mM DTT, 20 mM β-glycerophosphate, 100 μM sodium orthovanadate]. Phosphorylation reactions were performed in 30 μl volume containing JNK-KB, 20 μM ATP, 0.5 μCi [γ-^{32}P]ATP and 1 μg of GST-c-Jun 30°C for 20 min, stopped by the addition of 4×Laemli sample buffer and resolved on a 10% SDS-polyacrylamide gel. ERK immune complex assays were performed essentially in the same conditions but using an anti-ERK specific antibody (sc-154 from Santa Cruz), MBP as a substrate in the phosphorylation reaction, and samples were resolved onto a 12% SDS-polyacrylamide gel.

9.4.4 Immunocytochemistry

Cells were rinsed twice in PBS, fixed with 3.7% paraformaldehyde in PBS for 15 min at room temperature, permeabilized with 0.5% Triton X-100 for 15 min, and treated with 0.1 M glycine in PBS for 15 min. The non-specific sites were blocked by incubation with PBS containing 1% bovine serum albumin (BSA) or goat serum for 30 min at room temperature. Cells were then washed in PBS containing 0.05% Tween-20 for 5 min and incubated with the primary antibodies diluted in PBS for 1 h at room temperature or overnight at 4°C. The following primary antibodies were used: mouse monoclonal anti-c-Jun phosphorylated on serine-63 (Santa Cruz sc-822), mouse monoclonal against human JNK1 (Pharmingen 15701 A), and mouse monoclonal anti-JNK1 phosphorylated on threonine-183 and tyrosine-185 (Santa Cruz sc-6254). Cells were then incubated for 45 min with TxR-conjugated goat anti-mouse (Jackson Immunoresearch, West Grove, Pa., USA). To amplify the phospho-JNK and JNK1 staining, anti-mouse Ig-digoxigenin, F(ab')$_2$-fragment, followed by anti-digoxigenin-rhodamine (Boehringer Mannheim, Mannheim, Germany), or biotinylated anti-mouse, followed by streptavidin-rhodamine (Jackson Immunoresearch, West Grove, Pa.), secondary antibodies were used. Confocal microscopy was performed with a BioRad MRC-1024 laser-scanning microscope, equipped with an Axiovert 100 invert microscope (Zeiss, Jena, Germany).

9.4.5 Whole-Cell Extracts and Subcellular Fractionation

To prepare whole-cell extracts, the monolayers were washed twice in PBS and the cells were lysed by incubation in radioimmunoprecipitation assay (RIPA) buffer [150 mM NaCl, 1.5 mM MgCl$_2$, 10 mM NaF, 10% glycerol, 4 mM ethylenediaminetetraacetate (EDTA), 1% Triton X-100, 0.1% SDS, 1% sodium deoxycholate, 50 mM HEPES-NaOH pH 7.4], plus phosphatase- and protease-inhibitor mixture (PPIM: 25 mM β-glycerophosphate, 1 mM sodium orthovanadate, 1 mM PMSF, 10 µg/ml leupeptin, 10 µg/ml aprotinin) for 15 min on ice followed by centrifugation at 13,000 rpm for 10 min at 4°C. For subcellular fractionation, cells were lysed in nuclear precipitation buffer (NPB) (10 mM Tris-HCl, pH 7.4, 2 mM MgCl$_2$, 140 mM NaCl, plus PPIM) supplemented with 0.1% Triton X-100 by incubating on ice for 10 min. The lysate was layered onto 50% w/v sucrose/NPB and centrifuged at 13,000 rpm for 10 min. Supernatants were taken as cytosolic fraction. Pellets (nuclei) were washed with NPB and extracted with Dignam C buffer (20 mM HEPES-NaOH, pH 7.9, 25% glycerol, 0.42 M NaCl, 1.5 mM MgCl$_2$, 0.2 mM EDTA, 1 mM DTT, plus PPIM) for 30 min at 4°C. Nuclear extracts were cleared by centrifugation at 13,000 rpm for 10 min at 4°C.

Acknowledgements. We thank M.T. Berciano and M. Lafarga for their help with the confocal microscopy and T. Martínez and M. González for their technical assistance. A.B. and M.M. were supported by predoctoral fellowships from Ministerio de Educación y Cultura of Spain. This work was supported by grants SAF98-0060 and SAF2001-2291 to A.M. and 1FD97–0281-CO2–01 to A.M. and C.C. from the Comisión Interministerial de Ciencia y Tecnología, Plan Nacional de Investigación y Desarrollo and PM97–0115 to C.C. from the Plan General del Conocimiento, Ministerio de Educación y Cultura of Spain.

References

Adcock IM, Lane SJ, Brown CR, Lee TH, Barnes PJ (1995) Abnormal glucocorticoid receptor-activator protein 1 interaction in steroid-resistant asthma. J Exp Med 182:1951–1958

Arias J, Alberts AS, Brindle P, Claret FX, Smeal T, Karin M, Feramisco J, Montminy M (1994) Activation of cAMP and mitogen responsive genes relies on a common nuclear factor. Nature 370:226–229

Ausubel FM, Brent R, Kingston RE, Moore DD, Seidman JG, Smith JA, Struhl K (1994) Current protocols in molecular biology. Greene Publishing Associates and Wiley-Interscience, New York

Barnes PJ, Karin M (1998) Nuclear factor-kappaB: a pivotal transcription factor in chronic inflammatory disease. N Engl J Med 336:1066–1071

Beato M, Herrlich P, Schütz G (1995) Steroid hormone receptors: many actors in search of a plot Cell 83:851–857

Behrens A, Sibilia M, Wagner EF (1999) Amino-terminal phosphorylation of c-Jun regulates stress-induced apoptosis and cellular proliferation. Nat Genet 21:326–329

Caelles C, González-Sancho JM, Muñoz A (1997) Nuclear hormone receptor antagonism with AP-1 by inhibition of the JNK pathway. Genes Dev 11:3351–3364

Chinenov Y, Kerppola TK (2001) Close encounters of many kinds: Fos-Jun interactions that mediate transcription regulatory specificity. Oncogene 20:2438–2452

Cole TJ, Blendy JA, Monaghan AP, Krieglstein K, Schimd W, Aguzzi A, Fantuzzi G, Hummier E, Unsicker K, Schütz G (1995) Targeted disruption of the glucocorticoid receptor gene blocks adrenergic chromaffin cell development and severely retards lung formation. Genes Dev 9:1608–1621

Devary Y, Gottlieb RA, Smeal T, Karin M (1992) The mammalian ultraviolet response is triggered by activation of Src tyrosine kinases. Cell 71:1081–1091

Felig P, Baxter JD, Broadus AE, Frohman LA (1987) Endocrinology and metabolism. McGraw-Hill, New York

González MV, González-Sancho JM, Caelles C, Muñoz A, Jiménez B (1999) Hormone-activated nuclear receptors inhibit the stimulation of the JNK and ERK signalling pathways in endothelial cells. FEBS Lett 459:272–276

González MV, Jiménez B, Berciano MT, González-Sancho JM, Caelles C, Lafarga M, Muñoz M (2000) Glucocorticoids antagonize AP-1 by inhibiting the activation/phosphorylation of JNK without affecting its subcellular distribution. J Cell Biol 150:1199–1207

Göttlicher M, Heck S, Herrlich P (1998) Transcriptional cross-talk, the second mode of steroid hormone receptor action. J Mol Med 76:480–489

Heck S, Kullmann M, Gast A, Ponta H, Rahmsdorf HJ, Herrlich P, Cato AC (1994) A distinct modulating domain in glucocorticoid receptor monomers in the repression of activity of the transcription factor AP-1. EMBO J 13:4087–4095

Helmberg A, Auphan N, Caelles C, Karin M (1995) Glucocorticoid-induced apoptosis of human leukemic cells is caused by the repressive function of the glucocorticoid receptor. EMBO J 14:452–460

Herrlich P (2001) Cross-talk between glucocorticoid receptor and AP-1. Oncogene 20:2465–2475

Jonat C, Rahmsdorf HJ, Park KK, Cato ACB, Gebel S, Ponta H, Herrlich P (1990) Antitumor promotion and antiinflammation: down-modulation of AP-1 (Fos/Jun) activity by glucocorticoid hormone. Cell 62: 1189–1204

Kamei Y, Xu L, Heinzel T, Torchia J, Kurokawa R, Gloss B, Lin SC, Heyman RA, Rose D, Glass CK, Rosenfeld MG (1996) A CBP integrator complex mediates transcriptional activation and AP-1 inhibition by nuclear receptors. Cell 85:403–414

Karin M (1995) The regulation of AP-1 activity by mitogen-activated protein kinases. J Biol Chem 270:16483–16486

Karin M (1998) New twists in gene regulation by glucocorticoid receptor: is DNA binding dispensable? Cell 93:487–490

Karin M, Liu Z, Zandi E (1997) AP-1 function and regulation. Curr Opin Cell Biol 9:240–246

Krstic MD, Rogatsky I, Yamamoto KR, Garabedian MJ (1997) Mitogen-activated and cyclin-dependent protein kinases selectively and differentially modulate transcriptional enhancement by the glucocorticoid receptor. Mol Cell Biol 17:3947–3954

Lasa M, Brook M, Saklatvala J, Clark AR (2001) Dexamethasone destabilizes cyclooxygenase 2 mRNA by inhibiting mitogen-activated protein kinase p38. Mol Cell Biol 21:771–780

Leppa S, Saffrich R, Ansorge W, Bohmann D (1998) Differential regulation of c-Jun by ERK and JNK during PC12 cell differentiation. EMBO J 17:4404–4413

Mangelsdorf DJ, Thummel C, Beato M, Herrlich P, Schütz G, Umesono K, Blumberg B, Kastner P, Mark M, Chambon P, Evans RM (1995) The nuclear receptor superfamily: the second decade. Cell 83:835–839

Mechta-Grigoriou, Gerald D, Yaniv M (2001) The mammalian Jun proteins: redundancy and specificity. Oncogene 20:2378–2389

Minden A, Karin M (1997) Regulation and function of the JNK subgroup of MAP kinases. Biochim Biophys Acta 1333:F85-F104

Minden A, Lin A, Smeal T, Dérijard B, Cobb M, Davis R, Karin M (1994) c-Jun N-terminal phosphorylation correlates with activation of the JNK subgroup but not the ERK subgroup of mitogen-activated protein kinases. Mol Cell Biol 14:6683–6688

Reichardt HM, Kaestner KH, Tuckermann J, Kretz O, Wessely O, Bock R, Gass P, Schmid W, Herrlich P, Angel P, Schutz G (1998) DNA binding of the glucocorticoid receptor is not essential for survival. Cell 93:531–541

Resche-Rigon M, Gronemeyer H (1998) Therapeutic potential of selective modulators of nuclear receptor action. Curr Opin Chem Biol 2:501–507

Rogatsky I, Logan SK, Garabedian MJ (1998) Antagonism of glucocorticoid receptor transcriptional activation by the c-Jun N-terminal kinase. Proc Natl Acad Sci USA 95:2050–2055

Schena M, Freedman LP, Yamamoto KR (1989) Mutations in the glucocorticoid receptor zinc finger region that distinguish interdigitated DNA binding and transcriptional enhancement activities. Genes Dev 3:1590–1601

Schüle R, Rangarajan P, Kliewer S, Ransone LJ, Bolado J, Yang N, Verma IM, Evans RM (1990) Functional antagonism between oncoprotein c-Jun and the glucocorticoid receptor. Cell 62:1217–1226

Tuckermann JP, Reichardt HM, Arribas R, Richter KH, Schütz G, Angel P (1999) The DNA binding-independent function of the glucocorticoid receptor mediates repression of AP-1-dependent genes in skin. J Cell Biol 147:1365–1370

Vayssière BM, Dupont S, Choquart A, Petit F, Garcia T, Marchandeau C, Gronemeyer H, Resche-Rigon M (1997) Synthetic glucocorticoids that dissociate transactivation and AP-1 transrepression exhibit antiinflammatory activity in vivo. Mol Endocrinol 11:1245–1255

Wehle H, Moll J, Cato ACB (1995) Molecular identification of steroid analogs with dissociated antiprogestin activities. Steroids 60:368–374

Yang-Yen HF, Chambard JC, Sun YL, Smeal T, Schmidt TJ, Drouin J, Karin M (1990) Transcriptional interference between c-Jun and the glucocorticoid receptor: mutual inhibition of DNA binding due to direct protein-protein interaction. Cell 62:1205–1215

10 Mast Cells as Targets for Glucocorticoids in the Treatment of Allergic Disorders

O. Kassel, A.C.B. Cato

10.1	Introduction	153
10.2	Effect of Glucocorticoids on Mast Cell Biology and Function	154
10.3	Effect of Glucocorticoids on Signal Transduction in Mast Cells	160
10.4	Conclusion and Future Prospects	168
References		169

10.1 Introduction

Mast cells are immune inflammatory cells residing in peripheral tissues and acting as sentinels. They are involved in diverse physiological processes such as tissue repair or wound healing (reviewed in Artuc et al. 1999), and they participate in defence against parasitic infection and other aspects of innate and acquired immunity (reviewed in Mekori and Metcalfe 2000; Wedemeyer and Galli 2000). In susceptible allergic individuals, this protection system is unbalanced and mast cells play a deleterious role in the pathophysiology of allergic disorders (reviewed in Holgate 2000; Bingham and Austen 2000; Hart 2001). An increase in the number of infiltrated mast cells, often in an activated state, is indeed observed at sites of allergic inflammation such as in the bronchus of asthmatic patients (Laitinen et al. 1993; Pesci et al. 1993; Bradding et al. 1994; Kassel et al. 2001a). Mast cells express high affinity IgE receptors (FcεRI) on their surface, which when aggregated by an antigen (allergen) recognized by specific receptor-bound IgE, initiate biochemical events leading to the release of inflammatory mediators (Metzger 1992).

By far the most effective treatment for allergic diseases is the use of glucocorticoids (reviewed in Barnes et al. 1998). Even after a short time of treatment, glucocorticoids reduce both the number and the activation of mast cells that have infiltrated the bronchus of asthmatic patients (for a review see Barnes et al. 2000). The precise mechanisms of the therapeutic benefit of glucocorticoids, most probably resulting from multiple and integrated effects, have not been fully elucidated. Recent advances in the molecular mechanism of action of glucocorticoids offer the possibility for a better understanding of the control of mast cell biology and function.

10.2 Effect of Glucocorticoids on Mast Cell Biology and Function

10.2.1 Effect of Glucocorticoids on Mast Cell Development

Mast cells originate from bone marrow haematopoietic progenitors (reviewed in Austen and Boyce 2001). Their precursors express the surface marker CD34, but not the high-affinity receptor for IgE (FcεRI) (Rottem et al. 1994). Unlike cells from other haematopoietic lineages, mast cell progenitors leave the bone marrow in a non-differentiated state, via the blood stream (Agis et al. 1993) to the peripheral tissues where differentiation and maturation occur under the influence of the microenvironment (Hayashi et al. 1985). These processes are controlled by mast cell growth factors (reviewed in Austen and Boyce 2001) such as stem cell factor or nerve growth factor produced by structural cells and released in the tissue microenvironment (Kim et al. 1997; Virchow et al. 1998; Kassel et al. 2001a). Differentiated mature mast cells then become $CD34^-/Fc\varepsilon RI^+$. This process of differentiation/maturation can be reproduced in vitro by cultivation of haematopoietic progenitor cells from bone marrow, spleen or foetal liver, in the presence of mast cell growth factors, until pure mast cell population are obtained (reviewed in Austen and Boyce 2001).

In the in vitro culture models, glucocorticoids inhibit mast cell development from haematopoietic progenitors (Irani et al. 1995; Eklund et al.

1997). However, they affect only early precursors, since proliferation is not inhibited when they are added at a later stage (Irani et al. 1995). Glucocorticoids might also affect mast cell maturation. Treatment of cultured mast cells derived from bone marrow progenitors with glucocorticoids reduces their expression of granular chymases (mouse mast cell protease-1 and -2), which are markers for granule differentiation (Eklund et al. 1997). Thus, the inhibition of mast cell early development may account for the reported reduction of tissue mast cell number following glucocorticoid treatment.

10.2.2 Effect of Glucocorticoids on Mast Cell Survival

A further contribution of glucocorticoids to the reduction of tissue mast cell numbers is through a decreased survival of infiltrated mast cells. Although glucocorticoids do not induce mast cell apoptosis in vitro (Tchekneva and Serafin 1994; Finotto et al. 1997), they may do so in vivo through an indirect mechanism. Mast cells are critically dependent on growth factors for their survival, and withdrawal of factors, such as stem cell factor, induces apoptosis (Yee et al. 1994). Some of the growth factors needed for mast cell survival, for example stem cell factor or nerve growth factor, are transcriptionally downregulated by glucocorticoids (Finotto et al. 1997; Kim et al. 1997; Kassel et al. 1998; Olgart and Frossard 2001; Pons et al. 2001). This could indirectly result in increased mast cell apoptosis as demonstrated in in vivo experiments in mice. In this study, topical treatment of mice with glucocorticoids decreased the number of cutaneous mast cells as a result of apoptosis brought about by the inhibition of stem cell factor production (Finotto et al. 1997).

10.2.3 Effect of Glucocorticoids on Mast Cell Mediator Release

Mast cells exert their deleterious effect in allergic inflammatory disorders by releasing numerous mediators (reviewed in Bingham and Austen 2000; Hart 2001). This is initiated by an antigen (allergen) recognized by specific IgE bound on the FcεRI at the cell surface. This process is described as mast cell activation. The activated mast cells use three

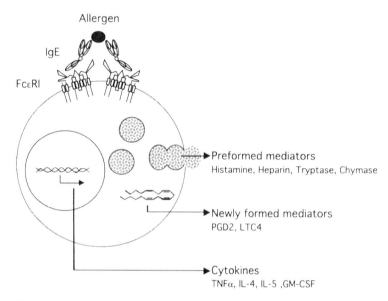

Fig. 1. Mast cell activation and mediator release. Allergen-mediated aggregation of specific IgE bound to the high-affinity receptor for IgE (*FcεRI*) promotes the release of three categories of mediators: preformed mediators stored in the granules and released by an exocytotic process, newly formed mediators derived from the arachidonic acid metabolism, and cytokines, through an enhanced transcription and release

different ways to release inflammatory mediators (Fig. 1). The first, which is an almost immediate reaction, is degranulation. This leads to the release of preformed mediators stored in the granules by an exocytotic process. Among these mediators are anaphylactic amines like histamine or serotonin as well as proteases such as chymase and tryptase, heparin and other hypersulphated proteoglycans. The second process of mediator release is the activation of the arachidonic acid metabolism leading to the release of arachidonic acid itself and its metabolites, which are in mast cells mainly prostaglandin D_2 and leukotriene C_4. The third process is the activation of the expression of cytokine genes, leading to the release of numerous cytokines such as tumour necrosis factor alpha (TNF-α), interleukins (IL)-4 and -6, transforming growth factor (TGF)-β or granulocyte-macrophage (GM)-CSF. These cytokines are involved in inflammation and tissue repair processes (reviewed in Kobayashi et al 2000).

In cultured mast cells, glucocorticoids inhibit all three processes: the degranulation process (Daeron et al. 1982; Heiman and Crews 1984; Collado-Escobar et al. 1990; Rider et al. 1996), the activation of arachidonic acid metabolism (Heiman and Crews 1984; Collado-Escobar et al. 1990; Rider et al. 1996), as well as the activation of the expression and release of cytokines (Wershil et al. 1995; Schmidt-Choudhury et al. 1996; Eklund et al. 1997; Kimata et al. 2001). However, these inhibitory effects occur with different kinetics. In RBL-2H3 rat mast cells, the FcεRI aggregation-induced degranulation, measured by the release of the granule marker β-hexosaminidase, is only partly inhibited after 5 h of pre-treatment with the synthetic glucocorticoid dexamethasone, and is fully inhibited after 24 h of treatment with this hormone (Fig. 2A). The activation of arachidonic acid metabolism is inhibited by dexamethasone only after prolonged treatment (Fig. 2B), and TNF-α release is totally abrogated both after 5 h and 24 h of treatment (Fig. 2C). The differences in the kinetics of action of glucocorticoids suggest that different mechanisms might be used for the inhibition of these processes.

The inhibition of mast cell activation by glucocorticoids has been observed in vitro but it occurs in vivo as well. In the bronchus of asthmatic patients, glucocorticoid treatment reduces the number of degranulated mast cells (for a review see Barnes et al. 2000) and the number of mast cells expressing proinflammatory cytokines such as IL-4 (Bradding et al. 1995). Thus, inhibition of mast cell activation and mediator release might be of substantial therapeutic benefit in allergic inflammation.

At the molecular level, glucocorticoids function by binding to a cytoplasmic receptor, the glucocorticoid receptor (GR), which is then transported into the nucleus where it regulates the expression of specific genes (Beato et al. 1995). Two different modes of action of the liganded GR have been described. (1) Transactivation – where the GR binds as a homodimer to specific sequences to enhance gene expression. (2) Transrepression – where the GR is thought to interact directly or indirectly with transcription factors already bound to DNA to downregulate their action (König et al. 1992; Nissen and Yamamoto 2000). Repression of the activity of transcription factors, such as activator protein (AP)-1

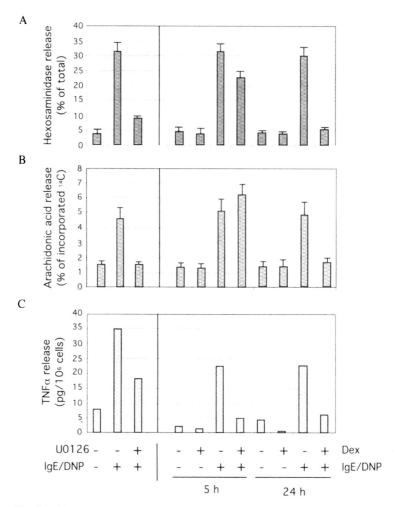

Fig. 2A–C. Legend see p. 159

or nuclear factor (NF)-κB, by the GR has been shown to be important for the anti-inflammatory action of glucocorticoids and might contribute to the inhibition of mast cell cytokine release (for reviews see Barnes 1998; Cato and Wade 1996; Herrlich 2001). The transcription factors AP-1 and NF-κB are activated following IgE receptor aggregation in mast cells (for a review see Nechushtan and Razin 1998), and are involved in the activation of cytokine genes expression (Hata et al. 1998; Novotny et al. 1998; Pelletier et al. 1998). The mechanisms of inhibition of degranulation and of arachidonic acid metabolism are less clear. Recent advances in the knowledge of glucocorticoid-mediated inhibition of signal transduction pathways may provide mechanistic explanations for this regulatory activity.

Fig. 2A–C. Involvement of Erk and effect of dexamethasone on degranulation, arachidonic acid metabolism, and TNF-α release by mast cells. **A** RBL-2H3 mast cells were treated for 30 min with solvent alone (dimethylsulphoxide) or with the mitogen-activated extracellular signal regulated kinase activating kinase (MEK) inhibitor U0126 (20 µM) (*left part*); or for 5 h or 24 h with solvent alone (ethanol) or dexamethasone (*Dex*, 0.1 µM) (*right part*), before sensitization for 1 h with monoclonal anti-2,4-dinitrophenyl (DNP) IgE. Cells were stimulated with DNP-bovine serum albumin (BSA) for 15 min, and degranulation was assessed by measurement of β-hexosaminidase release (Benhamou et al. 1986). Results are expressed as mean±standard error of mean (SEM) of three different experiments and presented as the percentage released from total cell content. **B** RBL-2H3 were labelled for 16 h with [^{14}C]arachidonate, and treated with U0126 or dexamethasone as in **A**. Cells were stimulated with DNP-BSA for 30 min, and arachidonic acid metabolism was assessed by measurement of ^{14}C release. Results are expressed as mean±SEM of three different experiments and presented as the percentage of ^{14}C released from total incorporated ^{14}C. **C** Cells were treated as in **A**, stimulated with DNP-BSA for 4 h, and TNF-α release was measured by enzyme-linked immunosorbent assay (ELISA). Results are means from two different experiments

10.3 Effect of Glucocorticoids on Signal Transduction in Mast Cells

10.3.1 Early Signalling Events

The binding of a specific antigen to FcεRI-bound IgE elicits the aggregation of the receptor. This promotes the recruitment and activation of the tyrosine kinases Syk, Lyn and Btk, the phosphorylation of the β and γ subunits of the receptor, and the initiation of several signalling cascades (for a review see Turner and Kinet 1999). A first potential level of glucocorticoid action in inhibiting mast cell activation might be a reduction of IgE binding to the FcεRI (Fig. 3). An early study showed a decrease in the number of IgE binding sites on mast cell membranes upon glucocorticoid treatment (Benhamou et al. 1986). This was confirmed by a more recent report describing a glucocorticoid-induced decrease in the expression of the receptor at the cell surface (Yamaguchi et al. 2001). These results are, however, in contradiction with previous reports showing a lack of glucocorticoid effect either at the receptor expression level or at the level of IgE binding to the receptor (Daeron et al. 1982; Rider et al. 1996). The effect of glucocorticoids on the expression of FcεRI in mast cells, therefore, remains to be clarified.

10.3.2 The Phospholipase C Pathway

Upon FcεRI aggregation, the phospholipase C (PLC)γ pathway is rapidly activated in mast cells (for a review, see Turner and Kinet 1999). This leads to the generation of inositol phosphates (IPs) including inositol 1,4,5-trisphosphate (IP3) responsible for the increase in intracellular free calcium $[Ca^{2+}]_i$, and of diacylglycerol responsible for the activation of protein kinase C (PKC) (Fig. 3). The activation of PKC and the increase in $[Ca^{2+}]_i$ are both necessary for antigen-mediated mast cell degranulation. The increase in $[Ca^{2+}]_i$ is also necessary to reach a full activation of phospholipase A2, the first enzyme in the arachidonic acid cascade (for reviews see Beaven and Baumgartner 1996; Turner and Kinet 1999). Glucocorticoids inhibit the generation of IPs (Collado-Escobar et al. 1990; Her et al. 1991; Rider et al. 1996), and the subsequent increase in $[Ca^{2+}]_i$ (Daeron et al. 1982; Heiman and Crews

Mast Cells as Targets for Glucocorticoids

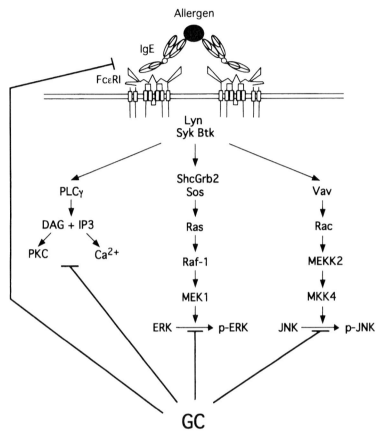

Fig. 3. Signalling pathways inhibited by glucocorticoids in mast cells. Allergen-mediated aggregation of specific IgE bound to the high-affinity receptor for IgE (*FcεRI*) promotes the recruitment and activation of the tyrosine kinases Lyn Syk and Btk, leading to the activation of several transduction pathways. The binding of IgE to the receptor, the phospholipase C (*PLC*)γ pathway, the extracellular regulated kinase (*Erk*) and the c-Jun NH$_2$-terminal kinase pathways are inhibited by glucocorticoids (*GC*)

1984; Collado-Escobar et al. 1990) through an unknown mechanism. This inhibition of the PLCγ pathway might contribute to the downregulation of antigen-mediated degranulation and activation of the arachidonic acid metabolism in mast cells.

10.3.3 The c-Jun NH2-Terminal Kinase Pathway

IgE receptor crosslinking activates the c-Jun NH_2-terminal kinase (JNK) pathway through the vav-rac-MEKK2-MKK4/7 cascade (Fig. 3) (for a review, see Turner and Kinet 1999). This pathway is essential for the expression of some of the cytokine genes turned on following mast cell antigenic activation, such as IL-4, IL-6 or TNF-α (Song et al. 1999; Garrington et al. 2000). In an effort to understand the effect of glucocorticoids on signalling in mast cells, we analysed the effect of the synthetic glucocorticoid dexamethasone on JNK activation in RBL-2H3 rat mast cells. Dexamethasone inhibited JNK phosphorylation, after 8 h of treatment, and the inhibition persisted up to 24 h (Fig. 4A). This result is in agreement with a previous report showing inhibition of JNK activity in antigen-activated mast cells as early as 6 h following glucocorticoid treatment (Hirasawa et al. 1998). We could also show that the negative regulation of JNK activity by glucocorticoid occurs independently of new protein synthesis. The addition of the protein synthesis inhibitor cycloheximide did not affect the inhibitory effect of dexamethasone on JNK phosphorylation in mast cells (Fig. 4B). This result demonstrates that glucocorticoid-mediated inhibition of JNK is a primary response of the receptor and does not require de novo protein synthesis. Similar conclusions were drawn from studies on the effect of the GR on the activity of JNK in other cell types (Caelles et al. 1997; Ventura et al. 1999; González et al. 2000). In HeLa cells, the effect of glucocorticoid on JNK phosphorylation has been further analysed. It has been shown to occur at a level downstream of mitogen-activated extracellular signal regulated kinase activating kinase kinase (MEKK) activation, and glucocorticoids inhibit already phosphorylated JNK, suggesting that JNK is the direct target of repression by the GR (Caelles et al. 1997). This inhibition of JNK was proposed as an alternative mechanism for AP-1 repression by glucocorticoids (Caelles et al. 1997; González et al. 2000) and might contribute to a repression of the expression of proinflamma-

Mast Cells as Targets for Glucocorticoids 163

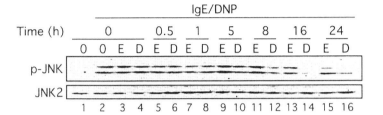

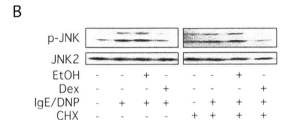

Fig. 4A, B. Effect of glucocorticoids on JNK activity in mast cells. RBL-2H3 mast cells were treated for the indicated time (**A**) or for 16 h (**B**) with either medium alone (*0*), ethanol (*E*), or dexamethasone (0.1 μM, *D*), in the presence or absence of the protein synthesis inhibitor cycloheximide (5 μg/ml, *CHX*) (**B**). Cells were sensitized for 1 h with monoclonal anti-DNP IgE, then stimulated with DNP-BSA for 15 min. Cell lysates were prepared and subjected to Western blotting, using an anti-phosphoJNK antibody. Membranes were stripped and reprobed with a phosphorylation state independent anti-JNK-2 antibody. Results are representative of three different experiments

tory cytokines (Swantek et al. 1997). Whether such a mechanism as described in HeLa cells occurs in mast cells remains to be determined. However, in an in vivo model of skin inflammation in asthmatic patients, glucocorticoid treatment induces a decrease in the number of inflammatory cells expressing phosphorylated JNK without affecting the number of cells expressing total JNK (Sousa et al. 1999). Thus, the inhibition by glucocorticoids of JNK activity might be relevant in the treatment of inflammatory disorders.

10.3.4 The Extracellular Regulated Kinase Pathway

FcεRI aggregation is also coupled to the activation the Erk-1and -2 pathway through a Grb2-sos-ras-Raf1-MEK1 cascade (Fig. 3) (Offermanns et al. 1994; Hirasawa et al. 1995a; 1995b). With the use of the MEK1 inhibitor U0126, we were able to strongly reduce the FcεRI aggregation-induced degranulation, measured by the release of the granule marker β-hexosaminidase in RBL-2H3 mast cells (Fig. 2A). The use of the MEK1 inhibitor also abrogated the activation of arachidonic acid metabolism (Fig. 2B), and partially inhibited TNF-α release (Fig. 2C). These results are all in agreement with earlier studies showing that activated Erk-1 and -2 are involved in the generation of arachidonic acid via phospholipase A_2 and in the production of the inflammatory cytokines TNF-α and GM-CSF in mast cells (Hirasawa et al. 1995b; Zhang et al. 1997; Kimata et al. 2000). Thus the Erk-1/2 pathway seems to be central in the activation of mast cell mediator release.

We and other investigators have shown that glucocorticoids inhibit Erk-1/2 activation following IgE receptor triggering in mast cells (Rider at al. 1996; Kassel et al. 2001b). This negative regulation of Erk-1/2 activity requires a prolonged treatment with glucocorticoids (16 to18 h) (Rider et al. 1996; Kassel et al. 2001b), suggesting the involvement of de novo protein synthesis. We confirmed this by the use of cycloheximide, which abrogated the glucocorticoid-mediated inhibition of Erk-1/2 activity (Kassel et al. 2001b). The negative regulation of Erk-1/2 phosphorylation by glucocorticoid was also observed in other cell types albeit not in every cell. In immune inflammatory cells, S49.1 murine thymoma cells, U937 monocytic cells and mouse primary splenocytes, Erk-1/2 phosphorylation induced by the phorbol ester 12-*O*-tetradecanoyl phorbol 13-acetate (TPA) was inhibited by dexamethasone after 16 h of treatment (Fig. 5). However, this effect was not observed in Cos-7 simian kidney epithelial cells, human lung primary fibroblasts, A549 human lung epithelial cells, and murine NIH-3T3 fibroblasts (Fig. 5 and unpublished results). The glucocorticoid-mediated inhibition of Erk-1/2 is therefore cell-type specific. This is in agreement with previous studies in different cell types showing either an inhibition (Hulley et al. 1998; González et al. 1999; Hirasawa et al. 2001) or lack of inhibition (Swantek et al. 1997; Fernandes et al. 1999; Gewert et al. 2000) of Erk-1/2 activity by glucocorticoids.

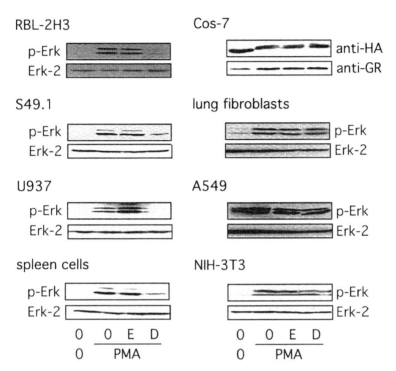

Fig. 5. Dexamethasone mediated inhibition of phosphorylation of Erk-1 and -2 is cell-type specific. Confluent serum starved cells were treated for 16 h with either medium alone (*0*), ethanol (*E*), or dexamethasone (0.1 μM, *D*). Cells were then stimulated with PMA (100 ng/ml) for 15 min. Cell lysates were prepared and subjected to Western blotting, using an anti-phosphoErk antibody. As a loading control, membranes were stripped and reprobed with a phosphorylation state independent anti-Erk-2 antibody. Cos-7 cells where transiently transfected with expression vectors encoding glucocorticoid receptor (*GR*) and haemagglutinin (*HA*)-tagged Erk-2. Cells where treated with dexamethasone and activated with PMA as described above. Cell lysates were prepared and subjected to Western blotting using an anti-HA antibody. This enabled a control of the transfection efficiency, and the phosphorylation of HA-Erk-2 was observed as a shift in electrophoretic mobility. Membranes were stripped and reprobed with an anti-GR antibody. Results are representative of two different experiments

The mechanism of glucocorticoid-mediated inhibition of the Erk-1/2 pathway has been studied at different levels by several groups. First, glucocorticoids were reported to inhibit the recruitment of the adaptor protein Grb2 to activated epidermal growth factor (EGF) receptor at the plasma membrane (Croxtall et al. 2000). Such an effect would inhibit the activation of the Erk-1/2 cascade within minutes (Croxtall et al. 2000). It is therefore unlikely to be related to the inhibition of the Erk-1/2 pathway in mast cells, which requires a prolonged treatment with the hormone (Rider et al. 1996; Kassel et al. 2001b). Second, a direct interaction of liganded GR and Raf-1, the upstream kinase in the Erk pathway, has been demonstrated in vitro (Widén et al. 2000), which might contribute mechanistically to the inhibition of the Erk-1/2 pathway. However, in living cells, such an interaction is also expected to occur rapidly, and would not account for the late inhibition of Erk-1/2 activity by glucocorticoids in mast cells. Third, an association of Raf-1 and Hsp90 required for the signalling activity of Raf-1 is disrupted following glucocorticoid treatment of mast cells (Cissel and Beaven 2000). This inhibition of Raf-1 activity may be responsible for the negative action of this hormone on Erk-1/2 activity since it is not a fast process (Cissel and Beaven 2000). However, it occurs with different kinetics than those for the inhibition of Erk-1/2 activity. Thus, it is uncertain whether they are related or whether they occur separately.

These uncertainties in the mechanism of negative regulation of Erk-1/2 activity prompted us to study the effect of glucocorticoid on the Erk signalling pathway. We have shown that Erk-1/2 is a direct target for the inhibitory effect of GR in mast cells (Kassel et al. 2001b). We showed in in vitro dephosphorylation studies that lysates from dexamethasone-treated RBL-2H3 mast cells contain a tyrosine phosphatase activity that dephosphorylates a preactivated Erk-2 substrate. This activity was later identified as mitogen-activated protein (MAP) kinase phosphatase-1 (MKP-1), a dual specificity tyrosine phosphatase that dephosphorylates and inactivates MAP kinases (Alessi et al. 1993; Sun et al. 1993). We could confirm that GR, through its transactivation function, induces MKP-1 expression at the promoter level. The subsequent increase in MKP-1 protein level is necessary for the glucocorticoid-induced inhibition of Erk-1/2 activity in mast cells. Other tyrosine phosphatases like HePTP and MKP-3, both known to inactivate Erk-1/2 (Groom et al. 1996; Pettiford and Herbst 2000), were not upregulated upon glucocor-

Mast Cells as Targets for Glucocorticoids 167

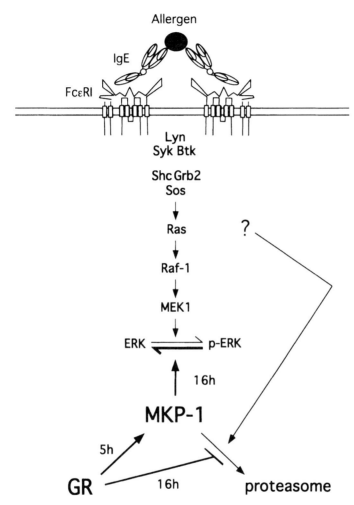

Fig. 6. Mechanism of MKP-1-dependent inhibition of Erk-1/2 activity by glucocorticoids in mast cells. Allergen-mediated aggregation of specific IgE bound to the high-affinity receptor for IgE (*FcεRI*) promotes the activation of the extracellular regulated kinase (*Erk*) pathway. Glucocorticoid receptor (*GR*) induces the expression of MAP kinase phosphatase-1 (*MKP-1*) after 5 h of glucocorticoid treatment. After 16 h of glucocorticoid treatment, the GR inhibits the degradation of MKP-1 by the proteasome, which is triggered upon IgE receptor activation. Both these effects of GR are required for the dephosphorylation of Erk observed after 16 h of glucocorticoid treatment

ticoid treatment of mast cells. This demonstrates that the effect of MKP-1 is rather specific (Kassel et al. 2001b). However, the dexamethasone-induced increase in MKP-1 protein levels is already maximal after 8 h of hormone treatment, a time at which Erk-1/2 activity is not yet inhibited. This apparent discrepancy is due to another regulation of MKP-1 levels. Following mast cell activation, MKP-1 is degraded via the proteasome pathway, a process inhibited by glucocorticoids only after prolonged treatment. Both the attenuation of proteasomal degradation and upregulation of MKP-1 expression are essential for glucocorticoid-mediated inhibition of Erk-1/2 activity in mast cells (Kassel et al. 2001b). This new mechanism of GR in the inhibition of Erk-1/2 activity in mast cells is illustrated in Fig. 6. In NIH-3T3 fibroblasts the regulation of MKP-1 levels is different. MKP-1 protein is constitutively degraded by the proteasome, and although glucocorticoids activate MKP-1 gene expression, they do not attenuate MKP-1 degradation. As a result Erk-1/2 activity is not inhibited in NIH 3T3 cells (Kassel et al. 2001b). This cell-type specificity of MKP-1 regulation by glucocorticoids might form the basis of the cell-type specificity of the inhibition of Erk-1/2 activity.

What might be the relevance of this MKP-1 dependent inhibition of the Erk-1/2 pathway by glucocorticoids in mast cell biology? Mast cell mediator release upon IgE receptor triggering is dependent on Erk-1/2 activation (Fig. 2). However, the effect of glucocorticoids on the three types of mediator release, i.e. degranulation, arachidonic acid metabolism and cytokine release, differs in term of kinetics (Fig. 2). Since MKP-1-dependent inhibition of the Erk-1/2 occurs only after a long pre-treatment with the hormone, it might account, at least in part, for the late repression of degranulation and arachidonic acid metabolism, but probably not for the early repression of TNF-α release.

10.4 Conclusion and Future Prospects

Glucocorticoids affect mast cell biology and function in many ways. They inhibit their development from haematopoietic progenitors, modulate their survival in the peripheral tissues, and inhibit the release of proinflammatory mediators upon activation of the IgE receptor. These effects, mostly studied in vitro, might contribute to the decrease in the number and activation of mast cells at sites of allergic inflammatory reactions following glucocorticoid treatment. The precise mechanisms by which the GR exerts these effects are gradually being elucidated. Increasing evidence point to a repressive role of glucocorticoids in the signal transduction events leading to IgE receptor-dependent mediator release. In particular, repression of the Erk-1/2 pathway through an MKP-1-dependent mechanism, relying on both MKP-1 induction and attenuation of its proteasomal degradation (Fig. 6), might participate in the anti-inflammatory effect of glucocorticoid treatment in mast cell-dependent allergic disorders. MKP-1 gene has been disrupted in mice. This disruption has not revealed any defects and the animals are viable and appear healthy (Dorfman et al. 1996). These mice remain to be analysed for their response to glucocorticoid treatment in models of allergic inflammation. These experiments will hopefully clarify the role of the inhibition of Erk-1/2 activity in the therapeutic effect of glucocorticoids in inflammation.

Another important aspect requiring further investigation is the cell-type specificity of the regulation of MKP-1 protein levels. Why is MKP-1 proteasomal degradation triggered upon activation of mast cells, whereas it is constitutive in fibroblasts? What are the mechanisms underlying these degradative pathways? Why and how is the attenuation of MKP-1 degradation by the GR effected in mast cells and not in fibroblasts? Answers to these questions will bring further insights into the molecular mechanism of GR in allergic disorders.

Acknowledgements. The authors thank Alessandra Sancono (Karlsruhe, Germany), Jörn Krätzschmar and Bertolt Kreft (Berlin, Germany) and Michael Stassen (Mainz, Germany) for their participation in the work presented here. We also thank the Dermatology Department in Schering AG (Berlin, Germany) for support. O. Kassel was partly supported by a fellowship from the French – German Science Foundation's exchange programme (Institut National de la Santé et de la Recherche Médicale – Deutsche Forschungsgemeinshaft).

References

Agis H, Willheim M, Sper WR, Wilfing A, Krömer E, Kabrna E, Spanblöchl E, Strobl H, Geissler K, Spittler A, Boltz-Nitulescu G, Madjic O, Lechner K, Valent P (1993) Monocytes do not make mast cells when cultured in the presence of SCF. Characterization of the circulating mast cell progenitor as a c-kit$^+$, CD34$^+$, Ly$^-$, CD14$^-$, CD17$^-$, colony forming cell. J Immunol 151:4221–4227

Alessi DR, Smythe C Keyse SM (1993) The human CL100 gene encodes a Tyr/Thr-protein phosphatase which potently and specifically inactivates MAP kinase and suppresses its activation by oncogenic ras in Xenopus oocyte extracts. Oncogene 8:2015–2020

Artuc M, Hermes B, Steckelings UM, Grutzkau A, Henz BM (1999) Mast cells and their mediators in cutaneous wound healing – active participants or innocent bystanders? Exp Dermatol 8:1–16

Austen KF, Boyce JA (2001) Mast cell lineage development and phenotypic regulation. Leuk Res 25:511–518

Barnes, PJ (1998) Antiinflammatory action of glucocorticoids: molecular mechanisms. Clin Sci 94:557–572

Barnes PJ, Pedersen S, Busse WW (1998) Efficacy and safety of inhaled corticosteroids. New developments. Am J Respir Crit Care Med 157:S1–S53

Barnes NC, Burke CM, Poulter LW, Schleimer RP (2000) The anti-inflammatory profile of inhaled corticosteroids: biopsy studies in asthmatic patients. Respir Med 94:S16–S21

Beato M, Herrlich P, Schutz G (1995) Steroid hormone receptors: many actors in search of a plot. Cell 83:851–857

Beaven MA, Baumgartner, RA (1996) Downstream signals initiated in mast cells by Fc epsilon RI and other receptors. Curr Opin Immunol 8:766–772

Benhamou M, Ninio E, Salem P, Hieblot C, Bessou G, Pitton C, Liu FT, Mencia-Huerta JM (1986) Decrease in IgE Fc receptor expression on mouse bone marrow-derived mast cells and inhibition of PAF-acether formation and of beta-hexosaminidase release by dexamethasone. J Immunol 136:1385–1392

Bingham CO 3rd, Austen KF (2000) Mast-cell responses in the development of asthma. J Allergy Clin Immunol 105:S527–S534

Bradding P, Roberts JA, Britten KM, Montefort S, Djukanovic R, Mueller R, Heusser CH, Howarth PH, Holgate ST (1994) Interleukin-4, interleukin-5, and interleukin-6 and tumor necrosis factor-alpha in normal and asthmatic airways – evidence for the human mast cell as a source of these cytokines. Am J Respir Cell Mol Biol 10:471–480

Bradding P Feather IH, Wilson S, Holgate ST, Howarth PH (1995) Cytokine immunoreactivity in seasonal rhinitis: regulation by a topical corticosteroid. Am J Respir Crit Care Med 151:1900–1906

Caelles C, González-Sancho JM, Muñoz A (1997) Nuclear hormone receptor antagonism with AP-1 by inhibition of the JNK pathway. Genes Dev 11:3351–3364

Cato AC, Wade E (1996) Molecular mechanisms of anti-inflammatory action of glucocorticoids. Bioessays 18:371–378

Cissel DS, Beaven, MA (2000) Disruption of Raf-1/Heat shock protein 90 complex and Raf signaling by dexamethasone in mast cells. J Biol Chem, 275:7066–7070

Collado-Escobar D, Cunha-Melo JR, Beaven MA (1990) Treatment with dexamethasone down-regulates IgE-receptor-mediated signals and up-regulates adenosine-receptor-mediated signals in a rat mast cell (RBL-2H3) line. J Immunol 144:244–250

Croxtall JD, Choudhury Q, Flower RJ (2000) Glucocorticoids act within minutes to inhibit recruitment of signalling factors to activated EGF receptors through a receptor-dependent, transcription-independent mechanism. Br J Pharmacol 130:289–298

Daeron M, Sterk AR, Hirata F, Ishizaka T (1982) Biochemical analysis of glucocorticoid-induced inhibition of IgE-mediated histamine release from mouse mast cells. J Immunol 129:1212–1218

Dorfman,K, Carrasco D, Gruda M, Ryan C, Lira S A, Bravo, R (1996) Disruption of the erp/mkp-1 gene does not affect mouse development: normal MAP kinase activity in ERP/MKP-1-deficient fibroblasts. Oncogene 13:925–931

Eklund KK, Humphries DE, Xia Z, Ghildyal N, Friend DS, Gross V, Stevens RL (1997) Glucocorticoids inhibit the cytokine-induced proliferation of mast cells, the high affinity IgE-receptor-mediated expression of TNF-alpha, and the IL-10-induced expression of chymases. J Immunol 158:4373–4380

Fernandes D, Guida E, Koutsoubos V, Harris T, Vadiveloo P, Wilson JW, Stewart AG (1999) Glucocorticoids inhibit proliferation, cyclin D1 expression, and retinoblastoma protein phosphorylation, but not activity of the extracellular-regulated kinases in human cultured airway smooth muscle. Am J Respir Cell Mol Biol 21:77–88

Finotto S, Mekori YA, Metcalfe DD (1997) Glucocorticoids decrease tissue mast cell number by reducing the production of the c-kit ligand, stem cell factor, by resident cells – in vitro and in vivo evidence in murine systems. J Clin Invest 99:1721–1728

Garrington TP, Ishizuka T, Papst PJ, Chayama K, Webb S, Yujiri T, Sun W, Sather S, Russell DM, Gibson SB, Keller G, Gelfand EW, Johnson GL (2000) MEKK2 gene disruption causes loss of cytokine production in response to IgE and c-Kit ligand stimulation of ES cell-derived mast cells. EMBO J 19:5387–5395

Gewert K, Hiller G, Sundler R (2000) Effects of dexamethasone on mitogen-activated protein kinases in mouse macrophages: implications for the regulation of 85 kDa cytosolic phospholipase A(2). Biochem Pharmacol 60:545–551

González MV, Gonzalez-Sancho JM, Caelles C, Munoz A, Jimenez B (1999) Hormone-activated nuclear receptors inhibit the stimulation of the JNK and ERK signalling pathways in endothelial cells. FEBS Lett 459:272–276

González MV, Jimenez B, Berciano MT, Gonzalez-Sancho JM, Caelles C, Lafarga M, Munoz A (2000) Glucocorticoids antagonize AP-1 by inhibiting the activation/phosphorylation of JNK without affecting its subcellular distribution. J Cell Biol 150:1199–1208

Groom LA, Sneddon AA, Alessi DR, Dowd S, Keyse SM (1996) Differential regulation of the MAP, SAP and RK/p38 kinases by Pyst1, a novel cytosolic dual-specificity phosphatase. EMBO J 15:3621–3632

Hart PH (2001) Regulation of the inflammatory response in asthma by mast cell products. Immunol Cell Biol 79:149–153

Hata D, Kitaura J, Hartman SE, Kawakami Y, Yokota T, Kawakami T (1998) Bruton's tyrosine kinase-mediated interleukin-2 gene activation in mast cells dependence on the c-Jun N-terminal kinase activation pathway. J Biol Chem 273:10979–10987

Hayashi C, Sonoda T, Nakano T, Nakayama H, kitamura Y (1985) Mast-cell precursors in the skin of mouse embryos and their deficiency in embryos of Sl/Sl[d] genotype. Dev Biol 109:234–241

Heiman AS, Crews FT (1984) Inhibition of immunoglobulin, but not polypeptide base-stimulated release of histamine and arachidonic acid by anti-inflammatory steroids. J Pharmacol Exp Ther 230:175–182

Her E, Reiss N, Braquet P, Zor U (1991) Characterization of glucocorticoid inhibition of antigen-induced inositolphosphate formation by rat basophilic leukemia cells: possible involvement of phosphatases. Biochim Biophys Acta 1133:63–72

Herrlich P (2001) Cross-talk between glucocorticoid receptor and AP-1. Oncogene 20:2465–2475

Hirasawa N, Scharenberg A, Yamamura H, Beaven MA, Kinet JP (1995a) A requirement for Syk in the activation of the microtubule-associated protein kinase/phospholipase A2 pathway by Fc epsilon R1 is not shared by a G protein-coupled receptor. J Biol Chem 270:10960–10967

Hirasawa N, Santini F, Beaven M A (1995b) Activation of the mitogen-activated protein kinase/cytosolic phospholipase A_2 pathway in a rat mast cell line. J Immunol 154:5391–5402

Hirasawa N, Sato Y, Fujita Y, Mue S, Ohuchi K (1998) Inhibition by dexamethasone of antigen-induced c-Jun N-terminal kinase activation in rat basophilic leukemia cells. J Immunol 161:4939–4943

Hirasawa N, Murakami A, Ohuchi K (2001) Expression of a 74-kDa histidine decarboxylase protein in a macrophage-like cell line RAW 2647 and inhibition by dexamethasone. Eur J Pharmacol 418:23–28

Holgate ST (2000) The role of mast cells and basophils in inflammation. Clin Exp Allergy 30:S28–S32

Hulley PA, Gordon F, Hough FS (1998) Inhibition of mitogen-activated protein kinase activity and proliferation of an early osteoblast cell line (MBA 154) by dexamethasone: role of protein phosphatases. Endocrinology 139:2423–2431

Irani AA, Nilsson G, Ashman LK, Schwartz LB (1995) Dexamethasone inhibits the development of mast cells from dispersed human fetal liver cells cultured in the presence of recombinant stem cell factor. Immunology 84:72–78

Kassel O, Schmidlin F, Duvernelle C, de Blay F, Frossard N (1998) Up- and down-regulation by glucocorticoids of the constitutive expression of the mast cell growth factor stem cell factor by human lung fibroblasts in culture. Mol Pharmacol 54:1073–1079

Kassel O, de Blay F, Duvernelle C, Olgart C, Israel-Biet D, Krieger P, Moreau L, Muller C, Pauli G, Frossard N (2001a) Local increase in the number of mast cells and expression of nerve growth factor in the bronchus of asthmatic patients after repeated inhalation of allergen at low-dose. Clin Exp Allergy 31:1432–1440

Kassel O, Sancono A, Krätzschmar J, Kreft B, Stassen M, Cato AC (2001b) Glucocorticoids inhibit MAP-kinase via increased expression and decreased degradation of MKP-1. EMBO J 20:7108–7116

Kim Y-K, Nakagawa N, Nakano K, Sulakvelidze I, Dolovich J, Denburg J (1997) Stem cell factor in nasal polyposis and allergic rhinitis: increased expression by structural cells is suppressed by in vivo topical corticosteroids. J Allergy Clin Immunol 100:389–399

Kimata M, Inagaki N, Kato T, Miura T, Serizawa I, Nagai H (2000) Roles of mitogen-activated protein kinase pathways for mediator release from human cultured mast cells. Biochem Pharmacol 60:589–594

Kimata M, Abe T, Yamaguchi I, Mito K, Tsunematsu M, Inagaki N, Nagai H (2001) Prednisolone inhibits an IgE-mediated late-phase allergic cutaneous reaction by interfering with the activation of mast cells in mice. Pharmacology 62:17–22

Kobayashi H, Ishizuka T, Okayama Y (2000) Human mast cells and basophils as sources of cytokines. Clin Exp Allergy 30:1205–1212

König H, Ponta H, Rahmsdorf HJ, Herrlich P (1992) Interference between pathway-specific transcription factors: glucocorticoids antagonize phorbol ester-induced AP-1 activity without altering AP-1 site occupancy in vivo. EMBO J 11:2241–2246

Laitinen LA, Laitinen A, Haahtela T (1993) Airway mucosal inflammation even in patients with newly diagnosed asthma. Am Rev Respir Dis 147:697–704

Mekori YA, Metcalfe DD (2000) Mast cells in innate immunity. Immunol Rev 173:131–140

Metzger H (1992) The receptor with high affinity for IgE. Immunol Rev 125:37–48

Nechushtan H, Razin E (1998) Deciphering the early-response transcription factor network in mast cells. Trends Immunol 19:441–444

Nissen RM, Yamamoto KR (2000) The glucocorticoid receptor inhibits NFκB by interfering with serine-2 phosphorylation of the RNA polymerase II carboxy-terminal domain. Genes and Dev 14:2314–2329

Novotny V, Prieschl EE, Csonga R, Fabjani G, Baumruker T (1998) Nrf1 in a complex with fosB, c-jun, junD and ATF2 forms the AP1 component at the TNF alpha promoter in stimulated mast cells. Nucleic Acids Res 26:5480–5485

Olgart C, Frossard N (2001) Human lung fibroblasts secrete nerve growth factor: effect of inflammatory cytokines and glucocorticoids. Eur Respir J 18:115–121

Offermanns S, Jones SV, Bombien E, Schultz G (1994) Stimulation of mitogen-activated protein kinase activity by different secretory stimuli in rat basophilic leukemia cells. J Immunol 152:250–261

Pelletier C, Varin-Blank N, Rivera J, Iannascoli B, Marchand F, David B, Weyer A, Blank U (1998) FcεRI-mediated induction of TNF-α gene expression in the RBL-2H3 mast cell line: regulation by a novel NF-κB-like nuclear binding complex. J Immunol 161:4768–4776

Pesci A, Foresi A, Bertorelli G, Chetta A, Oliveri D (1993) Histochemical characteristics and degranulation of mast cells in epithelium and lamina propria of bronchial biopsies from asthmatic and normal subjects. Am Rev Respir Dis 147:684–689

Pettiford SM, Herbst R (2000) The MAP-kinase ERK2 is a specific substrate of the protein tyrosine phosphatase HePTP. Oncogene 19:858–869

Pons F, Freund V, Kuissu H, Mathieu E, Olgart C, Frossad N (2001) Nerve growth factor secretion by human lung epithelial A549 cells in pro- and anti-inflammatory conditions. Eur J Pharmacol 428:365–369

Rider L G, Hirasawa N, Santini F, Beaven MA (1996) Activation of the mitogen-activated protein kinase cascade is suppressed by low concentrations of dexamethasone in mast cells. J Immunol 157:2374–2380

Rottem M, Okada T, Goff JP, Metcalfe DD (1994) Mast cells cultured from the peripheral blood of normal donors and patients with mastocytosis originate from a CD34+ Fc(epsilon)RI(–) cell population. Blood 84:2489–2496

Schmidt-Choudhury A, Furuta GT, Lavigne JA, Galli SJ, Wershil BK (1996) The regulation of tumor necrosis factor-α production in murine mast cells: pentoxifylline or dexamethasone inhibits IgE-dependent production of TNF-α by distinct mechanisms. Cell Immunol 171:140–146

Song JS, Haleem-Smith H, Arudchandran R, Gomez J, Scott PM, Mill JF, Tan TH, Rivera J (1999) Tyrosine phosphorylation of Vav stimulates IL-6 production in mast cells by a Rac/c-Jun N-terminal kinase-dependent pathway. J Immunol 163:802–810

Sousa AR, Lane SJ, Soh C, Lee TH (1999) In vivo resistance to corticosteroids in bronchial asthma is associated with enhanced phosphorylation of JUN N-terminal kinase and failure of prednisolone to inhibit JUN N-terminal kinase phosphorylation. J Allergy Clin Immunol 104:565–574

Sun H, Charles CH, Lau LF, Tonks NK (1993) MKP-1 (3CH134), an immediate early gene product, is a dual specificity phosphatase that dephosphorylates MAP kinase in vivo. Cell 75:487–493

Swantek JL, Cobb MH, Geppert TD (1997) Jun-N-terminal kinase/stress-activated protein kinase (JNK/SAPK) is required for lipopolysaccharide stimulation of tumor necrosis factor a (TNF-α) translation: glucocorticoids inhibit TNF-α translation by blocking JNK/SAPK. Mol Cell Biol 17:6274–6282

Tchekneva E, Serafin WE (1994) Kirsten sarcoma virus-immortalized mast cell lines Reversible inhibition of growth by dexamethasone and evidence for the presence of an autocrine growth factor. J Immunol 152:5912–5921

Turner H, Kinet J-P (1999) Signalling through the high-affinity IgE receptor FcεRI. Nature 402:B24–B30

Ventura JJ, Roncero C, Fabregat I, Benito M (1999) Glucocorticoid receptor down-regulates c-Jun amino terminal kinases induced by tumor necrosis factor alpha in fetal rat hepatocyte primary cultures. Hepatology 29:849–857

Virchow JC, Julius P, Lommatzsch M, Luttmann W, Renz H, Braun A (1998) Neurotrophins are increased in bronchoalveolar lavage fluid after segmental allergen provocation. Am J Respir Crit Care Med 158:2002–2005

Wedemeyer J, Galli SJ (2000) Mast cells and basophils in acquired immunity. Br Med Bull 56:936–955

Wershil BK, Furuta GT, Lavigne JA, Choudhury AR, Wang ZS, Galli SJ (1995) Dexamethasone or cyclosporin A suppress mast cell-leukocyte cytokine cascades. Multiple mechanisms of inhibition of IgE- and mast cell-dependent cutaneous inflammation in the mouse. J Immunol 154:1391–1398

Widén C, Zilliacus J, Gustafsson JA, Wikstrom AC (2000) Glucocorticoid receptor interaction with 14–3-3 and Raf-1, a proposed mechanism for crosstalk of two signal transduction pathways. J Biol Chem 275:39296–39301

Yamaguchi M, Hirai K, Komiya A, Miyamasu M, Furomoto Y, Teshima R, Ohta K, Morita Y, Galli SJ, Ra C, Yamamoto K (2001) Regulation of mouse mast cell surface Fc epsilon RI expression by dexamethasone. Int Immunol 13:843–851

Yee NS, Paek I, Besmer P (1994) Role of kit-ligand in proliferation and suppression of apoptosis in mast cells: basis for radiosensitivity of White spotting and Steel mutant mice. J Exp Med 179:1777–1787

Zhang C, Baumgartner RA, Yamada K, Beaven MA (1997) Mitogen-activated protein (MAP) kinase regulates production of tumor necrosis factor-alpha and release of arachidonic acid in mast cells. J Biol Chem 272:13397–13402

11 Cytosolic Glucocorticoid Receptor-Interacting Proteins

A.-C. Wikström, C. Widén, A. Erlandsson, E. Hedman, J. Zilliacus

11.1	Introduction and Background	177
11.2	GR and Chaperone Interaction	179
11.3	GR, NF-κB and Repression of Inflammation	181
11.4	GR and Cytoskeletal Protein Interaction	182
11.5	GR and Intracellular Localization	183
11.6	Other GR-Interacting Proteins	184
11.7	GR Interaction with 14-3-3 and Raf-I	185
11.8	Further Analysis of Cytosolic Proteins Interacting with GR	187
11.9	Summary	191
References		191

11.1 Introduction and Background

The classic view of glucocorticoid mechanism of action involves the concept of a non-liganded receptor residing in the cytoplasm of cells. The receptor is maintained in an inactive state, with regard to the ability to bind to DNA, by the interaction with a multi-protein chaperone complex that contains, among other proteins, a dimer of hsp90, hsp70, hsp40, FKBP52 and p23. Of these proteins, hsp90 has been shown to maintain glucocorticoid receptor (GR) in a conformation optimal for ligand binding, and it also serves to cover the two nuclear localization signals (NL) of GR. NL1 is mapped to certain basic amino acid sequences in the DNA-binding domain and hinge region, and NL2 is broadly mapped to the ligand-binding domain of GR. Upon ligand

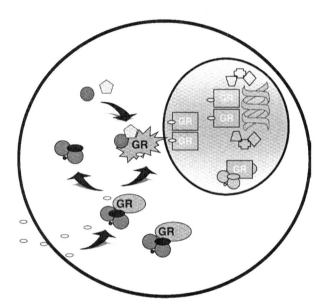

Fig. 1. Proposed model of GR action. Glucocorticoids bind intracellular glucocorticoid receptors (*GR*). Upon binding and activation, GR dissociate from some of the heat-shock proteins in the GR protein multi-complex, but also probably recruit new interaction partners and enter the nucleus. Not all of the receptor-associating proteins are identified and fully characterized. In the nucleus, GR also bind to adaptor proteins and coregulators

binding, GR dissociates from hsp90 and also from several of the other chaperone proteins, and translocates to the nucleus. In the nucleus, GR is present as a homodimer and acts as a ligand-dependent transcription factor binding specific DNA sequences, i.e., glucocorticoid response elements (GRE) of glucocorticoid-regulated genes. In this process, GR can associate with adaptor proteins and the basal transcription machinery, as well as with specific transcription factors and coregulators. The classical view of GR mechanism of action has been shown to be even more complex, and GR action most likely encompasses more protein interactions than described above. Not only the shedding of chaperone proteins but also the recruitment of novel interaction partners upon ligand binding may be of importance for glucocorticoid signalling as shown in a model in Fig. 1. Some of these interactions and possible functional consequences thereof will be outlined below, and this will also serve as a background to why further studies of cytosolic protein–protein interactions involving GR may be relevant for a more thorough understanding glucocorticoid mechanism of action.

11.2 GR and Chaperone Interaction

Steroid receptors are functionally dependent on interactions with chaperone proteins and the prerequisites, and functional implications of these interactions have been reviewed in detail, most comprehensively by Pratt and Toft (1997) and more recently updated in two reviews by Cheung and Smith (2000) as well as by Pratt and co-workers (Pratt et al. 1999). In brief, chaperone proteins such as heat-shock proteins (hsp) assist in folding processes and also in targeting of proteins for degradation. In the case of GR and progesterone receptor (PR), hsp also plays a role in intracellular transport of the receptors, dependent on the interaction with immunophilins and the cytoskeletal network. The involved chaperone hsp90 together with hsp70 seems to be necessary for a high-affinity ligand binding by PR and GR especially. Other steroid receptors like the oestrogen receptor (ER) and the androgen receptor (AR) have a more stable ligand-binding ability in the absence of chaperones, although ligand binding as well as ER stability is enhanced by chaperone interaction. The minimal chaperone complex for steroid receptors is reported to require hsp90 and hsp70 and also hsp40, a homologue of the bacterial DnaJ and yeast YDJ/1. Two additional co-chaperones have been shown to further functionally enhance steroid receptor activity, the hsp90-binding protein p23 and an hsp-organizing protein (hop) that brings together hsp90 and hsp70. In addition to this, several non-essential proteins take part in the chaperone-containing steroid-receptor multi-complex, hsp70 interacting protein (hip), BAG-1, the immunophilins FKBP52 and FKBP51, the cyclosporin A-binding immunophilin called cyclophilin 40 (Cyp40) and the protein phosphatase PP5. The interplay with the steroid receptors also involves ATP–adenosine diphosphate (ADP) binding of the chaperones and is dependent on the intrinsic weak ATPase activity of hsp90 and hsp70. There is a certain selectivity in which non-essential protein that is recruited to the multiprotein complex and various steroid receptors, as well as other signalling molecules reported to associate with hsp90 and hsp70, does show selectivity towards which proteins it associates with. For instance, GR preferentially interacts with FKBP52, whereas PR rather associates with FKBP51 and ER interacts with both Cyp-40 and FKBP52. The immunophilins are interesting from the point of view of intracellular receptor movement, and probably they serve to contact the molecular motors dynein and kinesin and thereby affect intracellular receptor transport, at least in the case of a preserved cytoskeletal network. Another interesting functional consequence of immunophilin action has been demonstrated in squirrel monkey. This new world primate (*Saimiri boliviensis boliviensis*) shows a marked resistance especially to glucocorticoids with a low glucocorticoid-binding affinity and also a higher circulating level of cortisol. It has been demonstrated that this low affinity to cortisol does not depend on the GR itself but rather on perturbed levels of FKBP51 and FKBP52, with a tenfold increase of the former and a 50% reduction of the latter as compared to what is present in human cells.

The possibility of using the knowledge about steroid receptor interactions with chaperones to modulate steroid dependent responses has been suggested. A recent report by Bagatell et al. shows some preliminary promising results in ligand-independent destabilization of ER and PR in severe combined immunodeficient (SCID) mice with breast cancer, by using the hsp90-binding drugs geldanamycin and radicicol (Bagatell et al. 2001). Enhancing the function of chaperones may in turn have positive effects in diseases that depend on aggregation of defect steroid receptors as in spinal and bulbar muscular atrophy (SBMA). This disease is caused by a polyglutamine tract expansion in the AR. By overexpression of a combination of hsp40 and hsp70 in a neuronal cell model of SBMA, aggregate formation could be reduced (Kobayashi et al. 2000). By learning more about the interplay of steroid hormone receptors, chaperones, co-chaperones and other cytosolic steroid receptor-interacting proteins, it may in future become possible to employ treatment strategies depending on these interactions.

11.3 GR, NF-κB and Repression of Inflammation

Studies of GR acting to repress inflammation have demonstrated that GR can interact with the pro-inflammatory mediator proteins p50 and p65 (RelA) in the NF-κB family as well as with c-fos and c-jun in the AP-1 family of proteins. These interactions regulate the involved signalling pathways and have, in the case of GR, been suggested to be a major mechanistic explanation for glucocorticoid-dependent inhibition of inflammation. The cross-talk between AP-1 and GR have recently been reviewed (Karin and Chang 2001) and will not be discussed further in this chapter.

The interaction and subsequent cross-talk of NF-κB and GR has been assumed to be ligand dependent and to occur in the nucleus of cells. The interaction is known to involve the Rel homology domain of p65 and the DNA-binding domain of GR (Ray and Prefontaine 1994; Caldenhoven et al. 1995; Scheinman et al. 1995; McKay and Cidlowski 1998). Two amino acids in the second zinc finger of GR DNA-binding domain (DBD), R488 and K490 have been shown to be crucial for p65 interaction (Liden et al. 1997). The precise mechanism for the mutual repression of NF-κB and GR signalling is not fully elucidated. In most cell types, the interaction seems to provide a mechanistic explanation for the repression. In some cell types, a GR-dependent up-regulation of IκBα has been suggested to contribute to glucocorticoid inhibition of NF-κB-mediated inflammation (Aljada et al. 1999). RelA and GR both depend upon the coactivators CREB-binding protein (CBP) and steroid-receptor coactivator-1 (SRC-1) for maximal activity. It has been suggested that cross-talk between the p65 and GR is due, at least in part, to nuclear competition for limiting amounts of the coactivators CBP and SRC-1 (Sheppard et al. 1998), but other reports speak against this possibility as a general mechanism (De Bosscher et al. 2000). Another possibility for an antagonistic action of p65 and GR may be that either signal molecule prevents the other from binding to its cognate response element. This possibility, at least for the binding of p65 to DNA, has been experimentally excluded in two separate experimental systems. In in vivo footprinting experiments, TNF-α induced binding of p65 to the intercellular adhesion molecule (ICAM)-1 promoter, even after the U397 cells are incubated with glucocorticoids (Liden et al. 2000). However, the p65-induced footprint is changed in a way that suggests modification of NF-κB binding by GR, when a glucocorticoid is added, suggesting the formation of a transcriptionally inert complex on the DNA. This would indicate that GR effects on NF-κB signalling involve mechanisms downstream of the actual DNA interaction, and such a mechanism has been suggested by work in the Yamamoto laboratory (Nissen and Yamamoto 2000). Experiments using chromatin immunoprecipitation have demonstrated that p65 and GR are associated simultaneously on the IL-8 as well as on the ICAM-1 promoter and that the GR binding prevents phosphorylation of a specific serine in the carboxy-terminal domain of RNA-polymerase II, which is necessary for transcriptional activation. The interaction of GR and p65 may inhibit p65 activity either by preventing the action of a kinase or by promoting the recruiting of a phosphatase that in turn leads to an inhibition of a necessary RNA-polymerase II phosphorylation. Other studies by Doucas et al. recently showed that RelA repression of GR is dependent on the protein kinase A (PKA) site at amino acid residue 276 of RelA and that GR-mediated inhibition of NF-κB transactivation is PKA-dependent (Doucas et al. 2000). In what way this relates to the lack of RNA-polymerase II phosphorylation reported by Nissen and Yamamoto (2000) is not clear at the present time.

11.4 GR and Cytoskeletal Protein Interaction

Several reports have shown that GR can interact with cytoskeletal proteins. Both tubulin and actin have been shown to interact directly or indirectly with GR using immuno-based methods (Pratt et al. 1989; Miyata and Yahara 1991). The microtubule interaction has been further corroborated using confocal laser microscopy, immunocytochemistry and drugs that specifically disorganize the various cytoskeletal proteins (Akner et al. 1990; Akner et al. 1991). The functional implications of these interactions are yet not studied in detail and experimental data do not to date clearly demonstrate that these proteins have an obvious role in intracellular transport of GR. However, some recent studies indicate the possibility of FKBP52 mediating GR interaction with dynein and thereby movement of GR to the nucleus (Galigniana et al. 2001). However, studies of the role of cytoskeletal proteins for the intracellular distribution of PR have not supported an obvious role of the cytoskeletal protein interaction for PR (Perrot-Applanat et al. 1992).

11.5 GR and Intracellular Localization

Within the nuclear receptor family of proteins and more specifically also in the steroid hormone receptor sub-family, GR in contrast to most of the other receptors has a distinct localization in both the cytoplasm and the nucleus. ERα and ERb have been reported to have a preferential localization in the nucleus (Hiroi et al. 1999; Choi et al. 2001; Shughrue and Merchenthaler 2001) and PR, which is more closely related to GR than ER, also has a more distinct nuclear localization than GR (Perrot-Applanat et al. 1985; Lim et al. 1999). This has been reported to be due to sequence differences in between the ligand-binding domain of GR and PR. More precisely, the helices involved in the two receptors have been mapped by the construction of a series of chimaeric receptors constructs linked to GFP, allowing in vivo intracellular localization studies (Wan et al. 2001). Unliganded GR is mainly localized in the cytoplasm and upon ligand binding translocates to the nucleus, whereas unliganded PR is localized in the nucleus and upon ligand binding maintains this localization but slightly changes appearance to a more punctuated pattern. In the case of GR, the short splicing variant GRβ has been reported to have a nuclear localization regardless of whether ligand is present or not (Oakley et al. 1996, 1997). For the more abundant, transcriptionally active, full-length form of GR, GRα, addition of ligand leads to a translocation of GR to the nucleus that occurs within minutes (Htun et al. 1996). Reports on ligand withdrawal further indicate that non-liganded GR is rapidly detached from DNA and only slowly, after several hours, returns to the cytoplasm (Qi et al. 1989). A fraction of GR does not seem to translocate and remains in the cytoplasm. The GR localization observed in experiments employing GFP-GR constructs as described above reflects the overall distribution of GR at any given time point. Another approach also involving GFP-GR expression but employing photobleaching and interaction with a tandem array of mouse mammary tumour GREs performed by Gordon Hager and collaborators has demonstrated that there is a rapid shuttling of GR in between the nucleoplasm and distinct site-specific response elements in chromatin (McNally et al. 2000).

The movement of GR most likely involves transport proteins, both to import and to export GR from the nucleus. A few reports are available on the mechanism of GR transport. Savory et al. (1999) suggest that pendulin/importin α is involved in the nuclear import of GR, depending on NL1. With regard to GR export from the nucleus, there are conflicting reports. Savory et al. (1999) claim that the transport of GR from nucleus to cytoplasm is inhibited by leptomycin B, which indicates that the exportin 1/CRM-1 pathway mediates this transport. In contrast to this, Liu and DeFranco (2000) show that other pathways also could be postulated to be involved in GR export. Elucidating the precise pattern of GR–protein interactions may not only shed light on GR signalling but could also contribute to the understanding of the transport mechanisms that are involved in GR nucleo-cytoplasmic shuttling.

11.6 Other GR-Interacting Proteins

GR has also been reported to interact with proteins involved in certain other signal transduction pathways, i.e. transcription factors signal transducer and activator of transcription 5 (Stat5), Ets-related transcription factors, and many more. Immunoprecipitation studies (Cella et al. 1998) revealed that in mammary cells Stat5a, Stat5b and GR are physically associated in vivo. This provides a mechanistic explanation for how glucocorticoids can affect milk protein gene expression by GR association with Stat5. Further studies by Wyszomierski et al. (1999) demonstrated that Stat5 and GR could affect the intracellular localization of each other. In general, the functional outcome of Stat5 and GR interaction has been shown to be dependent on promoter context, as well on levels and activational state of the involved transcription factors.

There are several reports in the literature that refer to co-operation between the Ets-family transcription factor PU.1 and nuclear receptors, especially GR. For instance, Ets sites have been found close to a GRE in the rat tyrosine aminotransferase (TAT) gene (Espinas et al. 1994). Additionally, GR activation of the natural Fcgamma RI promoter construct requires both Stat1 and the PU.1 transcription factor (Aittomaki et al. 2000).

The reported interactions of GR with adaptor proteins, coactivators, corepressors and the role of these interactions in chromatin remodelling and GR activation is beyond the scope of this presentation and the reader is referred to a recent review by Deroo and Archer (2001).

11.7 GR Interaction with 14-3-3 and Raf-I

One protein reported to interact with GR is the 14-3-3 eta protein. In a yeast two-hybrid screening and by glutathione S-transferase pull-down experiments, 14-3-3 has been shown to be a GR-binding protein (Wakui et al. 1997). This interaction occurred with the ligand-binding domain of GR and was glucocorticoid-ligand dependent. The study also indicated that hsp90 prevented GR–14-3-3 interaction. 14-3-3 proteins are highly conserved and widely distributed, and they have been suggested to serve as adaptor proteins (i.e. bind two different proteins at the same time) and to function in intracellular transport. Examples of proteins reported to interact with 14-3-3 are Cdc25, IRS-1 and Raf-1.

Because the methods used to investigate GR–14-3-3 interaction have relied on the interaction of overexpressed proteins, we found it of interest to investigate whether GR interacts with 14-3-3 under conditions where GR and its associated partners are present at concentrations comparable to the in vivo situation. We used a specific immunoaffinity chromatographic purification of GR in liver cytosol from adrenalectomized rats in which 14-3-3 was found to co-purify with GR in several functional states with regard to the presence of ligand and activation. Liganded/activated GR showed the strongest 14-3-3 interaction in a semi-quantitative analysis. The 14-3-3 interaction with non-liganded GR suggested that 14-3-3 and GR could associate when GR is complexed to hsp90 (Widén et al. 2000).

As glucocorticoids could affect glucose homeostasis as well as cellular growth rates, it was of interest to study whether we could observe possible 14-3-3-mediated GR interactions with proteins relevant for these processes. We found that Raf-1 was co-immunopurified with GR, especially when liganded and activated. We were not able to demonstrate the presence of IRS-1 in the GR-associated complex, although the commercially available antibodies used detected IRS-1 in rat liver cytosol clearly.

We do not know if GR interacts directly with Raf-1 or whether this interaction is mediated via 14-3-3. Studies of the effect of 14-3-3 on Raf-1 are so far not conclusive as 14-3-3 has been reported to either have no effect on Raf (Suen et al. 1995) or to activate Raf (Fantl et al. 1994; Irie et al. 1994) in in vitro experiments. One study implies that 14-3-3 is bound to Raf-1 in the cytosol and that this binding is necessary to shuttle

Raf-1 to the cell membrane and to activate Ras. This study also shows that 14-3-3 dissociates completely from Raf-1 at the plasma membrane (Roy et al. 1998). Because we have found several co-purifying proteins for liganded/activated GR in cytosolic extracts, any other protein present in the extract may actually mediate the GR-Raf-1-interaction. One candidate could be a chaperone protein such as hsp70 that was present in all cases where we saw an interaction. In contrast to this, we found that Raf-1 co-purified with GR when GR was dissociated from hsp90, so we were able to exclude that GR-Raf-1 interaction was mediated via hsp90 in this experimental system.

We also had indications that a low affinity interaction of GR, 14-3-3 and Raf-1 could occur regardless of ligand and GR activation state but that the conformational change of GR most likely induced by ligand binding as well as the concomitant detachment from hsp90 leads to an increased recruitment of 14-3-3 and Raf-1 to GR. In further studies of the interaction between 14-3-3 and GR, we also used 14-3-3-antibodies to immunoprecipitate 14-3-3 and looked for co-precipitating proteins. This experiment verified the 14-3-3 interaction with GR but also demonstrated that only a fraction of 14-3-3, being an abundant protein with several isoforms, interacts with GR. The observation of 14-3-3 complexed with liganded/activated GR may indicate the need for GR to be continuously chaperoned in all functional states. The functional relevance of GR-Raf-1 interaction could be of special importance in the regulation of cellular growth and differentiation. Glucocorticoids are reported to induce apoptosis in certain T-lymphocyte populations and growth arrest in G1 in lymphocytes (King and Cidlowski 1998), whereas in hepatoma cells apoptosis is suppressed by glucocorticoids (Evans-Storms and Cidlowski 2000). Raf-1 in turn is a key mediator of mitogen-activated growth responses and is recruited by Ras to further signal mitogen responses via the mitogen-activated protein kinase pathway. In many cells, Raf-1 and glucocorticoid signals would be expected to counterbalance one another, and a direct or indirect physical interaction of these two proteins may provide a route for cross-talk. Furthermore, 14-3-3 has been shown to transport Raf-1 both to and from the cell membrane where Ras is anchored (Campbell et al. 1998) and either or both of these transport steps of 14-3-3 may be affected by a GR/14-3-3 interaction (Roy et al. 1998). Glucocorticoids have been demonstrated by Croxtall et al. (2000) to exert inhibitory effects on the Ras/Raf signalling pathway in A549 human adenocarcinoma cells, an inhibition that could be reversed by the glucocorticoid antagonist, RU486.

Our observations that GR and Raf-1 are found within the same protein complex in the cytoplasm of rat liver cells provides a mechanistic explanation for effects of glucocorticoids on Ras–Raf signalling and vice versa. Furthermore, the unexpected finding of Raf-1 in the GR multiprotein complex indicates the usefulness of our immunoaffinity-based technique in native systems to discover yet other currently unknown GR–protein interaction. We suggest that cytosolic GR is present in one or several multiprotein complexes, putatively called receptosomes, and enables cross-talk between GR and other signalling pathways, one example of which is Ras/Raf-1. 14-3-3 proteins may serve as adaptors also in GR interactions with other proteins. Addition of glucocorticoids into this system might trigger changes in GR–protein interactions leading to rapid non-genomic effects. Ongoing studies are aimed at identifying other possible proteins participating in the formation of the GR-containing receptosome.

11.8 Further Analysis of Cytosolic Proteins Interacting with GR

11.8.1 Analysis of GR-Interacting Proteins by Western Blotting

Most studies of GR-interacting proteins are performed studying the interaction of overexpressed proteins in cell lines and/or in vitro experiments. This approach has the inherent risk of inducing artificial interactions that may not occur in vivo. The cellular distribution of GR may also become perturbed by overexpression of GR and this in itself may lead to interactions normally not occurring in vivo. We are currently performing an analysis of GR interactions in cytosols prepared from rat liver, rat hepatoma cells and human gingival fibroblasts. We found it of interest to investigate whether GR interacts with p65 under conditions where GR and its associated partners are present at concentrations comparable to the in vivo situation (C. Widén et al., unpublished results). Indeed, we found that p65 and its inhibitor protein, IκBα, co-purified with GR in a non-liganded/non-activated (i.e. in the presence of molybdate) and in a liganded/activated state (i.e. ligand was added to the cytosol prior to heat activation). We also tested the strength of the cytosolic GR-p65, and GR-IκBα interactions using a stringent washing procedure involving stepwise salt gradient washes from 50 mM to 2.4 M NaCl. Both GR-p65 and GR-IκBα interactions withstood 2.4 M NaCl thus indicating that the interactions were strong. We are now studying the involvement of p50 in this interaction, as well as if these interactions involve the total cellular pools of the involved proteins and how the interaction takes place in cytosol and nuclear extracts respectively.

11.8.2 The Role of GR and NF-κB in Herpes Simplex Type-1 Infection

Herpes simplex type 1 (HSV-1), similar to many other viruses, influences and regulates cellular proteins in order to promote its own propagation. Doing this process, HSV-1 has been shown to induce the pro-inflammatory mediator NF-κB (Rong et al. 1992; Patel et al. 1998; Amici et al. 2001). A reported persistent nuclear translocation of NF-κB in HSV-1-infected cells has also been shown to coincide with increased binding of NF-κB to specific DNA-response elements (Patel et al. 1998). Whether the increased DNA-bound NF-κB reflects increased activity and subsequent transcription of target genes is not clear, as some authors report decreased expression of a reporter-gene while others show increased reporter gene transcription (Rong et al. 1992; Amici et al. 2001). GR expression or activity has not previously been studied in response to HSV-1 infection, but other pathogens such as human immunodeficiency virus (HIV) has been shown to mediate increased GR activity. More specifically, the viral protein Vpr seems to act as a coactivator for GR and thereby up-regulates transcription of GR-dependent genes (Refaeli et al. 1995; Kino et al. 1999). As treatment with antiviral drugs is not sufficient to alleviate the painful virus associated inflammatory reaction, glucocorticoids have been suggested and used as a complementary treatment for HSV-1 lesions. However, this therapy is controversial as it has been suggested that glucocorticoid treatment may increase viral yield in vitro (Harrell and Sydiskis 1982; Dreyer et al. 1989) as well as in vivo (Awan et al. 1998), while other reports suggest a decreased viral yield (Notter and Docherty 1978).

As we are interested in the functional consequence of GR interaction with other signalling pathways, we studied if an HSV-1 infection, reported to affect NF-κB levels and signalling, also affected the levels of the NF-κB-interacting protein GR. The conflicting reports about virus production after glucocorticoid treatment also made us interested in how the regulation of the involved proteins affected HSV-1 production.

We established a cell system where dexamethasone treatment prior to infection, leads to increased viral production (A. Erlandsson et al., submitted for publication). Using this information, we studied how HSV-1 infection, with or without glucocorticoid treatment, affected GR and NF-κB levels and activity. We found as previously reported for HSV-1 infection in a variety of cell lines, that the expression of NF-κB in a primary gingival fibroblast culture was up-regulated in response to HSV-1 infection. We could observe that this was concomitant with increased p65 levels in both cytoplasm and nucleus. In addition to the increased nuclear NF-κB expression, we also showed a striking up-regulation of GR as well as an increased nuclear translocation. This increase of GR protein levels as a consequence of viral infection has to our knowledge not been reported previously, neither for HSV-1 nor for other virus infected cell cultures.

We also studied the effects of dexamethasone addition upon GR levels, NF-κB levels and virus production. We found that NF-κB up-regulation by HSV-1 infection is not further affected by glucocorticoids. In contrast to this, the endogenous, well-known glucocorticoid-induced down-regulation of GR (Dong et al. 1988), which was clearly detected after 24 h glucocorticoid treatment, lead to a much lower GR up-regulation after HSV-1 infection, when compared to cells not treated with glucocorticoids. In this case, a higher virus production was noticed. Dexamethasone treatment of the cells when performed simultaneously with HSV-1 infection resulted in "normal" up-regulation of GR and "normal" virus production. This leads us to suggest that up-regulation of GR may be an important factor in the defence against an HSV-1 infection, presumably by counteracting an NF-κB effect.

11.8.3 MALDI-TOF-Based Analysis of GR-Associated Proteins

Before using matrix-assisted laser desorption ionization-time of flight-based (MALDI-TOF) analysis, the proteins of the isolated GR complex are separated by two-dimensional (2D) electrophoresis and the reproducibility of the protein pattern is analysed. This approach allows not only the determination of the molecular weights and the isoelectric points of the proteins of the GR multi-complex but also a semi-quantitative analysis of the single protein spots. Thus, we are able to perform a thorough mapping of the proteins that belong to the GR receptosome and to monitor the changes in its protein composition, which take place under different experimental conditions. In the next step, the proteins that are studied by two-dimensional electrophoresis are digested with trypsin and analysed by high-resolving MALDI-TOF mass spectrometry, a process described in a recent review (Bonk and Humeny 2001). In those cases where the full-length sequence of a protein is available, a peptide mass fingerprint is usually sufficient to clearly identify a protein. If further information is required to find out the identity of a protein, a fragmentation analysis and structural TOF post-source decomposition (FAST-PSD) can be carried out to produce sequence tags for specific database searches. In certain situations de novo sequencing is required.

Using this approach, we have been able to obtain a reproducible pattern of GR-interacting proteins in 2D gels for both non-liganded non-activated GR and for cytosolic in vitro liganded, activated GR. The identification of these proteins using MALDI-TOF is ongoing. Preliminary results have made it possible to identify several novel GR-interacting proteins, among those could be noted HGPRT (hypoxanthine-guanine phosphoribosyltransferase), albumin and albumin precursor, cytokeratin, haemoglobin, and DNA/K-type molecular chaperone grp75 precursor, but also previously identified proteins such as hsp90, hsp70 and DNA/J have been possible to demonstrate by this method. However, these results needs to be confirmed by using other methodologies suited to study protein interactions and more importantly, these interactions also needs to be studied with regard to their biological and functional relevance.

The mass spectrometric analysis of the GR multi-complex is not only an efficient approach to produce its protein composition accurately under various experimental conditions. It can also be used to study whether certain subunits of the GR complex are phosphorylated and to specifically identify distinct phosphorylation sites.

11.9 Summary

Studies of GR-interacting proteins can provide valuable insights into the regulation of GR cellular signalling. The cytoplasmic localization of GR and reports of GR interaction with such a plethora of other cytoplasmic proteins may point to a unique role for GR in modulating and integrating other signalling pathways. A better insight into these interactions could serve as a tool when trying to understand and modify GR signalling.

Acknowledgements. We wish to thank Ms Marika Rönnholm for skilful technical assistance and professor Sam Okret for critical reading of the manuscript. This work was supported by the Foundation for Knowledge and Competence, Medivir AB, Åke Wibergs foundation, Research foundation of the Karolinska Institutet, Swedish Medical Research Council Grants12557 and KI 13X-2819.

References

Aittomaki S, Pesu M, Groner B, Janne OA, Palvimo JJ, Silvennoinen O (2000) Cooperation among Stat1, glucocorticoid receptor, and PU.1 in transcriptional activation of the high-affinity Fc gamma receptor I in monocytes. J Immunol 164:5689–5697

Akner G, Sundqvist KG, Denis M, Wikstrom AC, Gustafsson JA (1990) Immunocytochemical localization of glucocorticoid receptor in human gingival fibroblasts and evidence for a colocalization of glucocorticoid receptor with cytoplasmic microtubules. Eur J Cell Biol 53:390–401

Akner G, Mossberg K, Wikstrom AC, Sundqvist KG, Gustafsson JA (1991) Evidence for colocalization of glucocorticoid receptor with cytoplasmic microtubules in human gingival fibroblasts, using two different monoclonal anti-GR antibodies, confocal laser scanning microscopy and image analysis. J Steroid Biochem Mol Biol 39:419–432

Aljada A, Ghanim H, Assian E, Mohanty P, Hamouda W, Garg R, Dandona P (1999) Increased IkappaB expression and diminished nuclear NF-kappaB in human mononuclear cells following hydrocortisone injection. J Clin Endocrinol Metab 84:3386–3389

Amici C, Belardo G, Rossi A, Santoro MG (2001) Activation of I kappa b kinase by herpes simplex virus type 1. A novel target for anti-herpetic therapy. J Biol Chem 276:28759–28766

Awan AR, Harmenberg J, Flink O, Field HJ (1998) Combinations of antiviral and anti-inflammatory preparations for the topical treatment of herpes simplex virus assessed using a murine zosteriform infection model. Antivir Chem Chemother 9:19–24

Bagatell R, Khan O, Paine-Murrieta G, Taylor CW, Akinaga S, Whitesell L (2001) Destabilization of steroid receptors by heat shock protein 90-binding drugs: a ligand-independent approach to hormonal therapy of breast cancer. Clin Cancer Res 7:2076–2084

Bonk T, Humeny A (2001) MALDI-TOF-MS analysis of protein and DNA. Neuroscientist 7:6–12

Caldenhoven E, Liden J, Wissink S, Van de Stolpe A, Raaijmakers J, Koenderman L, Okret S, Gustafsson JA, Van der Saag PT (1995) Negative cross-talk between RelA and the glucocorticoid receptor: a possible mechanism for the antiinflammatory action of glucocorticoids. Mol Endocrinol 9:401–412

Campbell SL, Khosravi-Far R, Rossman KL, Clark GJ, Der CJ (1998) Increasing complexity of Ras signaling. Oncogene 17:1395–1413

Cella N, Groner B, Hynes NE (1998) Characterization of Stat5a and Stat5b homodimers and heterodimers and their association with the glucocorticoid receptor in mammary cells. Mol Cell Biol 18:1783–1792

Cheung J, Smith DF (2000) Molecular chaperone interactions with steroid receptors: an update. Mol Endocrinol 14:939–946

Choi I, Ko C, Park-Sarge OK, Nie R, Hess RA, Graves C, Katzenellenbogen BS (2001) Human estrogen receptor beta-specific monoclonal antibodies: characterization and use in studies of estrogen receptor beta protein expression in reproductive tissues. Mol Cell Endocrinol 181:139–150

Croxtall JD, Choudhury Q, Flower RJ (2000) Glucocorticoids act within minutes to inhibit recruitment of signalling factors to activated EGF receptors through a receptor-dependent, transcription-independent mechanism. Br J Pharmacol 130:289–298

De Bosscher K, Vanden Berghe W, Vermeulen L, Plaisance S, Boone E, Haegeman G (2000) Glucocorticoids repress NF-kappaB-driven genes by disturbing the interaction of p65 with the basal transcription machinery, irrespective of coactivator levels in the cell. Proc Natl Acad Sci U S A 97:3919–3924

Deroo BJ, Archer TK (2001) Glucocorticoid receptor-mediated chromatin remodeling in vivo. Oncogene 20:3039–3046

Dong Y, Poellinger L, Gustafsson JA, Okret S (1988) Regulation of glucocorticoid receptor expression: evidence for transcriptional and posttranslational mechanisms. Mol Endocrinol 2:1256–1264

Doucas V, Shi Y, Miyamoto S, West A, Verma I, Evans RM (2000) Cytoplasmic catalytic subunit of protein kinase A mediates cross-repression by NF-kappa B and the glucocorticoid receptor. Proc Natl Acad Sci U S A 97:11893–11898

Dreyer LL, Sydiskis RJ, Bashirelahi N (1989) Effect of dexamethasone on herpes simplex virus replication in mouse neuroblastoma cells (NB41A3): receptor characteristics. J Clin Lab Anal 3:236–243

Espinas ML, Roux J, Ghysdael J, Pictet R, Grange T (1994) Participation of Ets transcription factors in the glucocorticoid response of the rat tyrosine aminotransferase gene. Mol Cell Biol 14:4116–4125

Evans-Storms RB, Cidlowski JA (2000) Delineation of an antiapoptotic action of glucocorticoids in hepatoma cells: the role of nuclear factor-kappaB. Endocrinology 141:1854–1862

Fantl WJ, Muslin AJ, Kikuchi A, Martin JA, MacNicol AM, Gross RW, Williams LT (1994) Activation of Raf-1 by 14-3-3 proteins. Nature 371:612–614

Galigniana MD, Radanyi C, Renoir JM, Housley PR, Pratt WB (2001) Evidence that the peptidylprolyl isomerase domain of the hsp90-binding immunophilin FKBP52 is involved in both dynein interaction and glucocorticoid receptor movement to the nucleus. J Biol Chem 276:14884–14889

Harrell AJ, Sydiskis RH (1982) The effect of dexamethasone on the replication of herpes simplex virus in human gingival fibroblast cultures. J Baltimore Coll Dent Surg 35:9–13

Hiroi H, Inoue S, Watanabe T, Goto W, Orimo A, Momoeda M, Tsutsumi O, Taketani Y, Muramatsu M (1999) Differential immunolocalization of estrogen receptor alpha and beta in rat ovary and uterus. J Mol Endocrinol 22:37–44

Htun H, Barsony J, Renyi I, Gould DL, Hager GL (1996) Visualization of glucocorticoid receptor translocation and intranuclear organization in living cells with a green fluorescent protein chimera. Proc Natl Acad Sci USA 93:4845–4850

Irie K, Gotoh Y, Yashar BM, Errede B, Nishida E, Matsumoto K (1994) Stimulatory effects of yeast and mammalian 14-3-3 proteins on the Raf protein kinase. Science 265:1716–1719

Karin M, Chang L (2001) AP-1-glucocorticoid receptor crosstalk taken to a higher level. J Endocrinol 169:447–451

King KL, Cidlowski JA (1998) Cell cycle regulation and apoptosis. Annu Rev Physiol 60:601–617

Kino T, Gragerov A, Kopp JB, Stauber RH, Pavlakis GN, Chrousos GP (1999) The HIV-1 virion-associated protein vpr is a coactivator of the human glucocorticoid receptor. J Exp Med 189:51–62

Kobayashi Y, Kume A, Li M, Doyu M, Hata M, Ohtsuka K, Sobue G (2000) Chaperones Hsp70 and Hsp40 suppress aggregate formation and apoptosis in cultured neuronal cells expressing truncated androgen receptor protein with expanded polyglutamine tract. J Biol Chem 275:8772–8778

Liden J, Delaunay F, Rafter I, Gustafsson J, Okret S (1997) A new function for the C-terminal zinc finger of the glucocorticoid receptor. Repression of RelA transactivation. J Biol Chem 272:21467–1472

Liden J, Rafter I, Truss M, Gustafsson JA, Okret S (2000) Glucocorticoid effects on NF-kappaB binding in the transcription of the ICAM-1 gene. Biochem Biophys Res Commun 273:1008–1014

Lim CS, Baumann CT, Htun H, Xian W, Irie M, Smith CL, Hager GL (1999) Differential localization and activity of the A- and B-forms of the human progesterone receptor using green fluorescent protein chimeras. Mol Endocrinol 13:366–375

Liu J, DeFranco DB (2000) Protracted nuclear export of glucocorticoid receptor limits its turnover and does not require the exportin 1/CRM1-directed nuclear export pathway. Mol Endocrinol 14:40–51

McKay LI, Cidlowski JA (1998) Cross-talk between nuclear factor-kappa B and the steroid hormone receptors: mechanisms of mutual antagonism. Mol Endocrinol 12:45–56

McNally JG, Muller WG, Walker D, Wolford R, Hager GL (2000) The glucocorticoid receptor: rapid exchange with regulatory sites in living cells. Science 287:1262–1265

Miyata Y, Yahara I (1991) Cytoplasmic 8S glucocorticoid receptor binds to actin filaments through the 90-kDa heat shock protein moiety. J Biol Chem 266:8779–8783

Nissen RM, Yamamoto KR (2000) The glucocorticoid receptor inhibits NFkappaB by interfering with serine-2 phosphorylation of the RNA polymerase II carboxy-terminal domain. Genes Dev 14:2314–2329

Notter MF, Docherty JJ (1978) Steroid hormone alteration of herpes simplex virus type 1 replication. J Med Virol 2:247–252

Oakley RH, Sar M, Cidlowski JA (1996) The human glucocorticoid receptor beta isoform. Expression, biochemical properties, and putative function. J Biol Chem 271:9550–9559

Oakley RH, Webster JC, Sar M, Parker CR, Jr., Cidlowski JA (1997) Expression and subcellular distribution of the beta-isoform of the human glucocorticoid receptor. Endocrinology 138:5028–5038

Patel A, Hanson J, McLean TI, Olgiate J, Hilton M, Miller WE, Bachenheimer SL (1998) Herpes simplex type 1 induction of persistent NF-kappa B nuclear translocation increases the efficiency of virus replication. Virology 247:212–222

Perrot-Applanat M, Logeat F, Groyer-Picard MT, Milgrom E (1985) Immunocytochemical study of mammalian progesterone receptor using monoclonal antibodies. Endocrinology 116:1473–1484

Perrot-Applanat M, Lescop P, Milgrom E (1992) The cytoskeleton and the cellular traffic of the progesterone receptor. J Cell Biol 119:337–348

Pratt WB, Toft DO (1997) Steroid receptor interactions with heat shock protein and immunophilin chaperones. Endocr Rev 18:306–360

Pratt WB, Sanchez ER, Bresnick EH, Meshinchi S, Scherrer LC, Dalman FC, Welsh MJ (1989) Interaction of the glucocorticoid receptor with the Mr 90,000 heat shock protein: an evolving model of ligand-mediated receptor transformation and translocation. Cancer Res 49:2222s–2229s

Pratt WB, Silverstein AM, Galigniana MD (1999) A model for the cytoplasmic trafficking of signalling proteins involving the hsp90-binding immunophilins and p50cdc37. Cell Signal 11:839–851

Qi M, Hamilton BJ, DeFranco D (1989) v-mos oncoproteins affect the nuclear retention and reutilization of glucocorticoid receptors. Mol Endocrinol 3:1279–1288

Ray A, Prefontaine KE (1994) Physical association and functional antagonism between the p65 subunit of transcription factor NF-kappa B and the glucocorticoid receptor. Proc Natl Acad Sci U S A 91:752–756

Refaeli Y, Levy DN, Weiner DB (1995) The glucocorticoid receptor type II complex is a target of the HIV-1 vpr gene product. Proc Natl Acad Sci USA 92:3621–3625

Rong BL, Libermann TA, Kogawa K, Ghosh S, Cao LX, Pavan-Langston D, Dunkel EC (1992) HSV-1-inducible proteins bind to NF-kappa B-like sites in the HSV-1 genome. Virology 189:750–756

Roy S, McPherson RA, Apolloni A, Yan J, Lane A, Clyde-Smith J, Hancock JF (1998) 14-3-3 facilitates Ras-dependent Raf-1 activation in vitro and in vivo. Mol Cell Biol 18:3947–3955

Savory JG, Hsu B, Laquian IR, Giffin W, Reich T, Hache RJ, Lefebvre YA (1999) Discrimination between NL1- and NL2-mediated nuclear localization of the glucocorticoid receptor. Mol Cell Biol 19:1025–1037

Scheinman RI, Gualberto A, Jewell CM, Cidlowski JA, Baldwin AS Jr (1995) Characterization of mechanisms involved in transrepression of NF-kappa B by activated glucocorticoid receptors. Mol Cell Biol 15:943–953

Sheppard KA, Phelps KM, Williams AJ, Thanos D, Glass CK, Rosenfeld MG, Gerritsen ME, Collins T (1998) Nuclear integration of glucocorticoid receptor and nuclear factor-kappaB signaling by CREB-binding protein and steroid receptor coactivator-1. J Biol Chem 273:29291–29294

Shughrue PJ, Merchenthaler I (2001) Distribution of estrogen receptor beta immunoreactivity in the rat central nervous system. J Comp Neurol 436:64–81

Suen KL, Bustelo XR, Barbacid M (1995) Lack of evidence for the activation of the Ras/Raf mitogenic pathway by 14-3-3 proteins in mammalian cells. Oncogene 11:825–831

Wakui H, Wright AP, Gustafsson J, Zilliacus J (1997) Interaction of the ligand-activated glucocorticoid receptor with the 14-3-3 eta protein. J Biol Chem 272:8153–8156

Wan Y, Coxe KK, Thackray VG, Housley PR, Nordeen SK (2001) Separable features of the ligand-binding domain determine the differential subcellular localization and ligand-binding specificity of glucocorticoid receptor and progesterone receptor. Mol Endocrinol 15:17–31

Widén C, Zilliacus J, Gustafsson JA, Wikstrom AC (2000) Glucocorticoid receptor interaction with 14-3-3 and Raf-1, a proposed mechanism for crosstalk of two signal transduction pathways. J Biol Chem 275:39296–39301

Wyszomierski SL, Yeh J, Rosen JM (1999) Glucocorticoid receptor/signal transducer and activator of transcription 5 (STAT5) interactions enhance STAT5 activation by prolonging STAT5 DNA binding and tyrosine phosphorylation. Mol Endocrinol 13:330–343

12 The Glucocorticoid Receptor β-*Isoform: A Perspective on Its Relevance in Human Health and Disease*

M.J.M. Schaaf, J.A. Cidlowski

12.1	Introduction	197
12.2	Structure of the Human Glucocorticoid Receptor Gene, mRNA and Protein	198
12.3	The hGR α-Isoform	200
12.4	The hGR β-Isoform	201
12.5	Conclusions	205
References		206

12.1 Introduction

Glucocorticoids are steroid hormones that are secreted by the adrenal gland in a diurnal rhythm and after acute stress. They have diverse effects ranging from altering an organism's metabolism and behavior to the function of its immune system. Clinically, they are used to treat a wide variety of diseases, including allergic and autoimmune diseases like asthma and rheumatoid arthritis. They are generally effective therapies against these pathologies because of their well-known anti-inflammatory effects. The actions of glucocorticoids are mediated by an intracellular receptor, the glucocorticoid receptor (GR). The GR is a member of the steroid receptor subfamily that includes the mineralocorticoid, progesterone, androgen, and estrogen receptors. In turn, the steroid receptors belong to the nuclear receptor superfamily, which includes the thyroid, retinoid, and vitamin-D receptors. Upon activation by their ligand, these receptors can act as transcription factors, i.e., they can activate or repress the transcription of specific target genes.

In humans, two isoforms of the GR have been identified, hGRα and hGRβ. Since these two receptors were first cloned, considerable attention has been given to hGRα, and a significant amount of knowledge about its mechanism of action has been gathered. In contrast, for about a decade, hGRβ has been largely overlooked and was considered a cloning artifact. Furthermore, because it does not bind hormone and is transcriptionally inactive, it was considered unimportant in human physiology. Recently, hGRβ has become the focus of a variety of studies, and a new perspective on its function and relevance to human disease has arisen. In this review, we will provide insight into this new research.

12.2 Structure of the Human Glucocorticoid Receptor Gene, mRNA and Protein

The human glucocorticoid receptor (hGR) gene is located on chromosome 5 and consists of 9 exons (as shown in Fig. 1). Exon 1 and the first part of exon 2 contain the 5′ untranslated region (UTR), exons 2–9 contain the coding sequences, and exon 9 the 3′UTR (Encio and Detera-Wadleigh 1991; Oakley et al. 1996). Initially, two forms of human glucocorticoid receptor cDNAs were cloned: hGRα and hGRβ cDNA (Hollenberg et al. 1985; Weinberger et al. 1985a), which were later shown to encode the α- and β-isoform of the receptor. hGRα and hGRβ mRNA both contain exons 1–8, but contain different versions of exon 9 as a result of alternative splicing; hGRα mRNA (5.5 kb) contains exon 9α, whereas GRβ mRNA (4.3 kb) contains exon 9β. A third hGR mRNA has been shown to exist, and in most tissues this transcript has a higher expression level than the other two previously discovered messengers (Oakley et al. 1996). This messenger, which is 7.0 kb in length, contains exons 1–8 and the entire exon 9, including exon 9α, the "J region," and exon 9β. It is thought that this mRNA is translated primarily into hGRα, but it remains possible that during translation exon 9α is skipped and hGRβ is produced (see Fig. 1).

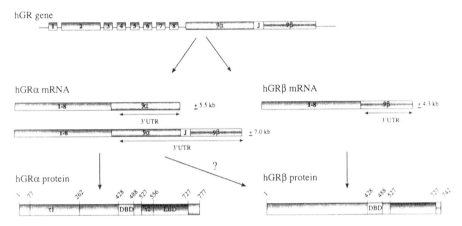

Fig. 1. The structure of the human glucocorticoid receptor (*hGR*) gene, mRNA, and protein. The hGR gene consists of nine exons. Exon 1 and the first part of exon 2 contain the 5'UTR, exons 2–9 contain the coding sequences, and exon 9 the 3'UTR. Three different messengers can be transcribed: they all contain exons 1–8, but contain different versions of exon 9 as a result of alternative splicing; hGRα mRNA (5.5 kb) contains exon 9α, whereas GRβ mRNA (4.3 kb) contains exon 9β. The third hGR mRNA (7.0 kb) contains the entire exon 9, including exon 9α, the "J region," and exon 9β. It is thought that this mRNA is translated into hGRα, but translation into hGRβ may be possible

Like other members of the steroid receptor family, the human glucocorticoid receptor contains three major domains, which are shown in Fig. 1 (Weinberger et al. 1985b; Giguere et al. 1986). Several functions of the receptor have been mapped within these domains, and these will be discussed in the following section. The most N-terminal domain is called the immunogenic domain, which in hGRα consists of amino acids 1–420. Amino acids 421–488 form the DNA-binding domain (DBD) of hGRα, and its C-terminal ligand-binding domain (LBD) consists of amino acids 527–777. The LBD of hGRβ is identical to that of hGRα until amino acid 727. After that hGRα and hGRβ diverge, and 15 unique amino acids form the most C-terminal part of hGRβ's LBD.

12.3 The hGR α-Isoform

The unliganded hGRα resides in the cytoplasm, where it forms a large heteromeric complex with several other proteins (Pratt and Toft 1997; Pratt and Dittmar 1998). One of these proteins, hsp90, plays a central role in the heterocomplex (Toft 1998) by associating with the LBD of the receptor and retaining the receptor in a conformation that can bind steroid but is inactive (Bresnick et al. 1989; Cadepond et al. 1991). Upon ligand binding, hGRα dissociates from the heterocomplex becomes hyperphosphorylated (Orti et al. 1989) and translocates to the nucleus. It is unclear what the function of the hyperphosphorylation is in hGRα signaling. Studies using a mutant mouse GR lacking all phosphorylation sites showed a modest decrease in transactivation properties for this mutant (Mason and Housley 1993; Webster et al. 1997), and reduced ligand-dependent downregulation at both the RNA and protein level (Webster et al. 1997). The mechanism by which the ligand-bound hGRα is transported into the nucleus has been studied extensively. Two domains in the hGRα have been suggested to be involved in this process. The first, nuclear localization signal (NL)1, is located in the C-terminal part of the DBD and extends into the hinge between DBD and LBD (Cadepond et al. 1992). The function of NL1 is inhibited by the LBD, and this inhibition can be abolished by ligand binding (Picard and Yamamoto 1987; Cadepond et al. 1992). The second nuclear localization signal, NL2, is located in the LBD (Picard and Yamamoto 1987), but its exact localization is unknown.

In the nucleus, hGRα binds – as a dimer (Tsai et al. 1988; Wrange et al. 1989) – to specific DNA sequences which are called GREs. Based on sequence analysis of GREs from several promoters, a consensus GRE sequence was determined. It consists of a palindromic 15-mer: 5′-GGTACAnnnTGTTCT-3′ (Beato et al. 1989). Binding of hGRα to GREs leads to transcription initiation through a complicated mechanism. The basic transcription machinery – consisting of RNA polymerase II and general transcription factors such as TATA box-binding protein (TBP) – is recruited to the promoter (Beato and Sanchez-Pacheco 1996). This process is facilitated by coactivators, which also enhance transcription initiation by their ability to remodel chromatin (Collingwood et al. 1999; McKenna et al. 1999; Glass and Rosenfeld 2000; Jenkins et al. 2001).

In addition to its transactivational properties, hGRα can repress gene transcription by one of several mechanisms. It can bind to negative GREs (nGREs) in the promoters of specific target genes (Sakai et al. 1988; Drouin et al. 1989; Cairns et al. 1993; Morrison and Eisman 1993). In addition, hGRα can also bind to composite GREs, which consist of a non-overlapping GRE and a binding site for a different transcription factor like AP-1 (Diamond et al. 1990; Pearce and Yamamoto 1993). Furthermore, hGRα can inhibit the activity of other transcription factors, like AP-1 and NF-κB, without binding to DNA. The latter mechanism probably requires protein–protein interaction between hGRα and these transcription factors (Jonat et al. 1990; Schule et al. 1990; Yang-Yen et al. 1990; Ray and Prefontaine 1994; Caldenhoven et al. 1995; McKay and Cidlowski 1998).

12.4 The hGR β-Isoform

The β-isoform of the glucocorticoid receptor, hGRβ, is a result of alternative splicing of the GR pre-mRNA, does not bind ligand, and cannot transactivate or transrepress gene transcription. hGRβ, therefore, did not garner much attention until it was discovered that it can act as a dominant negative inhibitor of hGRα in vitro. Expression of hGRβ in vivo may therefore contribute to glucocorticoid resistance. Bamberger et al. (1995) demonstrated that overexpression of hGRβ in COS cells inhibits hGRα-induced transactivation on a mouse mammary tumor virus (MMTV)-promoter-driven reporter assay. These data were confirmed by Oakley et al. (1996, 1999), who also showed that the transactivational activity of the endogenous hGRα in HeLa cells could be inhibited by hGRβ. However, several other labs could not detect this dominant negative inhibition of hGRβ (Hecht et al. 1997; Brogan et al. 1999; de Lange et al. 1999). What underlies this discrepancy is still unclear, although different experimental designs were used in these analyses.

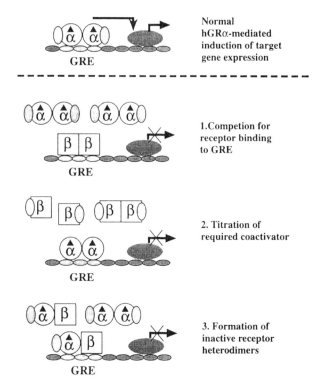

Fig. 2. Three possible mechanisms for hGRβ's dominant negative activity. First, it is hypothesized that hGRβ competes with hGRα for binding to GREs. Second, competition for coactivators has been proposed. Third, squelching of hGRα by forming inactive heterodimers may be hGRβ's mechanism of action

12.4.1 hGRβ's Mechanism of Action

The mechanism of hGRβ's dominant negative inhibition has not been elucidated yet. However, based on the available data, several theories exist, and these notions are presented in Fig. 2. First, it is hypothesized that hGRβ competes with hGRα for these binding sites, since it has been shown that hGRβ can bind to GREs (Leung et al. 1997; Oakley et al. 1999). In the absence of hormone, hGRβ's affinity for binding to a GRE-containing promoter is even higher than hGRα's (Oakley et al. 1999). However, this mechanism seems unlikely, since hGRβ does not inhibit androgen receptor (AR)- and progesterone receptor (PR)-induced transactivation on GRE-driven promoters (Oakley et al. 1999). Second, competition for coactivators has been proposed as a mechanism for hGRβ's dominant negative action. Studies in our lab have shown that hGRβ does not interact with glucocorticoid receptor-interacting protein (GRIP)-1 (M.R. Yudt et al., unpublished results), but interaction with other coactivators may be involved. Third, the formation of inactive heterodimers may likely reflect hGRβ's mechanism of action, since it has been shown by co-immunoprecipitation that hGRα and hGRβ form heterodimers in vivo and in vitro (Oakley et al. 1999; Strickland et al. 2001). This mechanism would also explain hGRβ's dominant negative inhibition of the transactivational properties of the human mineralocorticoid receptor – hMR (Bamberger et al. 1997) –, since it has been shown that hGRα and hMR are able to heterodimerize (Trapp and Holsboer 1996; Savory et al. 2001).

Recent studies have shown that the dominant negative activity is not the consequence of the lack of the 50 C-terminal amino acids of hGRα, but that the function is located in the 15 unique C-terminal amino acids in hGRβ. A truncation mutant of hGRα, containing amino acids 1–727, does not act as a dominant negative inhibitor (Oakley et al. 1999). However, mutation of the β-specific amino acids into hGRα results in a dominant negative form of the receptor (M.R. Yudt et al., in preparation).

12.4.2 Expression of hGRβ

In humans, expression of hGRβ has been found in a wide variety of tissues. In mice, a GRβ-isoform has not been observed, probably because the splice site is not conserved (Otto et al. 1997). By reverse transcriptase polymerase chain reaction (RT-PCR) and Northern blotting, hGRβ mRNA has been found in human brain, heart, liver, muscle, bone marrow, thymus, spleen, and lung (Bamberger et al. 1995; Oakley et al. 1996). hGRβ protein has been detected in several tissues including brain, heart, kidney, and lung (de Castro et al. 1996; Oakley et al. 1997). Surprisingly, immunocytochemistry revealed that hGRβ is predominantly localized in the nucleus, even in the absence of ligand. Expression levels of hGRβ are not homogenous in tissues where expression is observed. Specific cell types, mainly epithelial cells, show high expression levels of hGRβ. For example, epithelial cells lining the terminal bronchiole of the lung, forming the outer layer of Hassall's corpuscle in the thymus and lining the bile duct in the liver, show high expression levels of hGRβ (Oakley et al. 1996, 1997).

For most cell types, the ratio between their hGRα and hGRβ expression level is unclear yet. At the mRNA level, in lung and liver homogenates, a hGRα/-β ratio of approximately 300:1 and 500:1 was measured (Oakley et al. 1996), in peripheral blood mononuclear cells (PBMCs) the ratio was 600:1 (Honda et al. 2000), in pituitary it was 30–40:1 (Dahia et al. 1997; Mu et al. 1998), and in a human respiratory epithelial cell line, 3:1 (Pujols et al. 2001). At the protein level, the hGRα/-β ratio is mostly lower than it is at the mRNA level, which may be due to the longer half-life of the hGRβ protein (Webster et al. 2001). An immunocytochemistry study (using the synthetic peptide fused to bovine serum albumin as a standard) showed ratios between 1:1 and 1:5 for several human tissues (de Castro et al. 1996). By Western blotting, in PBMCs a ratio of 20:1 was measured, in neutrophils a ratio of 1:15 (Strickland et al. 2001), and in Epstein-Barr-virus transformed mononuclear cells 1:1.4 (Longui et al. 2000).

Apparently, the relative expression level of hGRβ at the protein level is remarkably higher than the mRNA expression level. Two explanations can be given for this discrepancy. First, the half-life of hGRβ protein is much longer than hGRα's (48 and 24 h, respectively), as measured after overexpressing the receptors in COS-1 cells (Webster et al. 2001). Second, in the mRNA studies, both the 5.5- and the 7.0-kb messengers are considered to encode hGRα, and the 4.5-kb messenger to encode hGRβ (Oakley et al. 1996). As previously mentioned, the 7.0-kb messenger may be translated into hGRβ, resulting in more hGRβ at the protein level.

Interestingly, the expression of hGRβ appears to be inducible by pro-inflammatory cytokines. A combination of interleukin (IL)-2 and IL-4 increases the amount of hGRβ-positive PBMCs almost threefold (Leung et al. 1997). Tumor necrosis factor (TNF)-α has been shown to increase hGRβ expression level in HeLa cells fourfold (Webster et al. 2001). Finally, IL-8 stimulation of neutrophils resulted in an almost twofold increase in hGRβ expression level (Strickland et al. 2001). Interestingly, these cytokine treatments result in decreased sensitivity to glucocorticoids in these cell types (Kam et al. 1993; Strickland et al. 2001; Webster et al. 2001), but whether this decreased sensitivity is due to the increased expression of hGRβ is unclear.

12.4.3 Relevance of hGRβ

Several recent studies have shown an association between increased expression of hGRβ and immune-related diseases. In ulcerative colitis patients, more hGRβ-positive lymphocytes were found as compared to the number in healthy controls (Honda et al. 2000). The hGRα/-β ratio in lymphocytes of patients with leukemia was observed to be lower in two studies (Shahidi et al. 1999; Longui et al. 2000). In glucocorticoid-resistant asthma patients, more hGRβ-positive cells (compared to healthy controls and glucocorticoid-responsive asthma patients) were found in their PBMCs (Leung et al. 1997; Hamid et al. 1999), bronchoalveolar cells (Hamid et al. 1999), and in cutaneous inflammatory lesions (Sousa et al. 2000). Patients with fatal asthma had more hGRβ-positive cells (compared to healthy controls) in inflammatory cells in their airways (Christodoulopoulos et al. 2000). Finally, DeRijk et al. (2001) have observed a polymorphism in the 3'UTR of hGRβ mRNA that is associated with rheumatoid arthritis. This polymorphism was shown to result in a more stable hGRβ mRNA in vitro, and is therefore believed to increase the hGRβ expression in patients carrying this polymorphism. Although the associations between increased expression of hGRβ and the immune-related diseases mentioned above are evident, it remains to be elucidated if the increase in hGRβ expression contributes to glucocorticoid resistance and/or the pathogenesis of these diseases.

12.5 Conclusions

The dominant negative inhibition of hGRα's transactivational properties by hGRβ, as seen by several groups (Bamberger et al. 1995; Oakley et al. 1996), is still being questioned by others (Hecht et al. 1997; Brogan et al. 1999; de Lange et al. 1999). However, recent studies show convincing evidence that this action of hGRβ is a very specific activity, which requires the presence of 15 β-specific amino acids in its C-terminus (Oakley et al. 1999; M.R. Yudt et al., in preparation). Another point often used to question hGRβ's relevance is its low expression level. However, since hGRβ-specific antibodies are available, it appears that the expression at the protein level (in contrast to the mRNA level) is in most tissues not much lower than that of hGRα (de Castro et al. 1996; Longui et al. 2000; Strickland et al. 2001). Additionally, the expression of hGRβ appears to be inducible by different pro-inflammatory cytokines (Leung et al. 1997; Strickland et al. 2001; Webster et al. 2001).

The latter result opens up a new perspective, especially since many inflammatory diseases are treated with glucocorticoids, and hGRβ may result in glucocorticoid resistance due to its dominant negative activity, which in turn seriously decreases the efficacy of the therapy. The relative expression levels of hGRα and hGRβ have been shown to be correlated with glucocorticoid resistance in asthma patients (Leung et al. 1997; Hamid et al. 1999; Sousa et al. 2000). It may therefore be useful to screen the hGRα/-β ratio prior to treatment with glucocorticoids, so the efficacy of glucocorticoid treatment or the optimal dosage can be estimated. Therefore, we suggest that further studies should focus on the validity of the α/β ratio as a parameter to determine an individual's glucocorticoid responsiveness. Real-time RT-PCR (DeRijk et al. 2001) or flow cytometric analysis of immunostained cells (Berki et al. 1998; Wiegers et al. 2001) are possible tools that can be used for a quick screening of this α/β ratio.

In summary, over the last decade the perspective on the relevance of hGRβ has changed dramatically. Its action both in vitro and in vivo has been shown to be associated with glucocorticoid-resistance and a variety of immune-related diseases. Further research will be necessary to elucidate to what extent hGRβ contributes to glucocorticoid resistance in vivo and the pathogenesis of immune-related diseases.

References

Bamberger CM, Bamberger AM, de Castro M, Chrousos GP (1995) Glucocorticoid receptor beta, a potential endogenous inhibitor of glucocorticoid action in humans. J Clin Invest 95:2435–2441

Bamberger CM, Bamberger AM, Wald M, Chrousos GP, Schulte HM (1997) Inhibition of mineralocorticoid activity by the beta-isoform of the human glucocorticoid receptor. J Steroid Biochem Mol Biol 60:43–50

Beato M, Sanchez-Pacheco A (1996) Interaction of steroid hormone receptors with the transcription initiation complex. Endocr Rev 17:587–609

Beato M, Chalepakis G, Schauer M, Slater EP (1989) DNA regulatory elements for steroid hormones. J Steroid Biochem 32:737–747

Berki T, Kumanovics G, Kumanovics A, Falus A, Ujhelyi E, Nemeth P (1998) Production and flow cytometric application of a monoclonal anti-glucocorticoid receptor antibody. J Immunol Methods 214:19–27

Bresnick EH, Dalman FC, Sanchez ER, Pratt WB (1989) Evidence that the 90-kDa heat shock protein is necessary for the steroid binding conformation of the L cell glucocorticoid receptor. J Biol Chem 264:4992–4997

Brogan IJ, Murray IA, Cerillo G, Needham M, White A, Davis JR (1999) Interaction of glucocorticoid receptor isoforms with transcription factors AP-1 and NF-kappaB: lack of effect of glucocorticoid receptor beta. Mol Cell Endocrinol 157:95–104

Cadepond F, Schweizer-Groyer G, Segard-Maurel I, Jibard N, Hollenberg SM, Giguere V, Evans RM, Baulieu EE (1991) Heat shock protein 90 as a critical factor in maintaining glucocorticosteroid receptor in a nonfunctional state. J Biol Chem 266:5834–5841

Cadepond F, Gasc JM, Delahaye F, Jibard N, Schweizer-Groyer G, Segard-Maurel I, Evans R, Baulieu EE (1992) Hormonal regulation of the nuclear localization signals of the human glucocorticosteroid receptor. Exp Cell Res 201:99–108

Cairns C, Cairns W, Okret S (1993) Inhibition of gene expression by steroid hormone receptors via a negative glucocorticoid response element: evidence for the involvement of DNA-binding and agonistic effects of the antiglucocorticoid/antiprogestin RU486. DNA Cell Biol 12:695–702

Caldenhoven E, Liden J, Wissink S, Van de Stolpe A, Raaijmakers J, Koenderman L, Okret S, Gustafsson JA, Van der Saag PT (1995) Negative cross-talk between RelA and the glucocorticoid receptor: a possible mechanism for the antiinflammatory action of glucocorticoids. Mol Endocrinol 9:401–412

Christodoulopoulos P, Leung DY, Elliott MW, Hogg JC, Muro S, Toda M, Laberge S, Hamid QA (2000) Increased number of glucocorticoid receptor-beta-expressing cells in the airways in fatal asthma. J Allergy Clin Immunol 106:479–484

Collingwood TN, Urnov FD, Wolffe AP (1999) Nuclear receptors: coactivators, corepressors and chromatin remodeling in the control of transcription. J Mol Endocrinol 23:255–275

Dahia PL, Honegger J, Reincke M, Jacobs RA, Mirtella A, Fahlbusch R, Besser GM, Chew SL, Grossman AB (1997) Expression of glucocorticoid receptor gene isoforms in corticotropin-secreting tumors. J Clin Endocrinol Metab 82:1088–1093 de Castro M, Elliot S, Kino T, Bamberger C, Karl M, Webster E, Chrousos GP (1996) The non-ligand binding beta-isoform of the human glucocorticoid receptor (hGR beta): tissue levels, mechanism of action, and potential physiologic role. Mol Med 2:597–607

de Lange P, Koper JW, Brinkmann AO, de Jong FH, Lamberts SW (1999) Natural variants of the beta isoform of the human glucocorticoid receptor do not alter sensitivity to glucocorticoids. Mol Cell Endocrinol 153:163–168

DeRijk RH, Schaaf MJM, Turner G, Datson NA, Vreugdenhil E, Cidlowski JA, De Kloet ER, Sternberg EM, Detera-Wadleigh SD (2001) A glucocorticoid receptor gene variant that increases the stability of the glucocorticoid receptor beta-isoform mRNA is associated with rheumatoid arthritis. J Rheumatol (in press)

Diamond MI, Miner JN, Yoshinaga SK, Yamamoto KR (1990) Transcription factor interactions: selectors of positive or negative regulation from a single DNA element. Science 249:1266–1272

Drouin J, Trifiro MA, Plante RK, Nemer M, Eriksson P, Wrange O (1989) Glucocorticoid receptor binding to a specific DNA sequence is required for hormone-dependent repression of pro-opiomelanocortin gene transcription. Mol Cell Biol 9:5305–5314

Encio IJ, Detera-Wadleigh SD (1991) The genomic structure of the human glucocorticoid receptor. J Biol Chem 266:7182–7188

Giguere V, Hollenberg SM, Rosenfeld MG, Evans RM (1986) Functional domains of the human glucocorticoid receptor. Cell 46:645–652

Glass CK, Rosenfeld MG (2000) The coregulator exchange in transcriptional functions of nuclear receptors. Genes Dev 14:121–141

Hamid QA, Wenzel SE, Hauk PJ, Tsicopoulos A, Wallaert B, Lafitte JJ, Chrousos GP, Szefler SJ, Leung DY (1999) Increased glucocorticoid receptor beta in airway cells of glucocorticoid-insensitive asthma. Am J Respir Crit Care Med 159:1600–1604

Hecht K, Carlstedt-Duke J, Stierna P, Gustafsson J, Bronnegard M, Wikstrom AC (1997) Evidence that the beta-isoform of the human glucocorticoid receptor does not act as a physiologically significant repressor. J Biol Chem 272:26659–26664

Hollenberg SM, Weinberger C, Ong ES, Cerelli G, Oro A, Lebo R, Thompson EB, Rosenfeld MG, Evans RM (1985) Primary structure and expression of a functional human glucocorticoid receptor cDNA. Nature 318:635–641

Honda M, Orii F, Ayabe T, Imai S, Ashida T, Obara T, Kohgo Y (2000) Expression of glucocorticoid receptor beta in lymphocytes of patients with glucocorticoid-resistant ulcerative colitis. Gastroenterology 118:859–866

Jenkins BD, Pullen CB, Darimont BD (2001) Novel glucocorticoid receptor coactivator effector mechanisms. Trends Endocrinol Metab 12:122–126

Jonat C, Rahmsdorf HJ, Park KK, Cato AC, Gebel S, Ponta H, Herrlich P (1990) Antitumor promotion and antiinflammation: down-modulation of AP-1 (Fos/Jun) activity by glucocorticoid hormone. Cell 62:1189–1204

Kam JC, Szefler SJ, Surs W, Sher ER, Leung DY (1993) Combination IL-2 and IL-4 reduces glucocorticoid receptor-binding affinity and T cell response to glucocorticoids. J Immunol 151:3460–3466

Leung DY, Hamid Q, Vottero A, Szefler SJ, Surs W, Minshall E, Chrousos GP, Klemm DJ (1997) Association of glucocorticoid insensitivity with increased expression of glucocorticoid receptor beta. J Exp Med 186:1567–1574

Longui CA, Vottero A, Adamson PC, Cole DE, Kino T, Monte O, Chrousos GP (2000) Low glucocorticoid receptor alpha/beta ratio in T-cell lymphoblastic leukemia. Horm Metab Res 32:401–406

Mason SA, Housley PR (1993) Site-directed mutagenesis of the phosphorylation sites in the mouse glucocorticoid receptor. J Biol Chem 268:21501–21504

McKay LI, Cidlowski JA (1998) Cross-talk between nuclear factor-kappa B and the steroid hormone receptors: mechanisms of mutual antagonism. Mol Endocrinol 12:45–56

McKenna NJ, Xu J, Nawaz Z, Tsai SY, Tsai MJ, O'Malley BW (1999) Nuclear receptor coactivators: multiple enzymes, multiple complexes, multiple functions. J Steroid Biochem Mol Biol 69:3–12

Morrison N, Eisman J (1993) Role of the negative glucocorticoid regulatory element in glucocorticoid repression of the human osteocalcin promoter. J Bone Miner Res 8:969–975

Mu YM, Takayanagi R, Imasaki K, Ohe K, Ikuyama S, Yanase T, Nawata H (1998) Low level of glucocorticoid receptor messenger ribonucleic acid in pituitary adenomas manifesting Cushing's disease with resistance to a high dose-dexamethasone suppression test. Clin Endocrinol (Oxf) 49:301–306

Oakley RH, Sar M, Cidlowski JA (1996) The human glucocorticoid receptor beta isoform. Expression, biochemical properties, and putative function. J Biol Chem 271:9550–9559

Oakley RH, Webster JC, Sar M, Parker CR, Jr., Cidlowski JA (1997) Expression and subcellular distribution of the beta-isoform of the human glucocorticoid receptor. Endocrinology 138:5028–5038

Oakley RH, Jewell CM, Yudt MR, Bofetiado DM, Cidlowski JA (1999) The dominant negative activity of the human glucocorticoid receptor beta isoform. Specificity and mechanisms of action. J Biol Chem 274:27857–27866

Orti E, Mendel DB, Smith LI, Munck A (1989) Agonist-dependent phosphorylation and nuclear dephosphorylation of glucocorticoid receptors in intact cells. J Biol Chem 264:9728–9731

Otto C, Reichardt HM, Schutz G (1997) Absence of glucocorticoid receptor-beta in mice. J Biol Chem 272:26665–26668

Pearce D, Yamamoto KR (1993) Mineralocorticoid and glucocorticoid receptor activities distinguished by nonreceptor factors at a composite response element. Science 259:1161–1165

Picard D, Yamamoto KR (1987) Two signals mediate hormone-dependent nuclear localization of the glucocorticoid receptor. EMBO J 6:3333–3340

Pratt WB, Dittmar KD (1998) Studies with purified chaperones advance the understanding of the mechanism of glucocorticoid receptor-hsp90 heterocomplex assembly. Trends Endocrinol Metab 9:244–252

Pratt WB, Toft DO (1997) Steroid receptor interactions with heat shock protein and immunophilin chaperones. Endocr Rev 18:306–360

Pujols L, Mullol J, Perez M, Roca-Ferrer J, Juan M, Xaubet A, Cidlowski JA, Picado C (2001) Expression of the human glucocorticoid receptor alpha and beta isoforms in human respiratory epithelial cells and their regulation by dexamethasone. Am J Respir Cell Mol Biol 24:49–57

Ray A, Prefontaine KE (1994) Physical association and functional antagonism between the p65 subunit of transcription factor NF-kappa B and the glucocorticoid receptor. Proc Natl Acad Sci USA 91:752–756

Sakai DD, Helms S, Carlstedt-Duke J, Gustafsson JA, Rottman FM, Yamamoto KR (1988) Hormone-mediated repression: a negative glucocorticoid response element from the bovine prolactin gene. Genes Dev 2:1144–1154

Savory JG, Prefontaine GG, Lamprecht C, Liao M, Walther RF, Lefebvre YA, Hache RJ (2001) Glucocorticoid receptor homodimers and glucocorticoid-mineralocorticoid receptor heterodimers form in the cytoplasm through alternative dimerization interfaces. Mol Cell Biol 21:781–793

Schule R, Rangarajan P, Kliewer S, Ransone LJ, Bolado J, Yang N, Verma IM, Evans RM (1990) Functional antagonism between oncoprotein c-Jun and the glucocorticoid receptor. Cell 62:1217–1226

Shahidi H, Vottero A, Stratakis CA, Taymans SE, Karl M, Longui CA, Chrousos GP, Daughaday WH, Gregory SA, Plate JM (1999) Imbalanced expression of the glucocorticoid receptor isoforms in cultured lymphocytes from a patient with systemic glucocorticoid resistance and chronic lymphocytic leukemia. Biochem Biophys Res Commun 254:559–565

Sousa AR, Lane SJ, Cidlowski JA, Staynov DZ, Lee TH (2000) Glucocorticoid resistance in asthma is associated with elevated in vivo expression of the glucocorticoid receptor beta-isoform. J Allergy Clin Immunol 105:943–950

Strickland I, Kisich K, Hauk PJ, Vottero A, Chrousos GP, Klemm DJ, Leung DY (2001) High constitutive glucocorticoid receptor beta in human neutrophils enables them to reduce their spontaneous rate of cell death in response to corticosteroids. J Exp Med 193:585–594

Toft DO (1998) Recent advances in the study of hsp90 structure and mechanism of action. Trends Endocrinol Metab 9:238–243

Trapp T, Holsboer F (1996) Heterodimerization between mineralocorticoid and glucocorticoid receptors increases the functional diversity of corticosteroid action. Trends Pharmacol Sci 17:145–149

Tsai SY, Carlstedt-Duke J, Weigel NL, Dahlman K, Gustafsson JA, Tsai MJ, O'Malley BW (1988) Molecular interactions of steroid hormone receptor with its enhancer element: evidence for receptor dimer formation. Cell 55:361–369

Webster JC, Jewell CM, Bodwell JE, Munck A, Sar M, Cidlowski JA (1997) Mouse glucocorticoid receptor phosphorylation status influences multiple functions of the receptor protein. J Biol Chem 272:9287–9293

Webster JC, Oakley RH, Jewell CM, Cidlowski JA (2001) Proinflammatory cytokines regulate human glucocorticoid receptor gene expression and lead to the accumulation of the dominant negative beta isoform: a mechanism for the generation of glucocorticoid resistance. Proc Natl Acad Sci USA 98:6865–6870

Weinberger C, Hollenberg SM, Ong ES, Harmon JM, Brower ST, Cidlowski J, Thompson EB, Rosenfeld MG, Evans RM (1985a) Identification of human glucocorticoid receptor complementary DNA clones by epitope selection. Science 228:740–742

Weinberger C, Hollenberg SM, Rosenfeld MG, Evans RM (1985b) Domain structure of human glucocorticoid receptor and its relationship to the v-erb-A oncogene product. Nature 318:670–672

Wiegers GJ, Knoflach M, Bock G, Niederegger H, Dietrich H, Falus A, Boyd R, Wick G (2001) CD4(+)CD8(+)TCR(low) thymocytes express low levels of glucocorticoid receptors while being sensitive to glucocorticoid-induced apoptosis. Eur J Immunol 31:2293–2301

Wrange O, Eriksson P, Perlmann T (1989) The purified activated glucocorticoid receptor is a homodimer. J Biol Chem 264:5253–5259

Yang-Yen HF, Chambard JC, Sun YL, Smeal T, Schmidt TJ, Drouin J, Karin M (1990) Transcriptional interference between c-Jun and the glucocorticoid receptor: mutual inhibition of DNA binding due to direct protein- protein interaction. Cell 62:1205–1215

13 Cooperation of Nuclear Transcription Factors Regulated by Steroid and Peptide Hormones

B. Groner, C. Shemanko

13.1	Models to Study the Sequential and Simultaneous Action of Steroid and Peptide Hormones	213
13.2	Extracellular Signals Mediate Their Effects Through the Activation of Transcription Factors	214
13.3	Stat5 and the Glucocorticoid Receptor Cooperate in the Transcription from the β-Casein Gene Promoter	217
13.4	Transcriptional Regulation by Nuclear Receptors as Ligand-Dependent Transcription Factors and as Cofactors of Stats	221
13.5	Conclusions	226
References		227

13.1 Models to Study the Sequential and Simultaneous Action of Steroid and Peptide Hormones

Cycles of growth, differentiation and apoptosis characterise the fate of mammary epithelial cells throughout the life of the individual. These cellular processes are under the control of steroid and peptide hormones (Topper and Freeman 1980). Growth and limited differentiation occurs during puberty and requires the action of oestrogen, prolactin and members of the transforming growth factor (TGF)-β family. The epithelial cells form ducts and terminal end buds, which grow and extend to the limits of the fat pad. Under the influence of the hormones secreted during pregnancy (oestrogen, prolactin and progesterone) the cells undergo a massive increase in numbers and further differentiate to form the functional lactating gland. Prolactin and its receptor are particularly important for mammary development. Prolactin plays a role in the morphological and biochemical differentiation of the epithelial cells during pregnancy and regulates milk protein synthesis during lactation. It enhances the growth of breast cancer cell lines and has also been found to be expressed in a paracrine fashion in human tumours.

The secretory epithelial cells secrete large amounts of milk into the ductal system during lactation, which is followed by a period of involution and apoptosis at weaning. The gland systematically reduces the number of terminally differentiated cells by programmed cell death. The renewal of cell numbers and their differentiation is characteristic for the cycle that the gland undergoes with each pregnancy and parturition.

13.2 Extracellular Signals Mediate Their Effects Through the Activation of Transcription Factors

Our understanding of the regulation of proliferative, differentiating and apoptotic processes has been greatly extended in the past few years. Extracellular hormones, growth factors or cytokines relay their effects on the transcription of genes through the recognition of specific receptors and intracellular signalling molecules. The consequence is a shift in the overall pattern of gene expression, involving quantitative and qualitative alterations. The regulation of casein gene expression by peptide and steroid hormones has been most valuable for understanding the mechanisms by which signalling pathways converge on gene expression. In addition to the basic insights into the linear pathways, the study of mammary epithelial cells and milk protein gene expression has allowed us to gain insights into hormone-mediated interactions in the nucleus. Prolactin and glucocorticoid hormones act synergistically to induce β-casein gene expression in mammary epithelial cells. The regulatory factors governing milk protein gene expression are not limited to Stat5 (signal transducer and activator of transcription) and the glucocorticoid receptor, but comprise additional signal-regulated components. These include C/EBPs, NFI, YYI and other transcription factors that are characterised in less detail (Altiok and Groner 1994; Robinson et al. 1998; Rosen et al. 1998, 1999).

The individual pathways utilised by prolactin and glucocorticoid hormones have been described in molecular detail for some time, it has recently been demonstrated that Stat5 and the glucocorticoid receptor (GR), the downstream effectors of prolactin and glucocorticoid hormones, interact and exhibit transcriptional synergy (Stöcklin et al. 1996; Cella et al. 1998) and transcriptional repression.

The classic model of transcriptional regulation by nuclear hormone receptors (reviewed by Aranda and Pascual 2001) involves binding of monomers, homodimers or retinoid X receptor (RXR) heterodimers to DNA regulatory sequences in the promoters of target genes. The DNA response element for monomers is a single hexameric site. Homodimers or heterodimers bind to two hexameric motifs arranged as palindromes, inverted palindromes or direct repeats. The arrangement and spacing determines binding specificity. Some hexameric response elements can mediate the transcriptional response from more than one ligand receptor complex, resulting in a transcriptional control of target genes from overlapping signals (for example, thyroid hormones and retinoic acid).

Dimerisation is mediated by the multifunctional ligand-binding domain (LBD) as well as the DNA-binding domain (DBD) of the nuclear hormone receptors (NHRs). The DBD zinc fingers mediate specific and high-affinity DNA binding as well as DNA-based receptor dimerisation. There are three types of heterodimeric complexes, unoccupied (unliganded) heterodimers, nonpermissive heterodimers and permissive heterodimers. Permissive heterodimers, such as peroxisome proliferator activated receptors (PPAR/RXR) can be activated by ligands of either RXR or its partner and are synergistically activated in the presence of both ligands. In non-permissive heterodimers, RXR often acts as a silent partner. Dimers can also bind negative response elements and cause gene repression. These sites have been found for the GR and thyroid hormone receptor (TR). Negative or positive regulation of gene transcription usually results from the interaction of NHR with corepressors or coactivators.

When glucocorticoid hormones, for example, are taken up into cells, they bind intracellularly to the latent form of the GR. Ligand binding induces a conformational change in GR and dissociation of heat shock proteins, which are associated with the transcriptionally inactive form of the receptor. This form of the GR resides primarily in the cytoplasm in the absence of ligand. Upon ligand binding, it translocates to the nucleus where it binds to palindromic glucocorticoid response elements (GREs) in the promoters of target genes (Beato et al. 1995; Bamberger et al. 1996). It also can interact with GRE half-sites, a DNA interaction which can be stabilised by other trans acting factors binding to adjacent sequences.

Prolactin binds to the extracellular domain of the prolactin receptor (PRL-R) and causes its dimerisation (Doppler 1994; Groner and Gouilleux 1995; Hynes et al. 1997). This complex initiates the Jak/Stat signal transduction cascade (Pellegrini and Dusanter-Fourt 1997). Jak2 is a cytoplasmic protein tyrosine kinase, which is associated with the intracellular domain of the prolactin receptor (Parganas et al. 1998). The enzymatic activity of this kinase is activated when two Jak2 molecules are brought in immediate proximity through the receptor dimerisation. The activated Jak2 tyrosine phosphorylates the PRL-R, resulting in phosphorylated tyrosine residues which in turn can serve as docking sites for the SH2 domains in Stat5. This latent transcription factor is subsequently tyrosine phosphorylated. Stat5 can now dimerise itself through a src homology (SH)2 tyrosine phosphate interaction and translocate to the nucleus (Ihle 1996; Berchtold et al. 1997; Darnell 1997). Activated Stat5 binds to DNA sites in the nucleus known as γ-interferon activated sites (GAS) elements and modulates the activity of target genes, e.g. the β-casein gene (Kazansky et al. 1995; John et al. 1999; Ihle 2001).

Stat5 was first identified as a DNA-binding protein in tissue extracts from lactating mammary gland (Wakao et al. 1992, 1994; Schmitt-Ney et al. 1991, 1992a,b). Molecular cloning and extensive molecular biological analysis of different tissues at various stages of differentiation revealed that Stat5 was not only expressed in mammary epithelial cells and activated by prolactin, but that many hormones, growth factors and cytokines use Stat5 as a signal transducer in other tissues (Gouilleux et al. 1994, 1995; Gobert et al. 1996; Mui et al. 1996; Teglund et al. 1998). Two clustered Stat5 genes, Stat5a and Stat5b, have been identified that are more than 90% identical and probably arose by gene duplication (Liu et al. 1995). Both Stat5a and Stat5b encode isoforms, some of which may arise by alternative splicing.

Particularly interesting are carboxy-truncated isoforms of Stat5. These naturally occurring carboxy-truncated Stat5 isoforms lack a functional transactivation domain and act as dominant negative inhibitors of Stat5-dependent transcription. They remain tyrosine phosphorylated and bound to GAS sites for longer periods of time than full-length Stat5 isoforms after PRL treatment, suggesting that the carboxy-terminal sequences may affect the interaction with a tyrosine phosphatase (Moriggl et al. 1996, 1997). In addition, these variants are able to efficiently recruit nuclear corepressors to gene promoters and thus actively participate in the formation of silent chromatin (Maurer et al. 2001).

13.3 Stat5 and the Glucocorticoid Receptor Cooperate in the Transcription from the β-Casein Gene Promoter

The hormonal synergism between prolactin and glucocorticoids in the induction of the β-casein milk protein gene promoter was investigated in molecular detail (Doppler et al. 1988; Stöcklin et al. 1996; Lechner et al. 1997). Each pathway involves the activation of latent transcription factors Stat5 and the glucocorticoid receptor. Initially it was cotransfection experiments with the wild-type versions of Stat5 and the glucocorticoid receptor as well as deletion and point mutants of these transcription factors which allowed us to gain further insights into their functional synergism. A single Stat5-binding site in the promoter region of a target gene is sufficient to confer the functional interaction. The synergism is enhanced when multimerised Stat5-binding sites are present. In contrast to the requirement for a specific Stat5-binding site, we find that no functional glucocorticoid response element is necessary. Synergistic activation of transcription needs Stat5 activation through tyrosine phosphorylation and the presence of the activation function (AF)-1 transactivation function in the glucocorticoid receptor. The DNA-binding domain of the glucocorticoid receptor and the transactivation domain of Stat5 are dispensable (Stöcklin et al. 1996, 1997, 1999).

Stat5 and the GR also synergise in the regulation of the whey acidic protein gene promoter (Mukhopadhyay et al. 2001), another milk protein gene. It is possible that a slightly different mechanism is involved. It depends upon the presence of the nuclear factor I response element, the Stat5 site and clustered GR half sites. NF1-B2 was found to participate in full synergism with Stat5 and GR. NF1-A4, NF1-B2 and NF1-X1 are developmentally regulated in the mammary gland, but interact with the DNA response element on this promoter with differential specificity.

GR appears to promote the active conformation of C/EBPbeta as well as prolong the activated state of Stat. p300/CREB-binding protein (CBP) is an important coactivator for Stat5 and GR, but was not rate limiting in the repression mediated by Stat5 on a GR-responsive promoter. The in vivo interaction between the GR and Stat5 was also detected in liver extracts and in NIH-3T3 cells. Mice homozygous for a GR mutation that disrupts dimerisation and DNA binding are viable, indicating that the mutant GR still functions in the mice (Reichardt et al. 1998). One possible explanation is that cross-talk with Stat5, and possibly other transcription factors, is responsible for their survival.

Proteins interacting with GR have complemented our conventional picture of GR action over the past several years. GR, as a specific DNA-binding protein, associates with transcriptional regulators through protein–protein interactions. These transcriptional coactivators (e.g. SRC-1 and CBP) function as mediators for the interaction of the receptor with components of the basal transcription apparatus and participate in the transcriptional induction. They recruit an enzymatic activity inherent in the coactivator molecules, i.e. histone acetyltransferase (Pfitzner et al. 1998). These enzymes cause a change in chromatin structure and change the accessibility for essential components of the transcriptional machinery (reviewed by Lee and Lee Kraus 2001).

Protein–protein interactions have also been described in situations in which GR exerts a negative role in transcription. Glucocorticoids can repress the transcription of genes which do not contain receptor-specific binding sites in their promoter, e.g. the collagenase gene. Repression is thought to be mediated by the interaction of the receptor with AP-1. The interaction is independent of the DNA binding of the receptor and does not interfere with the specific DNA binding of AP-1. A similar scenario has been suggested for the negative regulation of the gene of the α subunit of the human glycoprotein hormone where GR interferes with the positive action of the cAMP-response element binding protein, CREB. Interaction of the glucocorticoid receptor with the p65 component of the transcription factor NF-κB leads to a decreased binding activity and thus an interference with the induction of NF-κB regulated genes. The examples for non-conventional glucocorticoid receptor action comprise mostly cases of negative regulation. Only a few experiments have been reported in which the glucocorticoid interactions with AP-1 complexes of a particular composition lead to the enhancement of transcription.

Stat factors have also been found to associate with a number of proteins. In addition to the Stat homodimers, Stat1-Stat2 heterodimers and complexes of Stat1-Stat2 with p48, Stat1 with Sp1, Stat1 or Stat2 with CBP/p300, Stat3b with c-jun or CREB and Stat5 with GR. It is likely that these protein–protein interactions add to the functional diversity of Stat proteins. The associations with other transcription factors might specify target genes in particular cell types and modulate the strength of transcription. These interactions might complement regulatory mechanisms which are based on the composite structure of promoters with respect to transcription-factor binding sites. They bring into proximity binding sites for Stat factors and other transcription factors such as members of the ets family, GATA factors, nuclear factor 1, YY1, Sp1, CREB or C/EBP possibly to promote their functional interactions (Meier and Groner 1994).

While these experiments define a non-conventional function for GR, GR itself can act as a coactivator of Stat5. This coactivation function requires ligand activation of the wild-type form of the receptor. It is independent of the DNA-binding domain and can be observed when the ligand-binding domain of the receptor is deleted. The requirement for the proper ligand-dependent activation of the wild-type form of the receptor was corroborated in experiments with partial hormone antagonists. RU486 was not able to cause an increase in transcription of the β-casein promoter construct over that observed with prolactin activated Stat5 alone. Our experiments also indicate that the carboxyl-truncated variants of Stat5 might have a function which goes beyond their dominant negative phenotype observed in transfected cells. The truncated molecule is still able to cooperate with the glucocorticoid receptor in transcriptional induction. Since its DNA-binding activity is maintained over a much longer time upon cytokine induction, but the transactivation potential in conjunction with the glucocorticoid receptor is reduced when compared to the full-length molecule, it is possible that the truncated variants serve as a means to regulate extent and duration of the cytokine response.

Although our experiments indicate that the transactivation domain of Stat5 as well as the ligand-binding domain and the DNA-binding domain GR are not involved in the functional interaction, they do not allow conclusions about the physical interaction sites. The region of GR comprising the AF-1 function is required for the synergism, but might not directly participate in the interaction. We also cannot exclude an adaptor protein which interacts with both Stat5 and GR. Secondary modifications, phosphorylations on serine/threonine residues, have been detected in Stat5 as well as in GR. Stat5 is phosphorylated on serine/threonine residues prior to the lactogenic hormone induction of mammary epithelial cells. No increase in serine/threonine phosphorylation of Stat5 was observed upon hormone addition, although ERK2-MAPK activity was transiently induced through the activation of the prolactin receptor (Wartmann et al. 1996). Serine phosphorylation of Stat factors has been recognised as important for DNA binding and the transactivation potential. GR showed an increase in its phosphorylation upon hormone induction, but no functional significance could yet be assigned (Beisenherz et al. 2001). It is conceivable that these phosphorylation events could regulate specific factor interactions. Diverse, cell-type and cytokine-specific interactions between Stat factors and non-Stat transcription factors might contribute to the different phenotypic effects elicited by individual cytokines utilising the Jak-Stat pathway.

An inverse relationship was observed when the co-activation of Stat5 and the GR were studied and effects on GRE-dependent transcription was measured (Stöcklin et al. 1996; Biola et al. 2001). Stat5 suppresses the GR mediated activation of glucocorticoid responsive genes. Complex formation between Stat5 and GR diminishes the glucocorticoid response of a GRE-containing promoter and adds an additional level of regulation. Combinatorial regulation is observed in which the activation of two signals, prolactin and glucocorticoid hormones, can result in three different outcomes. Each signal by itself can determine the induction of promoter sequences with the appropriate response elements. The induction by both signals simultaneously causes a third outcome which is not the simple addition of the two signals but the provision of a new quality. Cytokine-dependent gene transcription is enhanced, glucocorticoid-dependent gene transcription is diminished.

13.4 Transcriptional Regulation by Nuclear Receptors as Ligand-Dependent Transcription Factors and as Cofactors of Stats

The non-conventional mode of nuclear hormone receptor function is based upon specific protein–protein interactions and allows the integration of seemingly unrelated signalling pathways (reviewed by Goettlicher et al. 1998). Cross-talk was initially suspected when the simultaneous activation of two transcription factors resulted in a different qualitative or quantitative regulation of genes compared to a situation in which only one of the factors was activated. The effect could even be observed in the regulation of genes that contained a DNA-response element in their promoter region for only one of the two transcription factors. The cross-talk of different families of transcription factors with NHRs has been investigated in detail. Examples for this regulatory principle include the functional interaction between the GR and AP-1, and also the GR and Stats. Other transcription factor families such as NF-κB/Rel, Oct and CCAAT/enhancer-binding protein (C/EBP) have also been shown to participate in cross-talk with members of the NHR family. The comprehensive molecular mechanisms for the negative or positive cross-talk that can ensue have not been fully elucidated, though many important aspects have been described (reviewed by Robyr et al. 2000).

Transcriptional cross-talk can be the consequence of different molecular interactions depicted schematically in Fig. 1. Two transcription factors with specific DNA-binding potential can interact, but that only one of them is bound to DNA via its DNA-binding domain (Fig. 1 part 1). The second factor is recruited to the promoter through protein–protein interactions. Each transcription factor can recognise different elements in the same promoter of the target gene (Fig. 1 part 2). Similar to the situation shown in Fig. 1 part 1, the recruitment of coactivators by both factors could lead to a higher local concentration of such activators and an increase in transcriptional strength. Transcription factor-binding sites are arranged so that they are neighbouring or overlapping with each other, and that proximity influences the activity or binding of the other protein (Fig. 1 part 3). It enables direct protein–protein interactions, but could also cause competition for DNA-binding sites or limiting amounts of cofactors. The effect of this third mode of

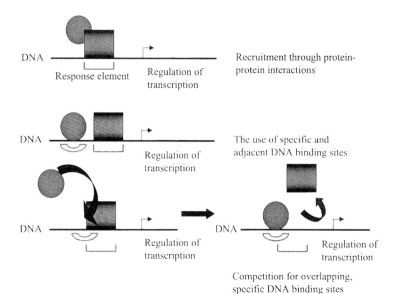

Fig. 1. Mechanisms of molecular crosstalk. *1*, Protein–protein interactions result in the recruitment of two transcription factors to the promoter sequence, only one of which is bound to a specific DNA sequence. *2*, Each transcription factor recognises its own element in the gene promoter with the possibility of direct or indirect contact. *3*, Specific DNA binding sites can involve overlapping sequences. The binding of factors is determined by relative affinity and can result in mutual interference. An example of this is the binding of YY1 and Stat5 to the β-casein promoter, where as long as YY1 is bound, the promoter is repressed. Upon Stat5 activation, YY1 is replaced by Stat5

cross-talk is often manifested in gene repression. The elucidation of the different modes of receptor action, classical and transcriptional cross-talk-mediated, have been aided by the utilisation of various ligands or gene mutation.

The Stat family of transcription factors are encoded by seven genes in mammalian cells, encoding Stat1, Stat2, Stat3, Stat4, Stat5a, Stat5b and Stat6. Diversity of function is further enhanced by the production of alternatively spliced variants and proteolytic posttranslational processing. Stat1, Stat3 and Stat4 mRNAs have been found to be alternatively spliced, and Stat5 can be proteolytically processed. These target genes

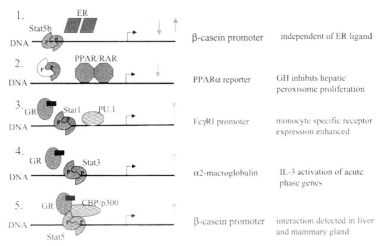

Fig. 2. Examples for transcriptional cross-talk between NHRs and Stats. *1*, Stat5 is able to affect gene regulation by its interaction with the ER. *2*, Stat5, in the absence of direct contact with PPARα, -γ or -δ, causes inhibition of transcription on a promoter with a PPRE. *3*, GR does not interact with Stat1 but promotes transcription, with PU.1 providing cell-type specificity. *4*, GR acts indirectly in concert with Stat3 to enhance the transcriptional response on the IL-6 response element of the rat α2-macroglobulin (*APRE*) promoter. *5*, GR and Stat5 interact to promote transcription on a promoter with Stat binding sites

are involved in a wide range of cellular processes including growth, differentiation, and apoptosis. The diversity of the signals which mediate their effects through Stat factors and the various functions affected by Stat factors in different cell types, make it reasonable to assume that combinatorial mechanisms are at work to mediate the cellular responses. Examples for such combinatorial effects have been found in addition to the well-studied interaction between Stat5 and GR and are summarised in Fig. 2.

Oestrogen receptors (ERs) ERα and ERβ bind the steroid hormone oestrogen, xenoestrogens and peptide growth factors to mediate effects on growth and development in a wide range of tissues. ER can cross-talk with other DNA-bound transcription factors in order to regulate gene expression. It can cause upregulation of genes with promoter-binding sites for Sp-1 sites and AP-1 (differential effects with the two ERs) and it can negatively affect genes with regulatory sites for NF-κB, GATA-1 and C/EBP. The receptors bind additional cofactors in order to integrate multiple signals.

Cross-talk has been observed between ER and Stat5. It causes oestrogen-dependent inhibition of Stat5-mediated transcriptional induction and is accompanied by decreased Stat5 DNA binding and tyrosine phosphorylation. The latter two events have also occurred independently of ER ligand. These observations, however, have not been consistently made in all cell types. Synergism between ERα or ERβ and Stat5b was also reported. Transcription induced by prolactin-activated Stat5b was enhanced by the expression of either ER, though with a larger quantitative effect through ERβ, in a mammary epithelial cell line. The synergism required activation of Stat5b, but not the ligand activation of the ERs. The DNA-binding domain of the ERs was essential for the functional interaction with Stat5b, though it was not determined if the effect is mediated by a direct physical interaction. Other members of the steroid hormone receptor family cross-talk with Stat5. The mineralocorticoid, progesterone, and glucocorticoid receptors synergise with Stat5 in the induction of transcription from a Stat-responsive promoter. Conversely, Stat5 negatively interfered with transcription from promoters carrying DNA-response elements for those receptors. The androgen receptor did not appear to interact with Stat5.

PPARs belong to a subfamily of NHR, consisting of three receptors, PPARαa, PPARδ and PPARγ, which are developmentally as well as tissue-specifically expressed. They are ligand-activated transcription factors which recognise PPAR response elements (PPRE) when heterodimerised with RXR. PPARs can repress gene transcription by negatively interfering with NF-κB, and AP-1. PPARα mediates hepatic peroxisome proliferation after induction by non-genotoxic carcinogens. This response is inhibited by growth hormone (GH), a peptide hormone that stimulates the activation of Stat1, -3, -5a and -5b. GH inhibition of PPARα reporter gene transcription was mediated specifically by Stat5b (Zhou et al. 1999). Stat5b tyrosine phosphorylation, its DBD, and its transcription activation domain, plus the amino-terminal AF-1 transactivation domain of PPARα, were essential to mediate the inhibition. Direct protein–protein interactions were not detected. It is possible that Stat5b competes with PPARα for an essential coactivator, though p300 and SRC-1 were shown not to be involved in this case. This negative cross-talk is most likely the molecular basis for the regulation of PPARα-dependent responses by GH. This could be important in the modulation of inflammatory responses by leukocytes, which have an intact GH-Jak-Stat pathway, or by inflammatory cells, which have the capacity to synthesise and secrete GH. GH may also inhibit tumour development by suppressing the PPARα response to peroxisome proliferators that have been linked to carcinogenesis. GH-activated Stat5b also inhibited the transcriptional activity of PPARγ, PPARδ and T3R.

GR functionally interacts with Stat1 and PU.1 in the mediation of inflammatory and immune responses (Aittomaki et al. 2000). interferon (IFN)-γ and Stat1 and GR signals converge on the Fcγ receptor I (FcγRI) gene promoter through cross-talk between Stat1 and GR. The upregulation of the high-affinity FcγRI in monocytes is important for endocytosis of immune complexes and in antibody-mediated cytotoxic reactions. Monocyte-specific transcription of this gene is mediated by a second response element which binds PU.1, an Ets-family transcription factor, and activation of this gene is dependent upon both Stat1 and PU.1. The action of GR results in an enhancement of Stat1-dependent gene activation. The cross-talk between Stat1 and GR does not appear to involve direct physical interaction between the two proteins. The use of GR variants showed that the functional cross-talk does not depend on the interaction of the GR with this promoter, though its DNA-binding domain was important, as was its hormone-dependent transactivation function. The DNA-binding function of active Stat1 was essential. It was shown that though the GR can activate transcription on the Stat1 responsive promoter, Stat1 did not affect GR-dependent transcription. The requirement of protein synthesis in the process indicates that GR may enhance transcription by inducing the synthesis of transcriptional coregulators. The recruitment of cofactors could not be ruled out, though CBP, an important coactivator for both GR and Stat5, was not a rate-limiting component.

Synergism of GR with Stat3 has been documented on the Stat3 responsive promoter, α_2-macroglobulin (Takeda et al. 1998). The synergism is dependent upon the binding of interleukin-6 (IL-6)/Jak-activated Stat3, to the DNA-response element, APRE. The testing of two other Stat3-responsive promoters containing a dual Stat3-response element plus a cAMP-responsive element did not result in synergism with the

GR, indicating a specificity for the APRE. There was no detectable physical interaction between the GR and Stat3, though the synergism appears to either involve an undefined coactivator process, and not secondary gene expression. The IL-6 activation of acute phase proteins, such as α_2-macroglobulin, is important in host defence mechanisms, and relies upon glucocorticoid for maximal expression.

13.5 Conclusions

The interaction of nuclear hormone receptors with Stat molecules is a common mechanism for specific gene regulation that is shared among many different family members, such as the ER, GR, PPAR (PPARα, -γ and -δ), mineralocorticoid, progesterone and T3R. It is likely that other members of the NHR family will be found to interact with one of the Stat family members in order to achieve transcription cross-talk. Although the cross-talk of NHR with Stats to regulate gene transcription is a common theme, the molecular mechanisms appear to depend upon the Stat, NHR and promoter sequences involved. Physical interaction of the NHR with the Stat is not always detectable nor is DNA binding of both partners commonly found. Another important question to be answered concerns the qualitative and quantitative effects of cross-talk. Two transcription factors which act positively in gene induction on their own can result in gene repression upon simultaneous activation and complex formation.

NHR mediate their transcriptional regulatory function through at least two qualitatively different classes of cofactors, coactivator and corepressor molecules and their associated chromatin remodelling components. The recruitment of these cofactors is tightly regulated and ordered and necessary for the transcription initiation and attenuation of hormone-regulated genes. How multiple cofactors are recruited into large transcription complexes and what determines the balance between positively and negatively acting factors is a focus of current research. The potential of Stats to recruit coactivators and corepressors with their associated chromatin remodelling functions makes them formally quite similar to TR and retinoic acid receptor (RAR), members of the NHR family. Through direct physical interactions or through adjacent DNA-binding sites Stats might be able to influence the timing of NHR binding to the DNA, the recruitment of chromatin remodelling complexes, assembly of the transcription preinitiation complex and the recruitment of cofactors and thus regulate the cellular response to nuclear hormones.

References

Aittomaki S, Pesu M, Groner B, Janne OA, Palvimo JJ, Silvennoinen O (2000) Cooperation among Stat1, glucocorticoid receptor, and PU.1 in transcriptional activation of the high-affinity Fc gamma receptor I in monocytes. J Immunol 164:5689–5697

Altiok S, Groner B (1994) Beta-casein mRNA sequesters a single stranded nucleic acid binding protein which negatively regulates the beta-casein gene promoter. Mol Cell Biol 14:6004–6012

Aranda A, Pascual A (2001) Nuclear hormone receptors and gene expression. Physiol Rev 81:1269–1304

Bamberger CM, Schulte HM, Chrousos GP (1996) Molecular determinants of glucocorticoid receptor function and tissue sensitivity to glucocorticoids. Endo Rev 17:245–261

Beato M, Herrlich P, Schuetz G (1995) Steroid hormone receptors: many actors in search of a plot. Cell 83:851–857

Beisenherz-Huss C, Mundt M, Herrala A, Vihko P, Schubert A, Groner B (2001) Specific DNA binding and transactivation potential of recombinant, purified Stat5. Mol Cell Endocrinol 183:101–112

Berchtold S, Morrigl R, Gouilleux F, Silvennoinen O, Groner B (1997) Cytokine receptor independent, constitutively active variants of MGF-Stat5. J Biol Chem 272:30237–30243

Biola A, Lefebvre P, Perrin-Wolff M, Sturm M, Bertoglio J, Pallardy M (2001) Interleukin-2 inhibits glucocorticoid receptor transcriptional activity through a mechanism involving STAT5 (signal transducer and activator of transcription 5) but Not AP-1. Mol Endocrinol 15:1062–1076

Cella N, Groner B, Hynes NE (1998) Characterisation of Stat5a and Stat5b homodimers and heterodimers and their association with the glucocorticoid receptor in mammary cells. Mol Cell Biol 18:1783–1792

Darnell JE Jr (1997) STATs and gene regulation. Science 277:1630–1635

Doppler W (1994) Regulation of gene expression by prolactin. Rev Physiol Biochem Pharmacol 124:93–130

Doppler W, Groner B, Ball RK (1988) Prolactin and glucocorticoid hormones synergistically induce expression of transfected rat β-casein gene promoter constructs in a mammary epithelial cell line. Proc Natl Acad Sci USA 86:104–108

Gobert S, Chretien S, Gouilleux F, Muller O, Pallard C, Dusanter-Fourt I, Groner B, Lacombe C, Gisselbrecht S, Mayeux P (1996) Identification of tyrosine residues in the intracellular domain of the erythropoietin receptor crucial for Stat5 activation. EMBO J 10:2434–2441

Goettlicher M, Heck S, Herrlich P (1998) Transcriptional cross-talk, the second mode of steroid hormone receptor action. J Mol Med 76:480–489

Gouilleux F, Wakao H, Mundt M, Groner B (1994) Prolactin induces phosphorylation of tyrosine 694 of mammary gland factor (MGF), a prerequisite for DNA binding and induction of transcription. EMBO J 13:4361–4369

Gouilleux F, Pallard C, Dusanter I, Wakao H, Haldosen LA, Norstedt G, Levy D, Groner B (1995) Prolactin, growth hormone, erythropoietin and granulocyte-macrophage colony stimulating factor induce MGF-Stat5 DNA binding activity. EMBO J 14:2005–2013

Groner B, Gouilleux F (1995) Prolactin mediated gene activation in mammary epithelial cells. Curr Opin Genet Develop 5:587–594

Hynes NE, Cella N, Wartmann M (1997) Prolactin stimulated pathways in mammary epithelial cells. J Mammary Gland Biol Neoplasia 1:207–214

Ihle JN (1996) Signaling by the cytokine superfamily in normal and transformed cells. Adv Cancer Res 68:23–65

Ihle JN (2001) The Stat family in cytokine signaling. Curr Opin Cell Biol 13:211–217

John S, Vinkemeier U, Soldaini E, Darnell JE Jr, Leonard WJ (1999) The significance of tetramerization in promoter recruitment by Stat5. Mol Cell Biol 19:1910–1918

Kazansky AV, Raught B, Lindsey SM, Wang YF, Rosen JM (1995) Regulation of mammary gland factor/Stat5a during mammary gland development. Mol Endocrinol 9:1598–1609

Lechner J, Welte T, Tomasi JK, Bruno P, Cairns C, Gustafsson J, Doppler W (1997) Promoter-dependent synergy between glucocorticoid receptor and Stat5 in the activation of beta-casein gene transcription. J Biol Chem 272:20954–20960

Lee KC, Lee Kraus W (2001) Nuclear receptors, coactivators and chromatin: new approaches, new insights. Trends Endocrinol Metab 12:191–197

Liu X, Robinson GW, Gouilleux F, Groner B, Hennighausen L (1995) Cloning and expression of two closely related homologues of MGF-Stat5 involved in prolactin signal transduction in mouse mammary tissue. Proc Natl Acad Sci USA 92:8831–8835

Maurer A, Wichmann C, Kunkel H, Heinzel T, Ruthardt M, Groner B, Grez M (2001) The Stat5-RARα fusion protein represses transcription and differentiation through interaction with a corepressor complex. Blood (in press)

Meier VS, Groner B (1994) The nuclear factor YY1 participates in repression of the beta-casein gene promoter in mammary epithelial cells and is counteracted by mammary gland factor during lactogenic hormone induction. Mol Cell Biol 14:128–137

Moriggl R, Gouilleux-Gruart V, Jähne R, Berchtold S, Gartmann C, Liu X, Hennighausen L, Sotiropoulos A, Groner B, Gouilleux F (1996) Deletion of the carboxyl terminal transactivation domain of MGF-Stat5 results in enhanced DNA binding and a dominant negative phenotype. Mol Cell Biol 16:5691–5700

Moriggl R, Berchtold S, Friedrich K, Standke G, Kammer W, Heim M, Wissler M, Stöcklin E, Gouilleux F, Groner B (1997) Comparison of the transactivation domains of Stat5 and Stat6 in lymphoid and in mammary epithelial cells. Mol Cell Biol 17:3663–3678

Mui AL-F, Wakao H, Kinoshita T, Kitamura T, Miyajima A (1996) Suppression of interleukin-3-induced gene expression by a C-terminal truncated Stat5: role of Stat5 in proliferation. EMBO J 15: 2425–2433

Mukhopadhyay SS, Wyszomierski SL, Gronostajski RM, Rosen JM (2001) Differential interactions of specific nuclear factor I isoforms with the glucocorticoid receptor and stat5 in the cooperative regulation of WAP gene transcription. Mol Cell Biol 21:6859–6869

Parganas E, Wang D, Stravopodis D, Topham DJ, Marine J-C, Teglund S, Vanin EF, Bodner S, Colamonici OR, van Deursen JM, Grosveld G, Ihle JN (1998) Jak2 is essential for signaling through a variety of cytokine receptors. Cell 93:385–395

Pellegrini S, Dusanter-Fourt I (1997) The structure regulation and function of the Janus kinases (JAKs) and the signal transducers and activators of transcription (STATs). Eur J Biochem 248:615–633

Pfitzner E, Jähne R, Wissler M, Stöcklin E, Groner B (1998) p300/CBP enhances the prolactin mediated transcriptional induction through direct interaction with the transactivation domain of Stat5, but does not participate in the Stat5 mediated suppression of the glucocorticoid response. Mol Endocrinol 12:1582–1593

Reichardt HM, Kaestner KH, Tuckermann J, Kretz O, Wessely O, Bock R, Gass P, Schmid W, Herrlich P, Angel P, Schutz G (1998) DNA binding of the glucocorticoid receptor is not essential for survival. Cell 93:531–541

Robinson GW, Johnson PF, Hennighausen L, Sterneck E (1998) The C/EBPβ transcription factor regulates epithelial cell proliferation and differentiation in the mammary gland. Genes Dev 12:1907–1916

Robyr D, Wolffe AP, Wahli W (2000) Nuclear hormone receptor coregulators in action: diversity for shared tasks. Mol Endocrinol 14:329–47

Rosen JM, Zahnow C, Kazansky A, Raught B (1998) Composite response elements mediate hormonal and developmental regulation of milk protein gene expression. Biochem Soc Symp 63:101–113

Rosen JM, Wyszomierski SL, Hadsell D (1999) Regulation of milk protein gene expression. Annu Rev Nutr 19:407–436

Schmitt-Ney M, Doppler W, Ball R, Groner B (1991) Beta-casein gene promoter activity is regulated by the hormone-mediated relief of transcriptional repression and a mammary gland specific nuclear factor. Mol Cell Biol 11:3745–3755

Schmitt-Ney M, Happ B, Ball R, Groner B (1992a) Developmental and environmental regulation of a mammary gland specific nuclear factor essential for transcription of the gene encoding beta-casein. Proc Natl Acad Sci USA 89:3130–3134

Schmitt-Ney M, Happ B, Hofer P, Hynes NE, Groner B (1992b) Mammary gland nuclear factor (MGF) activity is positively regulated by lactogenic hormones and negatively by milk stasis. Mol Endocrinol 6:1988–1997

Stöcklin E, Wissler M, Gouilleux F, Groner B (1996) Functional interactions between Stat5 and the glucocorticoid receptor. Nature (London) 383:726–728

Stöcklin E, Wissler M, Moriggl R, Groner B (1997) Specific DNA binding of Stat5, but not of glucocorticoid receptor, is required for the their functional cooperation in the regulation of gene transcription. Mol Cell Biol 17:6708–6716

Stöcklin E, Wissler M, Schaetzle D, Pfitzner E, Groner B (1999) Interactions in the transcriptional regulation exerted by Stat5 and by members of the steroid hormone receptor family. J Steroid Biochem Mol Biol 69:195–204

Takeda T, Kurachi H, Yamamoto T, Nishio Y, Nakatsuji Y, Morishige K, Miyake A, Murata Y (1998) Crosstalk between the interleukin-6 (IL-6)-JAK-STAT and the glucocorticoid-nuclear receptor pathway: synergistic activation of IL-6 response element by IL-6 and glucocorticoid. J Endocrinol 159:323–330

Teglund S, McKay C, Schuetz E, van Deursen J, Stravopodis D, Wang D, Brown M, Bodner S, Grosveld G, Ihle JN (1998) Stat5a and Stat5b proteins have essential and non-essential, or redundant, roles in cytokine responses. Cell 93:841–850

Topper YJ, Freeman CS (1980) Multiple hormone interactions in the developmental biology of the mammary gland. Physiol Rev 60:1049–1106

Wakao H, Schmitt-Ney M, Groner B (1992) Mammary gland specific nuclear factor (MGF) is present in lactating rodent and bovine mammary tissue and composed of a single polypeptide of 89 kDa. J Biol Chem 267:16365–16370

Wakao H, Gouilleux F, Groner B (1994) Mammary gland factor (MGF) is a novel member of the cytokine regulated transcription factor family and confers the prolactin response. EMBO J 13:2182–2191

Wartmann M, Cella N, Hofer P, Groner B, Liu X, Hennighausen L, Hynes NE (1996) Lactogenic hormone activation of Stat5 and transcription of the beta-casein gene in mammary epithelial cells is independent of p42 MAP kinase activation. J Biol Chem 271:31863–31868

Zhou YC, Waxman DJ (1999) STAT5b down-regulates peroxisome proliferator-activated receptor alpha transcription by inhibition of ligand-independent activation function region-1 trans-activation domain. J Biol Chem 274:29874–29882

14 Induction and Repression of NF-κB-Driven Inflammatory Genes

W. Vanden Berghe, K. De Bosscher*, L. Vermeulen*,
G. De Wilde*, G. Haegeman

14.1	Introduction	233
14.2	NF-κB	234
14.3	Molecular Mechanisms of NF-κB-Driven Gene Induction	237
14.4	Negative Modulation of NF-κB-Driven Gene Expression by NSAIDs	244
14.5	Negative Modulation of NF-κB-Driven Gene Expression by Glucocorticoids	246
14.6	Conclusions	255
References		259

14.1 Introduction

Aberrant gene expression is a primary cause of many disease-associated pathophysiologies. Therefore, the pharmacological modulation of transcription factor activity represents an attractive therapeutic approach to such disorders. Except for nuclear receptors, which are a direct target of pharmaceuticals, other known classes of transcription factors are largely regulated indirectly, i.e., by drugs, which impact upon signal transduction cascades and alter transcription factor modification (phosphorylation, acetylation, nitrosylation, hydroxylation) and/or their nuclear import (Emery et al. 2001; Pandolfi 2001).

*These authors contributed equally.

The inflammatory response is a highly regulated physiological process that is critically important to homeostasis. The precise physiological control of inflammation allows a timely reaction to invading pathogens or to other insults without an overreaction, that might itself cause damage to the host.

Over the last 10 years, the transcription factor NF-κB has been shown to be crucial for inducing genes involved in inflammation and in a wide range of diseases originating from chronic activation of the immune system, including asthma, atherosclerosis, inflammatory bowel disease, auto-immune diseases (such as multiple sclerosis and rheumatoid arthritis), late-life diseases, frailty, and dementia (Perkins 2000; Baldwin 2001; Silverman and Maniatis 2001; Tak and Firestein 2001). A plethora of immunoregulatory genes coding for cytokines, cytokine receptors, chemotactic proteins, or adhesion molecules, such as IL-1β, TNF-α, IL-2, IL-6, IL-8, MCP1, RANTES, IFN-β, GM-CSF, ICAM1, VCAM1 and E-selectin, contain NF-κB-responsive elements in their promoters or regulatory regions; this makes NF-κB an obvious target for immunosuppressive therapies and in inflammatory disorders (DiDonato et al. 1996; Baeuerle and Baichwal 1997; Handel 1997; Karin et al. 1997; Handel et al. 2000). Today, glucocorticoids and nonsteroidal anti-inflammatory drug (NSAIDs) are frequently used in the clinic as effective therapeutic agents to treat inflammatory diseases, cancer, and pain in all stages of malignancy, although the detailed molecular mechanisms of action are far from being understood.

14.2 NF-κB

Transcriptional regulators of the NF-κB/IκB family promote the expression of over 100 target genes, the majority of which participate in the host immune response (Ghosh et al. 1998). Gene knockout and other studies not only establish roles for NF-κB in the ontogeny of the immune system, but also demonstrate that NF-κB participates at multiple steps during oncogenesis and regulation of programmed cell death (Rayet and Gelinas 1999; Perkins 2000). In addition, the involvement of the ubiquitous transcription factor NF-κB in the pathogenesis of the inflammatory response has been well documented by experiments both in vitro and in vivo (Baeuerle and Baichwal 1997; Sha 1998; Perkins

2000; Baldwin 2001; Silverman and Maniatis 2001; Tak and Firestein 2001). NF-κB is latently present in the cytoplasm, under tight control of the associated protein IκBα (inhibitor-κB-α). NF-κB is a heterodimer, typically consisting of p50 and p65 monomeric proteins. A targeted disruption of the genes encoding p50 or p65 results in extreme immunodeficiencies, and even in lethality in the case of p65 knockout mice. The mammalian NF-κB/Rel family includes five members: p65 or RelA, RelB, c-Rel, NF-κB1 (p50/p105), and NF-κB2 (p52/p100). All members are characterized by a conserved stretch of 300 amino acids, designated as the rel homology domain (RHD). This domain is important for DNA binding and mutual interactions between the different Rel family members; it also serves as interaction surface for the inhibitory protein IκB. The IκB protein family comprises the following members: IκBα, IκBβ, IκBγ/p105, IκBδ/p100, IκBε, and Bcl3. They are characterized by several 30–33 amino acid motifs, called "ankyrin repeats." Potent inducers of NF-κB include IL-1, tumor necrosis factor (TNF), bacterial and viral products such as lipopolysaccharide (LPS), sphingomyelinase, dsRNA, Tax protein from human T-lymphotropic virus (HTLV), and pro-apoptotic and necrotic stimuli such as oxygen-free radicals, UV-irradiation and γ-irradiation. The first step in the activation process of NF-κB is a targeted phosphorylation of IκBα at serines 32 and 36. Subsequently, ubiquitinylation at lysines 21 and 22 takes place and predestine IκBα for degradation by the 26S proteasome complex (Yaron et al. 1998; Maniatis 1999; Spencer et al. 1999; Vuillard et al. 1999; Karin and Ben-Neriah 2000). This finally leads to release of the NF-κB protein, which migrates to the nucleus in order to exert its effects on gene regulation (Silverman and Maniatis 2001).

Many groups focused on the identification of the serine-specific IκB kinase complex (IKK), comprising multiple subunits and acting as an integrator of multiple NF-κB-activating stimuli (Karin 1999; Mercurio and Manning 1999; Israel 2000; Karin and Delhase 2000). The IKK complex consists of distinguished kinase subunits assembled in two IKKα/IKKβ heterodimers, and held together by one IKKγ monomer. The differential activity of the two IKK kinases on different IκB family members probably also results in a differentially regulated downstream NF-κB response and activity. Indeed, studies on these proteins confirmed the involvement of IKKβ in pro-inflammatory cytokine-induced activation of NF-κB. IKKα, on the other hand, was found to be essential for NF-κB activation in the limb and skin during embryogenesis, and to be crucial for B-cell maturation, formation of secondary lymphoid organs, increased expression of certain NF-κB target genes, and processing of the NF-κB2 (p100) precursor (Li et al. 1999; Takeda et al. 1999; Senftleben et al. 2001). Antagonistic effects of IKKα and IKKβ have recently been described in Wnt signaling depending on β-catenin phosphorylation and localization, thus integrating signaling events between the NF-κB and Wnt pathways (Lamberti et al. 2001). Results from IKKα and IKKβ double-deficient mice confirmed the importance of IKKs for NF-κB activation in vivo and demonstrated a neuroprotective role for these kinases during development (Li Q. et al. 2000). A third component, IKKγ (NEMO/IKKAP/FIP3), was designated as a scaffold platform for assembly of the IKK complex (reviewed in Cohen et al. 1998). Many proteins have been reported to activate the IKK complex, but so far there is no full understanding of specificity and redundancy. These proteins include protein kinase C (PKC) isozymes, mitogen-activated protein kinase kinase kinase (MAPKKK) family members, NIK, AKT/TPB, MEKK1, MEKK2, MEKK3, COT/TPL2, TAK1, and NAK (Israel 2000; Li X. et al. 2000; Tojima et al. 2000). Alternative IKK complexes causing NF-κB activation have also been identified (Pomerantz and Baltimore 1999; Peters et al. 2000). Furthermore, besides the classical IκB metabolism, variations have been described at the level of phosphorylation (Ser32, Ser36, Thr273, Tyr42) and degradation (non-proteasomal, lysosomal, or caspase-dependent) (Imbert et al. 1996; Barkett et al. 1997; Abu-amer et al. 1998; Bender et al. 1998; Cuervo et al. 1998; Li and Karin 1998; Liu L. et al. 1998; Miyamoto et al. 1998; White and Gilmore 1998; Beraud et al. 1999). The release and activity of NF-κB are subject to different control mechanisms. IκBα expression is controlled by NF-κB, which establishes an autoregulatory feedback loop and shuts down the activation of NF-κB (reviewed in Baeuerle and Baltimore 1996). Furthermore, NF-κB activation can be negatively regulated by a SUMO1 (small ubiquitin-like modifier-1 or sentrin) modification of unphosphorylated IκBα, leading to a degradation-resistant IκB molecule (Desterro et al. 1998), which may re-localize to particular subcellular compartments (Hodges et al. 1998).

Besides cytoplasmic regulation at the level of the IKK complex, various additional regulatory mechanisms have been described to affect nuclear activity of NF-κB and to require acetylation (Chen et al. 2001) or phosphorylation (Zhong et al. 1997; Wang and Baldwin 1998; Sakurai et al. 1999; Vanden Berghe et al. 2000; Wang et al. 2000; Schmitz et al. 2001; L. Vermeulen, submitted) at various amino acid residues in p65, most presumably affecting cofactor regulation (i.e., CBP/p300, p/CAF, SRC1, HDACs, Med, ARC) (Gerritsen et al. 1997; Zhong et al. 1998; Naar et al. 1999; Sheppard et al. 1999; Vanden Berghe et al. 1999a; Ashburner et al. 2001; Chen et al. 2001).

14.3 Molecular Mechanisms of NF-κB-Driven Gene Induction

Previously, we have focused on the transcriptional regulation of the interleukin-6 (IL-6) gene. The pleiotropic cytokine IL-6 affects inflammatory reactions, hematopoiesis, bone metabolism, reproduction, and aging. Aberrant IL-6 gene expression has been associated with multiple myeloma, neoplasia, rheumatoid arthritis, bowel disease, psoriasis, Alzheimer's disease, and post-menopausal osteoporosis (Akira et al. 1993; Erschler and Keller 2000). Serum IL-6 levels are currently considered a diagnostic marker for tumor progression and prognosis in various types of cancer (renal cell carcinoma, breast, lung, ovarian and gut cancer) (Wang et al. 1999; Zhang and Adachi 1999). Briefly, the IL-6 promoter behaves as a sophisticated biosensor for environmental stress, thus controlling immunological homeostasis (Vanden Berghe et al. 2000). This highly conserved promoter sequence revealed a complex control region that can be triggered by multiple activation pathways. In the case of TNF, the main transcriptional activator for IL-6 gene induction is NF-κB. We found that, apart from TNF-induced cytoplasmic NF-κB activation and nuclear DNA binding, the TNF-activated p38 and extracellular signal regulated kinase (ERK) MAPK pathways contribute to transcriptional activation of the IL-6 promoter by modulating the transactivation capacity of the NF-κB p65 subunit (Beyaert et al. 1996; Vanden Berghe et al. 1998). Upon further studying in vivo phosphorylation, we and others observed rapid phosphorylation of the p65 subunit in response to TNF and/or IL-1 in various cell lines (Li et al. 1994; Naumann and Scheidereit 1994) (Diehl et al. 1995; Beyaert et al. 1996; Bird et al. 1997; Zhong et al. 1998; Anrather et al. 1999; Egan et al. 1999; Sakurai et al. 1999; Shumilla et al. 1999; Sizemore et al. 1999;

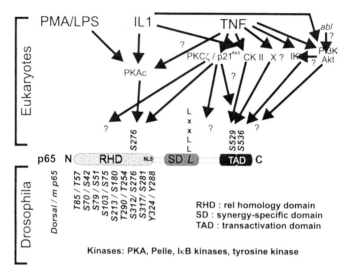

Fig. 1. Overview of the various TNF/IL-1 inducible pathways leading to p65/dorsal phosphorylations

Wang and Baldwin 1998; Madrid et al. 2000; L. Vermeulen, submitted). Besides inducible phosphorylation, p65 and p50 are also constitutively phosphorylated. Taken together, the current data indicate that TNF can induce phosphorylation of both the transactivation domain and the RHD of p65, but the contribution of each phosphorylation event in TNF-induced p65 transcriptional activity may be cell type-specific. Detailed phosphorylation studies of the NF-κB homologue dorsal in Drosophila and X-ray crystallography of p50/p65 dimer bound on DNA demonstrated that five conserved serines, although spread over the primary structure of the RHD, are rather close in the tertiary structure (near the PKA motif at position Ser276) so that they may form an endpoint of multiple kinase pathways (Chen et al. 1998; Huxford et al. 1998; Jacobs and Harrison 1998; Cramer and Muller 1999; Drier et al. 1999). Redundant kinases may act in concert with other enzymatic activities (acetylation) and could affect different stages of the transcription process (i.e., kinetics, translocation, initiation, elongation) or else may induce different target genes (Zhong et al. 1998; Drier et al. 1999; Mayr and Montminy 2001; Mayr et al. 2001; Saccani et al. 2001; West et al. 2001) An overview of the various inducible pathways leading to p65/dorsal phosphorylation is given in Fig. 1.

Another aspect of nuclear regulation of NF-κB activity relies on its interactions with chromatin-modifying cofactor complexes (Sheppard et al. 1999; Vanden Berghe et al. 1999a; Vanden Berghe et al. 2000; Saccani et al. 2001). One of the major problems in eukaryotic transcriptional regulation is the question of how the transcription machinery gains access to promoter DNA, wrapped in an amalgam of histones and proteins known as "chromatin" (Berger and Felsenfeld 2001). Current research indicates that reversible modifications of chromatin (acetylation, phosphorylation, methylation, ubiquitinylation, ADP ribosylation, glycosylation) affect relaxation or tightening of protein/DNA interactions (Strahl and Allis 2000; Berger and Felsenfeld 2001; Marmorstein 2001; Zhang and Reinberg 2001). This process originates from transcription factors which, upon DNA binding, recruit cofactor complexes having various enzymatic activities (i.e., acetyltransferase, deacetylase, methyltransferase, kinase, ubiquitin-ligase, etc.). In turn, the latter modify histones/nucleosomes and transcription factors, resulting in local and transient chromatin relaxation that attracts additional protein complexes (i.e., ATP-dependent chromatin-remodeling complexes, RNA polymerase holoenzyme, kinases, complexes involved in cell cycle control, proliferation or differentiation) (Marmorstein 2001). The complete pattern of chromatin "tags" residing around the promoter enhanceosome probably includes a unique message for further downstream signaling events (Strahl and Allis 2000; Jenuwein and Allis 2001; Wells 2001).

Unraveling this chromatin vocabulary is of crucial importance to understand many DNA-templated based processes as well as the origin of various diseases. Chromatin-modifying complexes have indeed been associated with control of cell growth and differentiation, whereas aberrant activities frequently result in transformation and tumorigenesis (Giles et al. 1998; Kouzarides 1999; Blobel 2000; Kung et al. 2000; Schiltz and Nakatani 2000; Wolffe 2001). Transcription factor-selective and signal-specific cofactor and/or HAT/HDAC recruitment have now become a prime focus of investigation (Hong et al. 1998; Ikeda et al. 1999; Kawasaki et al. 1998; Merika et al. 1998; Korzus et al. 1998; Utley et al. 1998; Xu et al. 1998; Perissi et al. 1999; Agalioti et al. 2000; Senger et al. 2000; Merika and Thanos 2001; Munshi et al. 2001; Urnov and Wolffe 2001). In this respect, our findings support a molecular model for synergistic transcription in which the co-integrator CBP/p300 may be recruited to the multi-responsive IL-6 promoter and stabilized by multiple protein–protein interactions with different DNA-binding transcription factors AP-1, CREB, C/EBP and NF-κB, sequentially arranged along the promoter sequence (Vanden Berghe et al. 1999a). The ultimate switch for gene induction is, however, achieved by the transcription factor NF-κB, e.g., in response to TNF, which engages the available CBP/p300 for transcriptional activation (Vanden Berghe et al. 1999a). Similar observations were recorded with other natural NF-κB-driven promoters, i.e., IL-8, E-selectin, and the HIV1 promoter activation (Van Lint et al. 1996; Sheridan et al. 1997; El Kharroubi et al. 1998; Vanden Berghe et al. 1999a). Whether this engagement relies on conformational changes of CBP by interaction with NF-κB (Parker et al. 1998) and/or is the result of concomitant phosphorylations of either NF-κB or CBP, e.g., by TNF induction, is at present not clear (Radhakrishnan et al. 1998; Wang and Baldwin 1998; Xu et al. 1998; Zhong et al. 1998). Zhong and colleagues (1998) have proposed an allosteric molecular model in which phosphorylation of the RHD induces intramolecular conformational changes of the N and C termini of p65, which may expose an entry platform for CBP interaction. p65-CBP interactions have now been mapped at the N and C termini, including a synergism-specific domain (p65 AA322–458) (Akimaru et al. 1997; Gerritsen et al. 1997; Perkins et al. 1997; Merika et al. 1998; Zhong et al. 1998). The latter domain revealed a single copy of the motif Leu-Gly-Ala-Leu-Leu and conforms to the consensus LxxLL, which is the signature motif present in CBP and CBP-interacting proteins; it forms a short amphipathic helix, providing the structural basis for protein–protein interactions (Torchia et al. 1997; Merika et al. 1998). An initial interaction between the LxxLL motifs in p65 (AA449–453) and CBP (AA358–362) may be further stabilized by simultaneous interactions of other CBP regions with neighboring domains. Furthermore, we cannot exclude that other factors corecruited to the IL-6 promoter also contribute to (part of) the acetylation process and/or chromatin-remodeling activities, since involvement of p/CAF, SRC, p160, bcl-3, ARC coactivators or HDAC1/2/3, Groucho, p202 corepressors has also been demonstrated in NF-κB transactivation (Min et al. 1996; Dubnicoff et al. 1997; Na et al. 1998; Valentine et al. 1998; Dechend et al. 1999; Heissmeyer et al. 1999; Naar et al. 1999; Sheppard et al. 1999; Tetsuka et al. 2000; Wen et al. 2000; Ashburner et al. 2001; Chen et al. 2001). These findings suggest that, by analogy with the coregulator exchange in nuclear receptor function (Glass and Rosenfeld 2000), transcriptional activity of NF-κB may also be regulated by a balance of corepressor and/or coactivator interactions, which might be modulated by a signal-induced modification of these proteins (Ito et al. 2000; Tetsuka et al. 2000; Ashburner et al. 2001).

Interference of NF-κB-dependent gene activity was observed by the repressor protein RBP-Jκ, which competes with NF-κB for DNA binding at the IL6-κB motif (Plaisance et al. 1997). One might postulate that the IL-6 gene promoter is dynamically regulated at the NF-κB site by an interplay of a coactivator complex including CBP/p300 (interacting with NF-κB), and a corepressor complex containing HDAC-1 (associated with RBP-Jκ) (Kao et al. 1998; Hsieh et al. 1999). In the absence of activated NF-κB, the IL-6-κB site is occupied by the RBP-Jκ corepressor complex, which could shield the RNA polymerase holoenzyme from CBP effects mediated by AP-1, CRE and C/EBP. However, deacetylase inhibitor experiments with trichostatin A (TSA) and studies with point-mutated IL-6 promoter variants, having lost RBP-Jκ binding without affecting NF-κB, only suggest a minor role of this repressor complex in promoter regulation in response to a sustained hypoxanthine-aminopterin-thymidin (HAT) activity (Vanden Berghe et al. 1999a).

At present, acetylases are known to modify, besides histones, a variety of other proteins including transcription factors (p53, E2F1, EKLF, TCF, GATA1, HMGI(Y), ACTR, TFIIF, TFIIEβ, HIV1-tat), nuclear acetylases (CBP/p300, p/CAF), and shuttling import factors (importin-α and α-tubulin), thus regulating many functions, such as DNA recognition, protein–protein interaction, and protein stability (Sakaguchi et al. 1998; Kouzarides 2000). In view of the synergy of NF-κB-driven gene transcription in response to sustained HAT activity (Vanden Berghe et al. 1999a), one might also expect acetylation of the NF-κB p50/p65. A role for inducible acetylation of NF-κB p65, affecting its association with IκB and cellular localization, has recently been demonstrated (Chen et al. 2001). Finally, by analogy with chromatin (acetylation)-dependent phosphorylation of CREB, a similar effect has been observed for NF-κB p65, in which acetylation prolongs its phosphorylation (Ashburner et al. 2001).

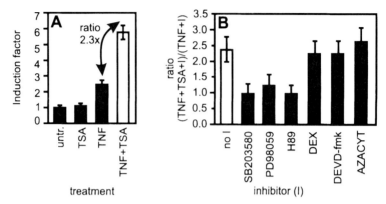

Fig. 2A, B. Stable cell pools transfected with an IL-6-luc promoter reporter gene construct (p1168hu.IL-6P-luc+) were left untreated, or were treated for 6 h with TNF alone or in combination with the deacetylase inhibitor trichostatin A (*TSA*). Lysates were assayed for reporter gene expression and normalized for protein concentration (**A**). Influence of inhibitors on the TSA effects are expressed as the ratio of TSA effect with or without the compound tested (**B**)

Besides MAPK-mediated effects on TNF-induced p65 phosphorylation, we also analyzed their possible interference in cofactor-dependent activities. Remarkably, MAPK (p38 and ERK) inhibitors were able to specifically revert p65-engaged HAT activity, whereas glucocorticoids, caspase (ZDEVD-fmk) or methylation (azacytidine) inhibitors were not (Vanden Berghe et al., unpublished results) (Fig. 2). Cross-coupling of MAPK signaling and histone acetylation has recently been further substantiated by the discovery of tandem cofactor complexes of acetylases (i.e., CBP, GCN5) and kinases (i.e., RSK2, MSK1, SNF1) (Alberts et al. 1998; Espinos et al. 1999; Witt et al. 2000; Lo et al. 2001; Merienne et al. 2001). In addition, as phosphorylation-dependent control of CBP transactivation and/or HAT capacity by PKA, PKC, ERK, p90rsk, CaMIV, and cyclin-dependent kinases is now well documented (Nakajima et al. 1996; Ait-si-ali et al. 1998; Chawla et al. 1998; Liu YZ et al. 1998; Snowden and Perkins 1998; Cohen et al. 1999; Hu et al. 1999; Liu et al. 1999; Zanger et al. 1999; Yuan and Gambee 2000; See et al. 2001), cytokine-induced phosphorylation of CBP has recently been observed (Ito et al. 2000). Other likely targets, sensitive to MAPK-dependent

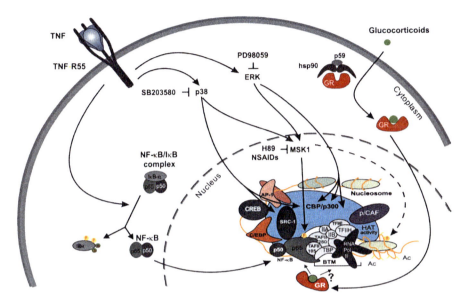

Fig. 3. Dual activation scheme for IL-6 gene expression in response to TNF. In this model, the IL-6 promoter recruits CBP which, upon phosphorylation of NF-κB p65, renders the NF-κB/CBP complex transcriptionally competent

regulation and which may contribute to gene transactivation, are the SWI/SNF chromatin-remodeling complex (Sif et al. 1998), p/CAF (Masumi et al. 1999), nucleosomes (Davie et al. 1999; Jin et al. 1999; Sassone-Corsi et al. 1999; Thomson et al. 1999b), locus control regions (Versaw et al. 1998), p21$^{Waf/Cip1}$ (Bottazzi et al. 1999; Magnaghi-Jaulin et al. 1999) and cell cycle components (Pumiglia and Decker 1997; Chau and Shibuya 1999; Roberts et al. 2000). Finally, the RNA polymerase carboxy-terminal domain (CTD), the TATA-box-binding protein (TBP), and the RNA synthesis process itself have also been found susceptible to MAPK modulation (Bensaude et al. 1999; Bonnet et al. 1999; Carter and Hunninghake 2000; Whitmarsh and Davis 2000; Zhu et al. 2000). The dual activation scheme for IL-6 gene expression in response to TNF is shown in Fig. 3.

In order to further understand IL-6 gene regulation, it will be required to determine how CBP/p300 can simultaneously integrate functions of various transcriptional activators present in the IL-6 promoter (Shikama et al. 1997; Torchia et al. 1998; Janknecht and Hunter 1999; Chan and La Thangue 2001; Munshi et al. 2001). Two valuable working models are currently available. Studies with the IFNβ enhanceosome show a strict stereospecific requirement for optimal CBP recruitment and transcriptional synergism, promoting a looping cofactor model with multiple transcription factor interactions (Carey 1998; Merika et al. 1998; Munshi et al. 1998; Wu and Hampsey 1999). Alternatively, a coactivator sequestration model has been proposed in which multiple transcription factors are competing for limiting amounts of CBP and become opponents to mediate CBP effects (Hottiger et al. 1998; Wadgaonkar et al. 1999). Along the same line, accessory proteins (e.g., SNIP1) may selectively regulate CBP access for specific transcription factors (Bhattacharya et al. 1999; O'Connor et al. 1999; Kim et al. 2001). So far, our experimental data for the IL-6 promoter model are in favor of the first model, since for optimal CBP synergistic activities all factors are required; nevertheless, the unique role of p65 in CBP engagement within the IL-6 promoter context remains to be further explored. To further unravel transcriptional activation of the IL-6 promoter in response to TNF at a detailed molecular level, it remains to be established how nucleosomes remodel at the IL-6 promoter upon inflammatory stimuli, and how this process is affected/driven by protein modification events (acetylation, phosphorylation) of various components of the IL-6 enhanceosome (Agarwal and Rao 1998a,b; Gribnau et al. 1998; Weinmann et al. 1999).

14.4 Negative Modulation of NF-κB-Driven Gene Expression by NSAIDs

NSAIDs comprise chemicals from a number of classes with anti-inflammatory as well as analgesic, antipyretic, and platelet-inhibitory actions (Brooks and Day 1991; Vane and Botting 1998). In addition, these agents induce regression of adenomatous polyps of the colon and prevent colon cancer development. The inhibition of cyclooxygenase (COX) activity to prevent prostaglandin synthesis is a well-characterized mechanism of action, but the anti-inflammatory effects require concentrations far above those needed for inhibition of prostaglandin

Induction and Repression of NF-κB-Driven Inflammatory Genes 245

synthesis; this indicates that, despite the ubiquitous use of NSAIDs in the treatment of inflammatory diseases, their entire mechanism of action is not completely understood (Vane et al. 1990; Vane and Botting 2000). For example, TNF-regulated transcriptional activation of NF-κB plays a key role in pathogenesis of inflammatory bowel disease (IBD) (Neurath et al. 1998; Schreiber et al. 1998). Increased NF-κB levels were observed in inflamed tissue, but not in adjacent uninvolved mucosa (Rogler et al. 1998). For this reason, the NF-κB system is an attractive target for therapeutic inhibition of chronic inflammatory conditions (Baeuerle and Baichwal 1997). Aspirin and sodium salicylate are examples of anti-inflammatory agents for which the molecular target is, at least partly, NF-κB. At concentrations measured in the serum of patients treated for chronic inflammatory conditions, both aspirin and salicylate inhibit activation of the NF-κB pathway (Pierce et al. 1996; Yin et al. 1998; Yamamoto and Gaynor 2001). These agents suppress TNF-induced mRNA synthesis of adhesion molecules and surface expression of VCAM1 and ICAM1 in endothelial cells (Pierce et al. 1996). The inhibition of the NF-κB pathway in endothelial cells prevents transendothelial migration of neutrophils. This suggests that the clinical importance of high-dose salicylates as anti-inflammatory agents is at least partially due to blocking of NF-κB activation, resulting in inhibition of leukocyte recruitment (Pierce et al. 1996). The observation that several NSAIDs can inhibit TNF-induced NF-κB activation might explain the beneficial effect of NSAIDs in anti-inflammatory therapy (Kopp and Ghosh 1994). Several reports indicated that NSAIDs can interfere with the NF-κB pathway by specific inhibition of ATP-binding to IKK: as IKK-dependent phosphorylation of IκB is markedly reduced, its degradation by the proteasome is prevented (Kopp and Ghosh 1994; Pierce et al. 1996; Kaiser et al. 1999; Yan and Polk 1999). This suggests that prevention of IκB degradation may underlie the in vivo anti-inflammatory effects of these drugs. However, Egan and colleagues (Egan et al. 1999) recently reported that inhibition of NF-κB-dependent transcription in intestinal epithelial cells and T lymphocytes can also occur without affecting IκB degradation, as the non-steroidal drug mesalamine interferes with NF-κB-dependent transcription by inhibiting IL-1-induced phosphorylation of the p65 subunit.

We found that NSAIDs from different classes are able to inhibit NF-κB-driven gene induction in response to TNF in the mouse fibrosarcoma cell line L929sA. Further experiments demonstrated that NSAIDs interfere with the NF-κB activity at the level of transactivation, probably by inhibition of MSK1-induced phosphorylation of the p65 subunit (G. De Wilde, unpublished data).

14.5 Negative Modulation of NF-κB-Driven Gene Expression by Glucocorticoids

Glucocorticoids (GCs) are the most widely used anti-inflammatory and immunomodulatory agents, and their mechanism of action is mainly based on interference with the activity of transcription factors, among them NF-κB. Understanding the precise molecular mechanism of gene repression by GCs is a controversial matter, due to many conflicting hypotheses. We discuss the three main mechanisms reported in the literature, namely the IκBα upregulatory model, the protein–protein interaction model, and the cofactor competition model.

GCs are steroid hormones, produced by the adrenal gland upon cytokine stimulation of the hypothalamic-pituitary-adrenal (HPA) axis, for instance after an immune challenge. They are involved in the regulation of immune responses, as well as in the control of stress responses and memory function (De Kloet et al. 1998, 2000). GCs are therapeutically useful agents to treat inflammation and various immune diseases. It is widely accepted that their potent anti-inflammatory and immunomodulatory actions result from the inhibition of the activity of transcription factors, such as AP-1 and NF-κB, which are involved in the activation of pro-inflammatory genes. Hence, GCs suppress the production and the effects of humoral factors involved in the inflammatory response; they inhibit leukocyte migration to the sites of inflammation and interfere with the functions of endothelial cells, leukocytes, and fibroblasts (Cato and Wade 1996). Due to their hormonal and lipophilic nature, GCs migrate freely through the cell membrane. At the level of gene expression, they exert their effects by binding to the glucocorticoid receptor (GR), a transcription factor that regulates several genes, either positively or negatively. GR belongs to a superfamily of thyroid/steroid nuclear hormone receptors and functionally comprises three domains: an N-terminal constitutive activation domain (Tau-1 or AF-1) (Beato et al. 1995), a central DNA-binding domain (DBD), and a C-terminal ligand-binding domain. The latter domain also contains activating functions Tau-2 and Tau-c or AF-2 (Giguere et al. 1986; Hollenberg and Evans 1988). In its unliganded resting state, this intracellular receptor protein is present in the cytoplasm as a complex with one p56, one hsp70, and two hsp90 immunophilin molecules. Ligand binding induces a conformational change of GR and releases GR from its chaperones, allowing its translocation to the nucleus. Genes positively regulated by GR are characterized by GRE-responsive elements in the promoter. However, these genes are not believed to contribute to the anti-inflammatory potential of GR, as the beneficial effects of activated GR are due to negative modulation of pro-inflammatory cytokines, whereas the side effects mainly result from the transactivating capacities of GR.

Gaining further insight into the exact mechanism and implications of the cross-talk between NF-κB and GR in different tissues is essential in order to be able to design appropriate anti-inflammatory drugs.

14.5.1 The IκBα Upregulatory Model

A first level at which GCs might act is the sequestration of NF-κB in the cytoplasm. In 1995, two independent groups proposed that a GC-induced increase in synthesis of IκBα caused retention of NF-κB within the cytoplasm by complexing this factor with newly synthesized IκBα protein (Auphan et al. 1995; Scheinman et al. 1995). A pulse chase experiment, followed by immunoprecipitation assays, showed that GCs did not interfere with the phosphorylation and subsequent breakdown of IκBα (Scheinman et al. 1995). According to these data, de novo synthesized IκBα protein interacts with pre-existing NF-κB, thus preventing it from migrating into the nucleus and activating its target genes. Furthermore, it was also suggested that IκBα is able to actively remove NF-κB from its promoter element in the nucleus, but this hypothesis needs further experimental evidence. In a number of cell lines, including HeLa cells, a murine T-cell hybridoma 2B4, and GR-expressing Jurkat T cells, gel shift analyses demonstrated that treatment with the synthetic glucocorticoid dexamethasone (DEX) affects the NF-κB DNA-binding activity (Auphan et al. 1995; Scheinman et al. 1995). In this way, the mechanism of gene repression by GCs was deduced to occur by squelching NF-κB from its response element. This mechanism was, however, not supported by data from other research groups, including ours (Brostjan et al. 1996; De Bosscher et al. 1997; Heck et al. 1997; Adcock et al. 1999; Lidén et al. 2000; Herrlich 2001). It was therefore assumed that the IκBα mechanism might be restricted to specific tissues or lymphoid cell types. In other cell lines, e.g., in L929sA fibroblasts, TC10 and BAEC endothelial cells, and A549 alveolar lung cells, the DNA-binding activity of NF-κB remained unaffected under conditions of gene repression (Brostjan et al. 1996; De Bosscher et al. 1997; Wissink et al. 1998). Additionally, also in monocytic U937 cells, GC repression of the intercellular adhesion molecule (ICAM) promoter was observed without a parallel inhibition of the DNA-binding capacity of NF-κB (Van de Stolpe et al. 1994). Apparently, the IκBα mechanism for repression of NF-κB cannot unequivocally be assigned to a specific subset of cell types. This is further documented by the fact that GCs inhibit NF-κB DNA-binding activity in cells of the cerebral cortex and hippocampus, without increasing IκBα levels. In contrast, in peripheral brain cells from the same animal, IκBα protein levels were enhanced after DEX treatment, suggesting that the effect of GCs is highly cell- and tissue-specific (Unlap and Jope 1997). Interestingly, there are GC analogues that no longer inhibit NF-κB activity, but do increase IκBα levels, indicating that GC-mediated transrepression and the event of IκBα upregulation are actually two separate and independent phenomena (Heck et al. 1997). Inversely, other GC analogues have been developed that exert transrepressive, but no transactivation effects (Vayssière et al. 1997; Vanden Berghe et al. 1999b). Furthermore, repressive effects by GR remain apparent, even in the presence of the protein synthesis inhibitor cycloheximide (Van de Stolpe et al. 1993; De Bosscher et al. 1997; Wissink et al. 1998), making it unlikely that de novo IκBα synthesis is a general phenomenon in transrepression. Along the same line, a study comparing the activity of various clinically important glucocorticoids showed that it is possible to prevent the TNF-induced degradation of IκBα to various extents without affecting the NF-κB DNA-binding activity (Hofmann et al. 1998). Finally, comparable GC repression of NF-κB has been observed in wild-type and IκBα$^{-/-}$ mouse embryonic fibroblasts (Beg et al. 1995; Doucas et al. 2000). These findings demonstrate that upregulation of IκBα and the phenomenon of glucocorticoid repression are two independent and separable processes which can be uncoupled. These results have now been confirmed by experiments using mice with a dimerization-defective GR$^{dim/dim}$ mutant (A458 T), which shows that GR DNA binding and IκB gene activation are dispensable for the anti-inflammatory activity of GR (Reichardt et al. 2001).

It should be noted that it is still unclear how GCs can stimulate the IκBα promoter, since the latter contains three κB-responsive elements, but no consensus-positive GC response element (GRE) element. The stimulatory capacity of GR on a stably transfected IκBα promoter construct is dependent on the cell type. While we have not observed any stimulatory effect of GCs on this construct in L929sA cells, but a distinct GC-mediated repression, Heck and coworkers found that GCs could indeed enhance the activity of the same construct in HeLa cells. In contrast, other nuclear receptors, such as androgen receptor (AR) and estrogen receptor (ER), are not able to enhance IκBα synthesis in LNCaP or MCF-7 cells, respectively, but do repress NF-κB-driven genes (Heck et al. 1997). After TNF or IL-1 treatment for about 30 min, the breakdown of IκBα is maximal; due to a positive feed-forward by NF-κB, the protein levels reappear after 1 or 2 h. It is puzzling how this NF-κB-dependent promoter displays a behavior distinct from other NF-κB-driven promoters in terms of its response to GCs. Perhaps a cell type-specific cofactor or positive cross-talk with other transcription factor(s) is involved in mediating the stimulatory effect of GCs in certain cell types. Synergistic activity of the IκBα promoter can also be observed under conditions of stimulated NF-κB and the nuclear receptor PPARα or of the orphan nuclear receptor RORα (Delerive et al. 2000; Delerive et al. 2001). Interestingly, PPARα ligand-dependent recruitment of the DRIP/TRAP complex onto a promoter configuration, in which Sp1 flanks NF-κB, lies on the basis of the observed transcriptional synergy (Delerive et al. 2002). Whether this mechanism can be generalized to GR and/or other cell types, needs to be further investigated (Rachez et al. 2000). Anyhow, these results suggest the existence of at least two independent pathways by which GCs mediate their inhibitory activity. Why a cell should need two ways to mediate transrepression by GCs and what the determining parameters for each of these pathways are, remains unsettled.

14.5.2 The Protein–Protein Interaction Model

A second model that has found support by many investigators over the last decade is the protein–protein interaction model (McEwan et al. 1997). Functional interactions between different protein families expand the range of possibilities with a given set of transcription factors. These mutual interferences may result in an inhibitory, additive, or synergistic effect on gene transcription, depending on the composition of the dimeric transcription factor and/or the cellular context. Some examples include the interaction between GR and Stat-5 (Stöcklin et al. 1996), GR and AP-1 (Jonat et al. 1990), p65 and C/EBP, and p65 and AP1 (Stein et al. 1993). As in most inflammatory gene promoters in which a so-called negative GRE has not been detected, transcriptional interference was discovered to result in most cases from cross-talk between GR and other transcription factors, such as NF-κB or AP-1. The protein–protein interaction model ascribes GC repression to a direct interaction between activated GR and NF-κB subunits, preferably in the nucleus. The p65 subunit of NF-κB and GR has been found associating in vitro (Ray and Prefontaine 1994; Scheinman et al. 1995) and also in vivo, by co-immunoprecipitation studies in A549 cells (Adcock et al. 1999). Our own results also support an exclusive, nuclear mechanism for the interference between NF-κB and GR. Using a chimeric protein linking the yeast Gal4 DNA-binding domain to the transactivator p65, we demonstrated activation of a Gal4-dependent promoter construct in an NF-κB promoter-independent and stimulus-independent way. The activity of this constitutively nuclear protein could be repressed by GCs, whereas the activity of a similar chimera containing the VP16 activator instead of p65 was unresponsive to the inhibitory effect of GCs (De Bosscher et al. 1997). Domain-mapping and transient transfection studies showed that the interaction between p65 and GR involved the N-terminal Rel homology domain of p65 and the zinc finger-containing DBD of GR (Stein and Yang 1995; Lidén et al. 1997; Wissink et al. 1997). However, a functional repression by GR also necessitates the C-terminal domain of p65, where the two transactivation functions, TA1 and TA2, are located (De Bosscher et al. 1997; Wissink et al. 1997). Several mechanistic models can be envisaged to explain the inhibitory protein–protein interferences between NF-κB and GR. The transcriptional inhibition by GR might be the result of a mutual masking of transactivating domains, the induction of post-translational modifications, conformational changes within the transcription initiation complex, or competition for coactivators (Dumont et al. 1998; see also Sect. 5.3, this chapter).

14.5.3 The Cofactor Competition Model

A huge number of coactivator proteins were recently identified (the majority by two-hybrid screenings) to interact with and potentiate the transactivating capacity of nuclear receptors. Coactivator proteins, such as CBP, p300 and SRC1, mediate the transcriptional activity, not only of nuclear receptors, but also of other transcription factors by means of their histone acetyltransferase functions (Horwitz et al. 1996; Glass et al. 1997). Acetylation at specific lysine residues of core histones is suggested to lead to a more relaxed configuration of chromatin, thereby facilitating transcription. Since these coactivators have been discovered to stimulate the activity of both nuclear receptors as well as of AP-1, it was first proposed by Kamei and coworkers that the mutual repression between retinoid acid receptor (RAR) or GR and AP-1 may result from a competition for limiting amounts of the essential coactivator CBP (Kamei et al. 1996). Aarnisalo and coworkers supported this competition mechanism for AR-mediated transrepression, but could not find evidence for this mechanism to explain GC repression of AP-1 activity (Aarnisalo et al. 1998). A similar mechanism, involving the coactivator SRC1, was suggested to also explain the GC-mediated repression of AP1-driven genes (Lee et al. 1998). After it was demonstrated that the activity of NF-κB can also be enhanced by the same coactivators (Gerritsen et al. 1997; Perkins 1997; Perkins et al. 1997), Sheppard and colleagues advanced that competition for limiting amounts of CBP might also account for the GR-mediated transrepression of NF-κB-dependent genes (Sheppard et al. 1999). While extra CBP indeed stimulates p65-mediated transactivation of several NF-κB-driven promoters, our experiments demonstrated that it did not, in any case, affect the relative levels of transrepression mediated by activated GR (De Bosscher et al. 2000a,b, 2001). Although the coactivators CBP, p300, and SRC-1 may be limiting for transactivation, our results show that this is not the case for transrepression. In contrast to Sheppard and Lee and

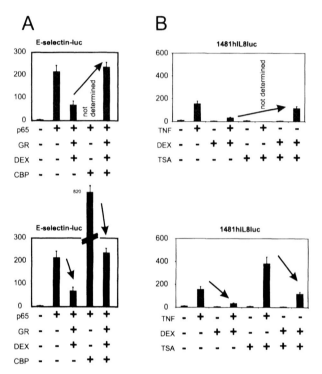

Fig. 4. A Rationale for the so-called competition model in which excess CBP is supposed to relieve GC-mediated gene repression. Overexpression of CBP results in higher expression levels (*bottom panel*); the level of gene "repression" is actually invariant, irrespective of the amount of CBP present in the cell. **B** Rationale for the HDAC model in which treatment with the deacetylase inhibitor trichostatin A (*TSA*) is supposed to relieve GC-mediated gene repression. Treatment with TSA, however, also results in higher expression levels of the reporter gene, whereas the level of gene repression is actually invariant

their colleagues (Lee et al. 1998; Sheppard et al. 1998), we measured the relative repression levels of the combined p65, CBP, and GR set-up as compared to the control, showing the synergism between p65 and CBP alone. The result shows that the relative repression mediated by GR remains unaffected in the presence of coactivating levels of CBP (Fig. 4). Furthermore, a coimmunoprecipation assay demonstrated that under conditions of repression by GCs, the interaction between p65 and

GR is not disrupted (De Bosscher et al. 2000a). In support of our conclusions, a number of arguments further counter the involvement of cofactor squelching in transrepression. Since various transcription factor families converge to the level of CBP/p300 for their transcriptional activities, the competition model struggles with a lack of specificity. In this scenario, one would expect that all factors sharing CBP are possible competitors. However, RAR, for instance does not mediate repression of NF-κB, but only represses AP-1. Furthermore, DNA binding-deficient mutants of p65, containing an intact predicted coactivator-recruiting transactivation domain, can no longer repress GC-mediated transactivation (Wissink et al. 1997). If a cell were to inactivate the entire cellular pool of a given co-activator or activator in response to one signal, such a mechanism would preclude responsiveness by other activators or cooperativity at other genes in response to additional signals. A number of recent observations are more consistent with the notion of territorial subdivision than the competition for factors (Doucas et al. 1999; Stenoien et al. 2000; Stewart and Crabtree 2000; Stenoien et al. 2001). If transcription factor complexes are assembled within segregated nuclear compartments, then cofactor effects may only affect the designated compartment without affecting the same factors in other compartments associated with different genes (Savory et al. 1999; Francastel 2000; Hager et al. 2000; Lemon and Tjian 2000; Wan et al. 2001).

Next to cofactor competition, an alternative cofactor mechanism involving histone deacetylases has been presented in which GR recruits HDAC to the p65-CBP complex, resulting in gene repression; in this configuration the histone deacetylase inhibitor TSA again relieves GR-mediated repression (Ito et al. 2000). However, similar to the CBP overexpression experiments, reporter gene activities in response to TNF+DEX+TSA should be compared to the response to TNF+TSA, demonstrating that relative repression is conserved under conditions of inhibited deacetylases (Fig. 4). In addition, promoter responsivity to TSA does not necessarily reflect sensitivity to GC (Fig. 5), whereas IL-8 and HIV promoter activity can be similarly increased with TSA, only the IL-8 promoter shows a strong repression in the presence of DEX. This proves that the dynamic balance of acetylation/deacetylation can be uncoupled from GR-mediated repression.

Since other nuclear receptors can also target NF-κB, and assuming that mutual antagonism occurs by different mechanisms, repression

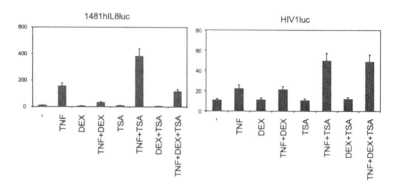

Fig. 5. Stable cell pools transfected with an IL-8 or HIV1 promoter reporter gene construct (1481IL-8-luc and HIV1-luc, respectively) were left untreated, pretreated with DEX, or treated for 6 h with TNF alone or in combination with the deacetylase inhibitor trichostatin A (*TSA*) and/or DEX. Lysates were assayed for reporter gene expression and normalized for protein concentration

might still result from targeting a specific p65-associated cofactor (e.g., by post-translational modifications) or from interference with the contact between p65 and the basal transcription machinery. We found preliminary evidence that the GR-mediated repression mechanism is dictated, in part, by the identity of the specific core TATA box context (De Bosscher et al. 2000a,b). This observation meets with the recent demonstration that GR inhibits phosphorylation of the carboxy-terminal domain of RNA polymerase II in the holoenzyme complex (Nissen and Yamamoto 2000). Moreover, it was previously reported that the TATA box context is also important for the response of certain promoters to the coactivating capacities of CBP (Martinez-Balbas et al. 1998). Basal transcription factors that have been identified to bind to p65 are TFIIB, TBP, and TAF$_{II}$105 (Schmitz et al. 1995; Yamit-Hezi and Dikstein 1998). The possibility exists that GCs repress NF-κB, or vice versa by a steric hindrance mechanism, i.e., by disrupting the interaction of p65 or GR, with one of these basal factors. The molecular mechanism has still to be investigated in more detail.

14.6 Conclusions

A central goal in modern molecular biology and biotechnology is to gain control, using small molecules, over biological pathways, among which the mRNA-synthesizing machinery is a particularly important target (Denison and Kodadek 1998; Young 1998). Cell fate is determined by extracellular signals (such as growth factors or pathogens), which are transmitted to the nucleus and culminate in the induction or repression of defined sets of genes. The transduction of signals from the cell surface to the nucleus often involves phosphorylation cascades that not only allow for rapid transmission but also serve to amplify the signal by activating multiple factors and genes. The fine-tuned combination of these signal-induced nuclear effects thus leads to integrated cellular responses, such as proliferation, differentiation, and apoptosis. Advances in our understanding of transcriptional regulation will allow us to develop new strategies to manipulate aberrant gene regulation in various pathologies.

14.6.1 Anti-inflammatory Therapy with MSK1 Inhibitors

As MSK1-induced phosphorylation of p65 is crucial for CBP engagement in the IL-6 enhanceosome, the p65-CBP interface and MSK1 may be attractive and potential targets for therapeutic inhibition in cancer (Waddick and Uckun 1999), viral replication (DeLuca et al. 1999; Tai et al. 2000), or inflammatory disorders (Baeuerle and Baichwal 1997; Chen et al. 1999), such as atopy (Beyaert 1999), bowel (Neurath et al. 1998; Jobin and Sartor 2000), or pulmonary diseases (Barnes and Adcock 1998; Barnes 1999), rheumatoid arthritis (Jue et al. 1999), and septic shock (Christman et al. 1998). The first pharmacological compounds which selectively interfere with NF-κB signaling have recently been reported and described to inhibit gut inflammation (Bork et al. 1997; Lyss et al. 1998; Egan et al. 1999). Chromium (VI) has been shown to interfere in nuclear competence of NF-κB in response to TNF by selectively inhibiting the p65/CBP interaction rather than DNA binding (Shumilla et al. 1999). NSAIDs are widely used as inhibitors of inflammation, although the precise mechanism of their therapeutic action is unknown. In contrast to previous reports dealing with the effects of conventional salicylates and sulfasalazines on IκB degradation (O'Neill 1998; Yin et al. 1998), we found a specific inhibitory role of NSAIDS in nuclear NF-κB function by preventing MSK1-induced phosphorylation and transactivation (G. De Wilde, unpublished data). The recent observation of tandem factor complexes combining (histone/factor) kinase and acetylase activities (i.e., CBP-RKS2 or CBP-MSK1) raises intriguing questions on the connection of histone language (H3/H4 phosphorylation/acetylation) and transcription factor code (CREB/p65 phosphorylation/acetylation) and may establish the basis for specific gene induction and a selective target for intervention (Thomson et al. 1999a,b; Clayton et al. 2000; Jenuwein and Allis 2001; Marmorstein 2001; Mayr and Montminy 2001; Mayr et al. 2001; Merienne et al. 2001).

14.6.2 Dissociated Glucocorticoid Hormones as Anti-inflammatory Drugs

GCs so far remain the most effective therapy for inflammatory disorders. For asthma, topical steroids are the mainstay for controlling the inflammatory component of the disease. However, their use is limited by the constellation of adverse effects associated with chronic, oral steroid use and the long-term adverse effects associated with inhaled steroid use. These include suppression of hypothalamic-pituitary axis, osteoporosis, reduced bone growth in the young, opportunistic infections, behavioral alterations, disorders of lipid metabolism, diabetes, impaired wound healing, skin atrophy, muscle atrophy, cataracts, peptic ulcers, hypertension (due to activation of the mineralocorticoid receptor), decreased carbohydrate tolerance, and disturbed water household balance of the body due to retention of water and sodium and excretion of potassium (Barnes and Adcock 1995; Karin 1998). To date physicians attempt to minimize these side effects using local therapy, intervals in the therapy, supplementation with calcium, vitamin D3, and estrogens, and using specific GCs with as few mineralocorticoid agonistic effects as possible.

Most of these detrimental side effects may be attributable to the endocrine activity of steroids and are largely identical to the syndromes of endogenous corticosteroid excess (Cushing's syndrome). Thus, the

Holy Grail of steroid pharmacology is the development of agents which have a better therapeutic ratio than the current steroids, especially on systemic administration. This may be achieved by the identification of molecules which elicit marked anti-inflammatory effects, but have a minor impact on endocrine responses. "Dissociated" corticosteroids are thus ligands for the GR that may offer the potential for a more selective anti-inflammatory profile (Belvisi et al. 2001a,b).

The search for new anti-inflammatory drugs that act via the GR, therefore, focuses on the separation of the transactivation and transrepressive potential of this nuclear receptor. While the positive activating properties are mainly associated with undesirable side effects, its negative interference with the activity of transcription factors, such as NF-κB and AP-1, greatly contributes to its anti-inflammatory and immune suppressive capacities. Previously, the compound RU24858 was demonstrated to dissociate between transactivation and transrepression in vitro and to exert strong AP-1 and NF-κB inhibition (transrepression), but little or no transactivation (Vayssiere et al. 1997; Vanden Berghe et al. 1999b). However, in vitro separation of transrepression from transactivation activity did not translate to an increased therapeutic ratio for GCs in vivo (Belvisi et al. 2001b).

Recently, we identified a new ligand, compound A (CpdA), which downmodulates TNF-induced IL-6 gene expression, but does not activate GRE-driven promoters [K. De Bosscher, unpublished, patent WO01/45693 (28.06.2001) – PCT/EP00/13347]. CpdA is a phenyl aziridine precursor with contraceptive actions, but which does not belong to the steroidal class of GR-binding ligands. However, it has been shown before to interfere with the HPA/human pituitary gonadotropin (HPG) axis, suggesting that it acts via the GR. We further found that the mechanism by which CpdA inhibits IL-6 and other pro-inflammatory genes, such as E-selectin and ICAM1, works via the GR and does not involve interference with the DNA-binding activity of NF-κB. Furthermore, CpdA also downmodulates AP1-driven gene expression, but in contrast with DEX, the JNK kinase pathway is not affected. Taken together, this compound represents a novel class of anti-inflammatory agents with dissociated properties and might hold potentials for therapeutic use. Whether CpdA or a derivative thereof could be useful as a therapeutic anti-inflammatory agent with fewer side effects than glucocorticoids depends not only on its ability to repress genes involved in the inflammatory response, but also on whether and to what extent it affects expression of the genes involved in reproduction. It will be instructive to compare the effective in vivo concentration that elicits anti-inflammatory actions with the concentration that elicits contraceptive effects. Further research on CpdA or other natural analogues as novel and more directed, effective anti-inflammatory drugs may offer realistic possibilities that could contribute to human welfare in the near future.

14.6.3 Final Conclusion

Because of its central role in various physiopathological processes (i.e., inflammatory diseases, bone diseases, oncogenesis, and neurodegeneration), NF-κB has become the first target for innovative, anti-inflammatory and/or anti-tumoral therapies in various pharmaceutical companies in Europe and the USA. The identification of new regulatory target molecules involved in NF-κB signaling, which are useful for pharmacological applications, has been a hot topic for many years.

Our observation that NF-κB activity is susceptible to various important modulations adds a new perspective to therapeutical applications, as several pharmaceutical companies which currently develop selective NF-κB inhibitors focus on the cytoplasmic regulatory event in which NF-κB is released from its physiologic inhibitor IκB (O'Neill 2001). As NF-κB is also a crucial component of the immune system, the latter strategy has the important disadvantage that blocking IκB degradation is a very drastic event with detrimental side effects. In this respect, innovative anti-inflammatory strategies may focus on less aggressive therapies, which interfere with NF-κB activity without affecting its DNA binding. To this purpose, the main scientific interest may shift from the cytoplasm to the NF-κB–chromatin interface in the nucleus. Furthermore, future drug compounds should interfere selectively with interactions of NF-κB with particular cofactors or defined nucleosomes, which could restrict drug activity to a limited set of disease-related genes, leaving homeostatic gene expression unaffected. Unraveling the relationship between chromatin and NF-κB regulation will definitely shed a new light on the design of innovative anti-inflammatory and/or anti-cancer drugs (Merienne et al. 2001; Saccani et al. 2001). A working model displaying this relationship is presented in Fig. 6. As chromatin-embed-

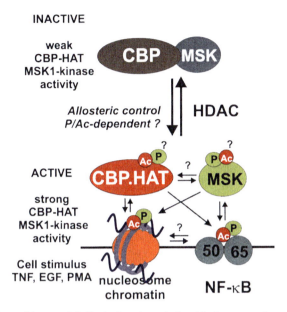

Fig. 6. A working model displaying the relationship between chromatin and NF-κB regulation

ded promoter enhanceosomes nowadays behave like sophisticated protein modules receptive to various signals, future therapies may benefit from combined structural (selective ligand) and signaling (selective inhibitors) approaches to establish effective but harmless treatments (Webster-J et al. 1997; Michael et al. 2000; Schmitz et al. 2001; Weinmann et al. 2001; L. Vermeulen, submitted).

References

Aarnisalo P, Palvimo JJ, Jänne OA (1998) CREB-binding protein in androgen receptor-mediated signaling. Proc Natl Acad Sci USA 95:2122–2127

Abu-amer Y, Ross FR, McHugh KP, Livolsi A, Peyron JF, Teitelbaum SL (1998) Tumor necrosis factor-alfa activation of nuclear transcription factor-kappaB in marrow macrophages is mediated by c-src tyrosine phosphorylation of IkappaBalfa. J Biol Chem 273:29417–29423

Adcock IM, Nasuhara Y, Stevens DA, Barnes PJ (1999) Ligand-induced differentiation of glucocorticoid receptor (GR) trans-repression and transactivation: preferential targetting of NF-kappaB and lack of I-kappaB involvement. Br J Pharmacol 127:1003–1011

Agalioti T, Lomvardas S, Parekh B, Yie J, Maniatis T, Thanos D (2000) Ordered recruitment of chromatin modifying and general transcription factors to the IFN-beta promoter. Cell 103:667–78

Agarwal S, Rao A (1998a) Long-range transcriptional regulation of cytokine gene expression. Curr Opin Immunol 10:345–52

Agarwal S, Rao A (1998b) Modulation of chromatin structure regulates cytokine gene expression during T cell differentiation. Immunity 9:765–775

Ait-si-ali S, Ramirez S, Barre FX, Dkhissi F, Magnaghi-Jaulin L, Girault JA, Robin P, Knibiehler M, Pritchard LL, Ducommun B, Trouche D, Harel-Bellan A (1998) Histone acetyltransferase activity of CBP is controlled by cycle-dependent kinases and oncoprotein E1 A. Nature 396:184–186

Akimaru H, Hou-Dx, Ishii S (1997) Drosophila CBP is required for dorsal dependent twist gene expression. Nat Genet 17:211–214

Akira S, Taga T, Kishimoto T (1993) Interleukin-6 in biology and medicine. Adv Immunol 54:1–78

Alberts AS, Geneste O, Treisman R (1998) Activation of SRF-regulated chromosomal templates by rho-family GTPases requires a signal that also induces H4 hyperacetylation. Cell 92:475–487

Anrather J, Csizmadia V, Soares MP, Winkler H (1999) Regulation of NF-kappaB RelA phosphorylation and transcriptional activity by p21(ras) and protein kinase Czeta in primary endothelial cells. J Biol Chem 274:13594–13603

Ashburner BP, Westerheide SD, Baldwin AS Jr (2001) The p65 (RelA) Subunit of NF-kappaB interacts with the histone deacetylase (HDAC) corepressors HDAC1 and HDAC2 to negatively regulate gene expression. Mol Cell Biol 21:7065–7077

Auphan N, DiDonato JA, Rosette C, Helmberg A, Karin M (1995) Immunosuppression by glucocorticoids: inhibition of NF-kappa B activity through induction of I kappa B synthesis. Science 270:286–290

Baeuerle PA, Baichwal VR (1997) NF-κB as a frequent target for immunosuppressive and anti-inflammatory molecules. Adv Immunol 65:111–137

Baeuerle PA, Baltimore D (1996) NF-kappa B: ten years after. Cell 87:13–20

Baldwin AS Jr (2001) Series introduction: the transcription factor NF-kappaB and human disease. J Clin Invest 107:3–6

Barkett M, Xue D, Horvitz HR, Gilmore TD (1997) Phosphorylation of IkappaB-alpha inhibits its cleavage by caspase CPP32 in vitro. J Biol Chem 272:29419–29422

Barnes PJ (1999) Novel approaches and targets for treatment of chronic obstructive pulmonary disease. Am J Respir Crit Care Med 160:S72–S79

Barnes PJ, Adcock IM (1995) Steroid resistance in asthma. QJM 88:455–468

Barnes PJ, Adcock IM (1998) Transcription factors and asthma. Eur Respir J 12:221–234

Beato M, Herrlich P, Schutz G (1995) Steroid hormone receptors: many actors in search of a plot. Cell 83:851–7

Beg AA, Sha WC, Bronson RT, Baltimore D (1995) Constitutive NF-kappaB activation, enhanced granulopoiesis, and neonatal lethality in IkappaBalfa-deficient mice. Genes Dev 9:2736–2746

Belvisi MG, Brown TJ, Wicks S, Foster ML (2001a) New glucocorticosteroids with an improved therapeutic ratio? Pulm Pharmacol Ther 14:221–227

Belvisi MG, Wicks SL, Battram CH, Bottoms SE, Redford JE, Woodman P, Brown TJ, Webber SE, Foster ML (2001b) Therapeutic benefit of a dissociated glucocorticoid and the relevance of in vitro separation of transrepression from transactivation activity. J Immunol 166:1975–1982

Bender K, Göttlicher M, Whiteside S, Rahmsdorf HJ, Herrlich P (1998) Sequential DNA damage-independent and -dependent activation of NF-kappaB by UV. EMBO J 17:5170–5181

Bensaude O, Bonnet F, Casse C, Dubois MF, Nguyen VT, Palancade B (1999) Regulated phosphorylation of the RNA polymerase II C-terminal domain (CTD). Biochem Cell Biol 77:249–255

Beraud C, Henzel WJ, Baeuerle PA (1999) Involvement of regulatory and catalytic subunits of phosphoinositide 3-kinase in NF-kappaB activation. Proc Natl Acad Sci USA 96:429–434

Berger SL, Felsenfeld G (2001) Chromatin goes global. Mol Cell 8:263–268

Beyaert R (1999) NF-kappaB as an emerging target in atopy. Emerg Ther Targets 3:1–16

Beyaert R, Cuenda A, Vanden Berghe W, Plaisance S, Lee JC, Haegeman G, Cohen P, Fiers W (1996) The p38/RK mitogen-activated protein kinase pathway regulates interleukin-6 synthesis in response to tumour necrosis factor. EMBO J 15:1914–1923

Bhattacharya S, Michels CL, Leung MK, Arany ZP, Kung AL, Livingston DM (1999) Functional role of p35srj, a novel p300/CBP binding protein, during transactivation by HIF-1. Genes Dev 13:64–75

Bird TA, Schooley K, Dower SK, Hagen H, Virca GD (1997) Activation of nuclear transcription factor NF-kappaB by interleukin-1 is accompanied by casein kinase II-mediated phosphorylation of the p65 subunit. J Biol Chem 272:32606–32612

Blobel GA (2000) CREB-binding protein and p300: molecular integrators of hematopoietic transcription. Blood 95:745–755

Bonnet F, Vigneron M, Bensaude O, Dubois MF (1999) Transcription-independent phosphorylation of the RNA polymerase II C-terminal domain (CTD) involves ERK kinases (MEK1/2). Nucleic Acids Res 27:4399–43404

Bork PM, Schmitz ML, Kuhnt M, Escher C, Heinrich M (1997) Sesquiterpene lactone containing Mexican Indian medicinal plants and pure sesquiterpene lactones as potent inhibitors of transcription factor NF-kappaB. FEBS Lett 402:85–90

Bottazzi ME, Zhu X, Bohmer RM, Assoian RK (1999) Regulation of p21(cip1) expression by growth factors and the extracellular matrix reveals a role for transient ERK activity in G1 phase. J Cell Biol 146:1255–1264

Brooks PM, Day RO (1991) Nonsteroidal anti-inflammatory drugs – differences and similarities. N Engl J Med 324:1716–1725

Brostjan C, Anrather J, Csizmadia V, Stroka D, Soares M, Bach FH, Winkler H (1996) Glucocorticoid-mediated repression of NFkappaB activity in endothelial cells does not involve induction of IkappaBalpha synthesis. J Biol Chem 271:19612–19616

Carey M (1998) The enhanceosome and transcriptional synergy. Cell 92:5–8

Carter AB, Hunninghake GW (2000) A constitutive active MEK –ERK pathway negatively regulates NF-kappa B-dependent gene expression by modulating TATA-binding protein phosphorylation. J Biol Chem 275:27858–27864

Cato AC, Wade E (1996) Molecular mechanisms of anti-inflammatory action of glucocorticoids. Bioessays 18:371–378

Chan HM, La Thangue NB (2001) p300/CBP proteins: HATs for transcriptional bridges and scaffolds. J Cell Sci 114:2363–2373

Chau AS, Shibuya EK (1999) Inactivation of p42 mitogen-activated protein kinase is required for exit from M-phase after cyclin destruction. J Biol Chem 274:32085–32090

Chawla S, Hardingham GE, Quinn DR, Bading H (1998) CBP: A signal-regulated transcriptional coactivator controlled by nuclear calcium and CaM kinase IV. Science 281:1505–1509

Chen FE, Huang DB, Chen YQ, Ghosh G (1998) Crystal structure of p50/p65 heterodimer of transcription factor NF-kappaB bound to DNA. Nature 391:410–413

Chen F, Castranova V, Shi X, Demers LM (1999) New insights into the role of nuclear factor-kappaB, a ubiquitous transcription factor in the initiation of diseases. Clin Chem 45:7–17

Chen L, Fischle W, Verdin E, Greene WC (2001) Duration of nuclear NF-kappaB action regulated by reversible acetylation. Science 293:1653–1657

Christman JW, Lancaster LH, Blackwell TS (1998) Nuclear factor kappa B: a pivotal role in the systemic inflammatory response syndrome and new target for therapy [see comments]. Intensive Care Med 24:1131–1138

Clayton AL, Rose S, Barratt MJ, Mahadevan LC (2000) Phosphoacetylation of histone H3 on c-fos- and c-jun-associated nucleosomes upon gene activation. EMBO J 19:3714–3726

Cohen L, Henzel WJ, Baeuerle PA (1998) IKAP is a scaffold protein of the IkappaB kinase complex. Nature 395:292–296

Cohen LE, Hashimoto Y, Zanger K, Wondisford F, Radovick S (1999) CREB-independent regulation by CBP is a novel mechanism of human growth hormone gene expression. J Clin Invest 104:1123–1130

Cramer P, Muller CW (1999) A firm hand on NFkappaB: structures of the IkappaBalpha-NFkappaB complex. Structure Fold Des 7:R1–R6

Cuervo AM, Hu W, Lim B, Dice FJ (1998) IkappaB is a substrate for a selective pathway of lysosomal proteolysis. Mol Biol Cell 9:1995–2010

Davie JR, Samuel SK, Spencer VA, Holth LT, Chadee DN, Peltier CP, Sun JM, Chen HY, Wright JA (1999) Organization of chromatin in cancer cells: role of signalling pathways. Biochem Cell Biol 77:265–275

De Bosscher K, Schmitz ML, Vanden Berghe W, Plaisance S, Fiers W, Haegeman G (1997) Glucocorticoid-mediated repression of nuclear factor-kappaB-dependent transcription involves direct interference with transactivation. Proc Natl Acad Sci USA 94:13504–1359

De Bosscher K, Vanden Berghe W, Haegeman G (2000a) Mechanisms of anti-inflammatory action and of immunosuppression by glucocorticoids: negative interference of activated glucocorticoid receptor with transcription factors. J Neuroimmunol 109:16–22

De Bosscher K, Vanden Berghe W, Vermeulen L, Plaisance S, Boone E, Haegeman, G (2000b) Glucocorticoids repress NF-kappaB-driven genes by disturbing the interaction of p65 with the basal transcription machinery, irrespective of coactivator levels in the cell. Proc Natl Acad Sci 97:3919–3924

De Bosscher K, Vanden Berghe W, Haegeman G (2001) Glucocorticoid repression of AP-1 is not mediated by competition for nuclear coactivators. Mol Endocrinol 15:219–227

De Kloet ER, Vreugdenhil E, Oitzl MS, Joels M (1998) Brain corticosteroid receptor balance in health and disease. Endocr Rev 19:269–301

De Kloet ER, Meijer OC, Vreugdenhil E, Joels M (2000) The Yin and Yang of nuclear receptors: symposium on nuclear receptors in brain, Oegstgeest, The Netherlands 13–14 April 2000. Trends Endocrinol Metab 11:245–248

Dechend R, Hirano F, Lehmann K, Heissmeyer V, Ansieau S, Wulczyn FG, Scheidereit C, Leutz A (1999) The Bcl-3 oncoprotein acts as a bridging factor between NF-kappaB/Rel and nuclear co-regulators. Oncogene 18:3316–3323

Delerive P, Gervois P, Fruchart JC, Staels B (2000) Induction of ikappa balpha expression as a mechanism contributing to the anti-inflammatory activities of peroxisome proliferator-activated receptor-alpha activators. J Biol Chem 275:36703–36707

Delerive P, Monté D, Dubois G, Trottein F, Fruchart-Najib Mariani J, Fruchart J-C, Staels B (2001) The orphan nuclear receptor RORalfa is a negative regulator of the inflammatory response. EMBO Reports 21:42–48

Delerive P, De Bosscher K, Vanden Berghe W, Fruchart JC, Haegeman G, Staels B (2002) DNA binding-independent induction of IkappaBalpha gene transcription by PPARalpha. Mol Endocrinol 16:1029–1039

DeLuca C, Kwon H, Lin R, Wainberg M, Hiscott J (1999) NF-kappaB activation and HIV-1 induced apoptosis. Cytokine Growth Factor Rev 10:235–253

Denison C, Kodadek T (1998) Small-molecule-based strategies for controlling gene expression. Chem Biol 5:R129–R145

Desterro J, Rodriguez M, Hay RT (1998) SUMO-1 modification of IkBalfa inhibits NF-kappaB activation. Mol Cell 2:233–239

DiDonato JA, Saatcioglu F, Karin M (1996) Molecular mechanisms of immunosuppression and anti-inflammatory activities by glucocorticoids. Am J Respir Crit Care Med 154:S11–S15

Diehl JA, Tong W, Sun G, Hannink M (1995) Tumor necrosis factor-alpha-dependent activation of a RelA homodimer in astrocytes. Increased phosphorylation of RelA and MAD-3 precede activation of RelA. J Biol Chem 270:2703–2707

Doucas V, Tini M, Egan DA, Evans RM (1999) Modulation of CREB binding protein function by the promyelocytic (PML) oncoprotein suggests a role for nuclear bodies in hormone signaling. Proc Natl Acad Sci USA 96:2627–2632

Doucas V, Shi Y, Miyamoto S, West A, Verma I, Evans RM (2000) Cytoplasmic catalytic subunit of protein kinase A mediates cross-repression by NF-kappaB and the glucocorticoid receptor. Proc Natl Acad Sci USA 97:11893–11898

Drier EA, Huang LH, Steward R (1999) Nuclear import of the Drosophila Rel protein Dorsal is regulated by phosphorylation. Genes Dev 13:556–568

Dubnicoff T, Valentine SA, Chen G, Shi T, Lengyel JA, Paroush Z, Courey AJ (1997) Conversion of dorsal from an activator to a repressor by the global corepressor Groucho. Genes Dev 11:2952–2957

Dumont A, Hehner SP, Schmitz ML, Gustafsson JA, Lidén J, Okret S, van der Saag PT, Wissink S, van der Burg B, Herrlich P, Haegeman G, De Bosscher K, Fiers W (1998) Cross-talk between steroids and NF-kappa B: what language? Trends Biochem Sci 23:233–235

Egan LJ, Mays DC, Huntoon CJ, Bell MP, Pike MG, Sandborn WJ, Lipsky JJ, McKean DJ (1999) Inhibition of interleukin-1-stimulated NF-kappaB RelA/p65 phosphorylation by mesalamine is accompanied by decreased transcriptional activity. J Biol Chem 274:26448–26453

El Kharroubi A, Piras G, Zensen R, Martin MA (1998) Transcriptional activation of the integrated chromatin-associated human immunodeficiency virus type 1 promoter. Mol Cell Biol 18:2535–2544

Emery JG, Ohlstein EH, Jaye M (2001) Therapeutic modulation of transcription factor activity. Trends Pharmacol Sci 22:233–240

Erschler WB, Keller ET (2000) Age-associated increased interleukin-6 gene expression, late-life diseases, and frailty. Annu Rev Med 51:245–270

Espinos E, Le Van Thai A, Pomies C, Weber MJ (1999) Cooperation between phosphorylation and acetylation processes in transcriptional control. Mol Cell Biol 19:3474–3484

Francastel C, Schubeler D, Martin DIK, Groudine M (2000) Nuclear compartmentalization and gene activity. Nat Rev Mol Cell Biol 1:137–143

Gerritsen ME, Williams AJ, Neish AS, Moore S, Shi Y (1997) CREB-binding protein/p300 are transcriptional coactivators of p65. Proc Natl Acad Sci USA 94:2927–2932

Ghosh S, May MJ, Kopp EB (1998) NF-kappaB and rel proteins: evolutionarily conserved mediators of immune responses. Annu Rev Immunol 16:225–260

Giguere V, Hollenberg SM, Rosenfeld MG, Evans RM (1986) Functional domains of the human glucocorticoid receptor. Cell 46:645–652

Giles RH, Peters DJM, Breuning MH (1998) Conjunction dysfunction: CBP/p300 in human disease. TIG 14:178–183

Glass CK, Rosenfeld MG (2000) The coregulator exchange in transcriptional functions of nuclear receptors. Genes Dev 14:121–141

Glass CK, Rose DW, Rosenfeld MG (1997) Nuclear receptor coactivators. Curr Opin Cell Biol 9:222–232

Gribnau J, de Boer E, Trimborn T, Wijgerde M, Milot E, Grosveld F, Fraser P (1998) Chromatin interaction mechanism of transcriptional control in vivo. EMBO J 17:6020–6027

Hager GL, Lim CS, Elbi C, Baumann CT (2000) Trafficking of nuclear receptors in living cells. J Steroid Biochem Mol Biol 74:249–254

Handel ML (1997) Transcription factors AP-1 and NF-kappa B: where steroids meet the gold standard of anti-rheumatic drugs. Inflamm Res 46:282–286

Handel ML, Nguyen LQ, Lehmann TP (2000) Inhibition of transcription factors by anti-inflammatory and anti-rheumatic drugs: can variability in response be overcome? Clin Exp Pharmacol Physiol 27:139–144

Heck S, Bender K, Kullmann M, Gottlicher M, Herrlich P, Cato AC (1997) I kappaB alpha-independent downregulation of NF-kappaB activity by glucocorticoid receptor. EMBO J 16:4698–4707

Heissmeyer V, Krappmann D, Wulczyn FG, Scheidereit C (1999) NF-kappaB p105 is a target of IkappaB kinases and controls signal induction of Bcl-3-p50 complexes. EMBO J 18:4766–4778

Herrlich P (2001) Cross-talk between glucocorticoid receptor and AP-1. Oncogene 20:2465–2475

Hodges M, Tissot C, Freemont PS (1998) Protein regulation: tag wrestling with relatives of ubiquitin. Curr Biol 8:R749–R752

Hofmann TG, Hehner SP, Bacher S, Droge W, Schmitz ML (1998) Various glucocorticoids differ in their ability to induce gene expression, apoptosis and to repress NF-kappaB-dependent transcription. FEBS Lett 441:441–446

Hollenberg SM, Evans RM (1988) Multiple and cooperative trans-activation domains of the human glucocorticoid receptor. Cell 55:899–906

Hong SH, Wong CW, Privalsky ML (1998) Signaling by tyrosine kinases negatively regulates the interaction between transcription factors and SMRT (silencing mediator of retinoic acid and thyroid hormone receptor) corepressor. Mol Endocrinol 12:1161–1171

Horwitz KB, Jackson TA, Bain DL, Richer JK, Takimoto GS, Tung L (1996) Nuclear receptor coactivators and corepressors. Mol Endocrinol 10:1167–1177

Hottiger MO, Felzien LK, Nabel GJ (1998) Modulation of cytokine-induced HIV gene expression by competitive binding of transcription factors to the coactivator p300. EMBO J 17:3124–3134

Hsieh JJD, Zhou S, Chen L, Young DB, Hayward SD (1999) CIR, a corepressor linking the DNA binding factor CBF1 to the histone deacetylase complex. Proc Natl Acad Sci USA 1:23–28

Hu SC, Chrivia J, Ghosh A (1999) Regulation of CBP-mediated transcription by neuronal calcium signaling. Neuron 22:799–808

Huxford T, Huang DB, Malek S, Ghosh G (1998) The crystal structure of the IkappaBalfa/NF-kappaB complex reveals mechanisms of NF-kappaB inactivation. Cell 95:759–770

Ikeda K, Steger DJ, Eberharter A, Workman JL (1999) Activation domain-specific and general transcription stimulation by native histone acetyltransferase complexes. Mol Cell Biol 19:855–863

Imbert V, Rupec RA, Livolsi A, Pahl HL, Traenckner EB, Mueller Dieckmann C, Farahifar D, Rossi B, Auberger P, Baeuerle PA, Peyron JF (1996) Tyrosine phosphorylation of I kappa B-alpha activates NF-kappa B without proteolytic degradation of IkappaB-alpha. Cell 86:787–798

Israel A (2000) The IKK complex: an integrator of all signals that activate NF-kappaB? Trends Cell Biol 10:129–133

Ito K, Barnes PJ, Adcock IM (2000) Glucocorticoid receptor recruitment of histone deacetylase 2 inhibits interleukin-1beta-induced histone H4 acetylation on lysines 8 and 12. Mol Cell Biol 20:6891–6903

Jacobs MD, Harrison SC (1998) Structure of an IkappaBalpha/NF-kappaB complex. Cell 95:749–758

Janknecht R, Hunter T (1999) Nuclear fusion of signaling pathways. Science 284:443–444

Jenuwein T, Allis CD (2001) Translating the histone code. Science 293:1074–1080

Jin Y, Wang Y, Walker DL, Dong H, Conley C, Johansen J, Johansen KM (1999) JIL-1: a novel chromosomal tandem kinase implicated in transcriptional regulation in Drosophila. Mol Cell 4:129–135

Jobin C, Sartor RB (2000) The IkappaB/NF-kappaB system: a key determinant of mucosal inflammation and protection. Am J Physiol Cell Physiol 278:C451–C462

Jonat C, Rahmsdorf HJ, Park KK, Cato AC, Gebel S, Ponta H, Herrlich P (1990) Antitumor promotion and anti-inflammation: down-modulation of AP-1 (Fos/Jun) activity by glucocorticoid hormone. Cell 62:1189–1204

Jue DM, Jeon KI, Jeong JY (1999) Nuclear factor kappaB (NF-kappaB) pathway as a therapeutic target in rheumatoid arthritis. J Korean Med Sci 14:231–238

Kaiser GC, Yan F, Polk DB (1999) Mesalamine blocks tumor necrosis factor growth inhibition and nuclear factor kappaB activation in mouse colonocytes. Gastroenterology 116:602–609

Kamei Y, Xy L, Heinzel T, Torchia J, Kurokawa R, Gloss B, Lin S-C, Heyman RA, Rose DW, Glass CK, Rosenfeld MG (1996) A CBP integrator complex mediates transcriptional activation and AP-1 inhibition by nuclear receptors. Cell 85:403–414

Kao HY, Ordentlich P, Koyano-Nakagawa N, Tang Z, Downes M, Kintner CR, Evans RM, Kadesch T (1998) A histone deacetylase corepressor complex regulates the Notch signal transduction pathway. Genes Dev 12:2269–2277

Karin M (1998) New twists in gene regulation by glucocorticoid receptor: is DNA binding dispensable? Cell 93:487–490

Karin M (1999) How NF-kappa B is activated: the role of the I kappa B kinase (IKK) complex. Oncogene 18:6867–6874

Karin M, Ben-Neriah Y (2000) Phosphorylation meets ubiquitination: the control of NF-[kappa]B activity. Annu Rev Immunol 18:621–663

Karin M, Delhase M (2000) The I kappa B kinase (IKK) and NF-kappa B: key elements of proinflammatory signalling. Semin Immunol 12:85–98

Karin M, Liu-Zg, Zandi E (1997) Ap-1 function and regulation. Curr Opin Cell Biol 9:240–246

Kawasaki H, Eckner R, Yao TP, Taira K, Chiu R, Livingston DM, Yokoyama KK (1998) Distinct roles of the co-activators p300 and CBP in retinoic-acid-induced F9-cell differentiation. Nature 393:284–289

Kim RH, Flanders KC, Reffey SB, Anderson LA, Duckett CS, Perkins ND, Roberts AB (2001) SNIP1 inhibits NF-kappa B signaling by competing for its binding to the C/H1 domain of CBP/p300 transcriptional co-activators. J Biol Chem 20:20

Kopp E, Ghosh S (1994) Inhibition of NF-kappa B by sodium salicylate and aspirin. Science 265:956–959

Korzus E, Torchia J, Rose DW, Xu L, Kurokawa R, McInerney EM, Mullen TM, Glass CK, Rosenfeld MG (1998) Transcription factor-specific requirements for coactivators and their acetyltransferase functions. Science 279:703–707

Kouzarides T (1999) Histone acetylases and deacetylases in cell proliferation. Curr Opin Genet Dev 9:40–48

Kouzarides T (2000) Acetylation: a regulatory modification to rival phosphorylation? EMBO J 19:1176–1179

Kung AL, Rebel VI, Bronson RT, Ch'ng LE, Sieff CA, Livingston DM, Yao TP (2000) Gene dose-dependent control of hematopoiesis and hematologic tumor suppression by CBP. Genes Dev 14:272–277

Lamberti C, Lin KM, Yamamoto Y, Verma U, Verma IM, Byers S, Gaynor RB (2001) Regulation of beta-catenin function by the I kappa B kinases. J Biol Chem 29:29

Lee SK, Kim HJ, Na SY, Kim TS, Choi HS, Im SY, Lee JW (1998) Steroid Receptor Coactivator-1 coactivates Activating Protein-1 mediated transactivations through interaction with the c-Jun and c-Fos subunits. J Biol Chem 273:16651–16654

Lemon B, Tjian T (2000) Orchestrated response: a symphony of transcription factors for gene control. Genes Dev 14:2551–2569. Review

Li N, Karin M (1998) Ionizing radiation and short wavelength UV activate NF-kappaB through two distinct mechanisms. Proc Natl Acad Sci USA 95:13012–13017

Li CC, Korner M, Ferris DK, Chen E, Dai RM, Longo DL (1994) NF-kappa B/Rel family members are physically associated phosphoproteins. Biochem J 303:499–506

Li ZW, Chu W, Hu Y, Delhase M, Deerinck T, Ellisman M, Johnson R, Karin M (1999) The IKKbeta subunit of IkappaB kinase (IKK) is essential for nuclear factor kappaB activation and prevention of apoptosis. J Exp Med 189:1839–1845

Li Q, Estepa G, Memet S, Israel A, Verma I (2000) Complete lack of NF-kappa B activity in IKK1 and IKK2 double-deficient mice: additional defects in neurulation. Genes Dev 14:1729–1733

Li X, Commane M, Nie H, Hua X, Chatterjee-Kishore M, Wald D, Haag M, Stark G (2000) Act1, an NF-kappa B-activating protein. Proc Natl Acad Sci USA 97:10489–10493

Lidén J, Delaunay F, Rafter I, Gustafsson J, Okret S (1997) A new function for the C-terminal zinc finger of the glucocorticoid receptor. Repression of RelA transactivation. J Biol Chem 272:21467–21472

Lidén J, Rafter I, Truss M, Gustafsson JA, Okret S (2000) Glucocorticoid effects on NF-kappaB binding in the transcription of the ICAM-1 gene. Biochem Biophys Res Commun 273:1008–1014

Liu L, Kwak YT, Bex F, Garcia-Martinez LF, Li XH, Meek K, Lane WS, Gaynor RB (1998) DNA-dependent protein kinase phosphorylation of IkappaBalfa and IkBbeta regulates NF-kappaB DNA binding properties. Mol Cell Biol 18:4221–4234

Liu YZ, Chrivia JC, Latchman DS (1998) Nerve growth factor up-regulates the transcriptional activity of CBP through activation of the p42/p44(MAPK) cascade. J Biol Chem 273:32400–32407

Liu YZ, Thomas NS, Latchman DS (1999) CBP associates with the p42/p44 MAPK enzymes and is phosphorylated following NGF treatment. Neuroreport 10:1239–1243

Lo WS, Duggan L, Tolga NC, Emre Belotserkovskya R, Lane WS, Shiekhattar R, Berger SL (2001) Snf1-a histone kinase that works in concert with the histone acetyltransferase Gcn5 to regulate transcription. Science 293:1142–1146

Lyss G, Knorre A, Schmidt TJ, Pahl HL, Merfort I (1998) The anti-inflammatory sesquiterpene lactone helenalin inhibits the transcription factor NF-kappaB by directly targeting p65. J Biol Chem 273:33508–33516

Madrid LV, Wang CY, Guttridge DC, Schottelius AJ, Baldwin AS Jr, Mayo MW (2000) Akt suppresses apoptosis by stimulating the transactivation potential of the RelA/p65 subunit of NF-kappaB. Mol Cell Biol 20:1626–1638

Magnaghi-Jaulin L, Ait-Si-Ali S, Harel-Bellan A (1999) Histone acetylation in signal transduction by growth regulatory signals. Semin Cell Dev Biol 10:197–203

Maniatis T (1999) A ubiquitin ligase complex essential for the NF-kappaB, Wnt/Wingless, and Hedgehog signaling pathways. Genes Dev 13:505–510

Marmorstein R (2001) Protein modules that manipulate histone tails for chromatin regulation. Nat Rev Mol Cell Biol 2:422–432

Martinez-Balbas MA, Bannister AJ, Martin K, Haus-Seuffert P, Meisterernst M, Kouzarides T (1998) The acetyltransferase activity of CBP stimulates transcription. EMBO J 17:2886–2893

Masumi A, Wang IM, Lefebvre B, YANG XJ, Nakatani Y, Ozato K (1999) The histone acetylase PCAF is a phorbol-ester-inducible coactivator of the IRF family that confers enhanced interferon responsiveness. Mol Cell Biol 19:1810–1820

Mayr B, Montminy M (2001) Transcriptional regulation by the phosphorylation-dependent factor CREB. Nat Rev Mol Cell Biol 2:599–609

Mayr BM, Canettieri G, Montminy MR (2001) Distinct effects of cAMP and mitogenic signals on CREB-binding protein recruitment impart specificity to target gene activation via CREB. Proc Natl Acad Sci USA 98:10936–10941

McEwan IJ, Wright PH, Gustafsson JA (1997) Mechanisms of gene expression by the glucocorticoid receptor: role of protein-protein interactions. Bioessays 19:153–160

Mercurio F, Manning AM (1999) Multiple signals converging on NF-kappa B. Curr Op Cell Biol 11:226–232

Merienne K, Pannetier S, Harel-Bellan A, Sassone-Corsi P (2001) Mitogen-regulated rsk2-cbp interaction controls their kinase and acetylase activities. Mol Cell Biol 21:7089–7096

Merika M, Thanos D (2001) Enhanceosomes. Curr Opin Genet Dev 11:205–208

Merika M, Williams AJ, Chen G, Collins T, Thanos D (1998) Recruitment of CBP/p300 by the IFN beta enhanceosome is required for synergistic activation of transcription. Mol Cell 1:277–287

Michael LF, Asahara H, Shulman AI, Kraus WL, Montminy M (2000) The phosphorylation status of a cyclic AMP-responsive activator is modulated via a chromatin-dependent mechanism. Mol Cell Biol 20:1596–1603

Min W, Ghosh S, Lengyel P (1996) The interferon-inducible p202 protein as a modulator of transcription: inhibition of NF-kappa B, c-Fos, and c-Jun activities. Mol Cell Biol 16:359–368

Miyamoto S, Seufzer BJ, Shumway SD (1998) Novel IkappaB alpha proteolytic pathway in WEHI231 immature B cells. Mol Cell Biol 18:19–29

Munshi N, Merika M, Yie J, Senger K, Chen G, Thanos D (1998) Acetylation of HMG I(Y) by CBP turns off IFNbeta expression by disrupting the enhanceosome. Mol Cell 2:457–468

Munshi N, Agalioti T, Lomvardas S, Merika M, Chen G, Thanos D (2001) Co-ordination of a transcriptional switch by HMGI(Y) acetylation. Science 293:1133–1136

Na S, Lee S, Han S, Choi H, Im S, Lee J (1998) Steroid receptor coactivator-1 interacts with the p50 subunit and coactivates nuclear factor kB-mediated transactivations. J Biol Chem 273:10831–10834

Naar AM, Beaurang PA, Zhou S, Abraham S, Solomon W, Tjian R (1999) Composite co-activator ARC mediates chromatin-directed transcriptional activation. Nature 398:828–832

Nakajima T, Fukimizu A, Takahashi J, Gage FH, Fisher T, Blenis J, Montminy MR (1996) The signal-dependent coactivator CBP is a nuclear target for pp90rsk. Cell 86:465–474

Naumann M, Scheidereit C (1994) Activation of NF-kappa B in vivo is regulated by multiple phosphorylations. EMBO J 13:4597–4607

Neurath MF, Becker C, Barbulescu K (1998) Role of NF-kappaB in immune and inflammatory responses in the gut. Gut 43:856–860

Nissen R, Yamamoto K (2000) The glucocorticoid receptor inhibits NF-kappa B by interfering with serine-2 phosphorylation of the RNA polymerase II carboxy-terminal domain. Genes Dev 14:2314–2329

O'Connor MJ, Zimmermann H, Nielsen S, Bernard HU, Kouzarides T (1999) Characterization of an E1A-CBP interaction defines a novel transcriptional adapter motif (TRAM) in CBP/p300. J Virol 73:3574–3581

O'Neill E (1998) A new target for aspirin. Nature 396:15–17

O'Neill L (2001) Inhibiting NF-κB. Trends Immunol 22:478

Pandolfi PP (2001) Transcription therapy for cancer. Oncogene 20:3116–127

Parker D, Jhala US, Radhakrishnan I, Yaffe MB, Reyes C, Shulman AI, Cantley LC, Wright PE, Montminy M (1998) Analysis of an activator:coactivator complex reveals an essential role for secondary structure in transcriptional activation. Molecular Cell 2:353–359

Perissi V, Dasen JS, Kurokawa R, Wang ZY, Korzus E, Rose DW, Glass CK, Rosenfeld MG (1999) Factor-specific modulation of CREB-binding protein acetyltransferase activity. Proc. Natl. Acad. Sci. USA 96:3652–3657

Perkins ND (1997) Achieving transcriptional specificity with NF-kappaB. Int J Biochem Cell Biol 29:1433–1448

Perkins ND (2000) The Rel/NF-kappaB family: friend and foe. Trends Biochem Sci 25:434–440

Perkins ND, Felzien LK, Betts JC, Leung K, Beach DH, Nabel GJ (1997) Regulation of NFkB by cyclin-dependent kinases associated with the p300 coactivator. Science 275:523–527

Peters RT, Liao SM, Maniatis T (2000) IKKepsilon is part of a novel PMA-inducible IkappaB kinase complex. Mol Cell 5:513–522

Pierce JW, Read MA, Ding H, Luscinskas FW, Collins T (1996) Salicylates inhibit I kappa B-alpha phosphorylation, endothelial-leukocyte adhesion molecule expression, and neutrophil transmigration. J Immunol 156:3961–3969

Plaisance S, Vanden Berghe W, Boone E, Fiers W, Haegeman G (1997) Recombination signal sequence binding protein Jkappa is constitutively bound to the NF-κB site of the interleukin-6 promoter and is a negative regulatory factor. Molecular and Cell Biol 17:3733–3743

Pomerantz JL, Baltimore D (1999) NF-kappaB activation by a signaling complex containing TRAF2, TANK and TBK1, a novel IKK-related kinase. EMBO J 18:6694–6704

Pumiglia KM, Decker SJ (1997) Cell cycle arrest mediated by the MEK/mitogen-activated protein kinase pathway. Proc Natl Acad Sci USA 94:448–452

Rachez C, Gamble M, Chang CP, Atkins GB, Lazar MA, Freedman LP (2000) The DRIP complex and SRC-1/p160 coactivators share similar nuclear receptor binding determinants but constitute functionally distinct complexes. Mol Cell Biol 20:2718–2726

Radhakrishnan I, Perez-alvarado GC, Parker D, Dyson HJ, Montminy MR, Wright PE (1998) Solution structure of the KIX domain of CBP bound to the transactivation domain of CREB; a model for activator:coactivator interactions. Cell 91:741–752

Ray A, Prefontaine KE (1994) Physical association and functional antagonism between the p65 subunit of transcription factor NF-kappa B and the glucocorticoid receptor. Proc Natl Acad Sci USA 91:752–756

Rayet B, Gelinas C (1999) Aberrant rel/nfkb genes and activity in human cancer. Oncogene 18:6938–6947

Reichardt HM, Tuckermann JP, Gottlicher M, Vujic M, Weih F, Angel P, Herrlich P, Schutz G (2001) Repression of inflammatory responses in the absence of DNA binding by the glucocorticoid receptor. EMBO J 20:7168–7173

Roberts CJ, Nelson B, Marton MJ, Stoughton R, Meyer MR, Bennett HA, He YD, Dai H, Walker WL, Hughes TR, Tyers M, Boone C, Friend SH (2000) Signaling and circuitry of multiple MAPK pathways revealed by a matrix of global gene expression profiles. Science 287:873–880

Rogler G, Brand K, Vogl D, Page S, Hofmeister R, Andus T, Knuechel R, Baeuerle PA, Scholmerich J, Gross V (1998) Nuclear factor kappaB is activated in macrophages and epithelial cells of inflamed intestinal mucosa. Gastroenterology 115:357–369

Saccani S, Pantano S, Natoli G (2001) Two waves of nuclear factor kappaB recruitment to target promoters. J Exp Med 193:1351–1359

Sakaguchi K, Herrera JE, Saito S, Miki T, Bustin M, Vassilev A, Anderson CW, Appella E (1998) DNA damage activates p53 through a phosphorylation-acetylation cascade. Genes Dev 12:2831–2841

Sakurai H, Chiba H, Miyoshi H, Sugita T, Toriumi W (1999) IkappaB Kinases Phosphorylate NF-kappa B p65 Subunit on Serine 536 in the Transactivation Domain. J Biol Chem 274:30353–30356

Sassone-Corsi P, Mizzen CA, Cheung P, Crosio C, Monaco L, Jacquot S, Hanauer A, Allis CD (1999) Requirement of Rsk-2 for epidermal growth factor-activated phosphorylation of histone H3. Science 285:886–891

Savory JG, Hsu B, Laquian IR, Giffin W, Reich T, Hache RJ, Lefebvre YA (1999) Discrimination between NL1- and NL2-mediated nuclear localization of the glucocorticoid receptor. Mol Cell Biol 19:1025–1037

Scheinman RI, Gualberto A, Jewell CM, Cidlowski JA, Baldwin AS, Jr (1995) Characterization of mechanisms involved in transrepression of NF-kappa B by activated glucocorticoid receptors. Mol Cell Biol 15:943–953

Schiltz RL, Nakatani Y (2000) The P/CAF acetylase complex as a potential tumor suppressor. Biochim Biophys Acta 1470:M37–M53

Schmitz ML, Stelzer G, Altmann H, Meisterernst M, Baeuerle PA (1995) Interaction of the COOH-terminal transactivation domain of p65 NF-kappa B with TATA-binding protein, transcription factor IIB, and coactivators. J Biol Chem 270:7219–7226

Schmitz ML, Bacher S, Kracht M (2001) IkB-independent control of NF-kappa B activity by modulatory phosphorylations. Trends Biochem Sci 26:187–191

Schreiber S, Nikolaus S, Hampe J (1998) Activation of nuclear factor kappa B inflammatory bowel disease. Gut 42:477–484

See RH, Calvo D, Shi Y, Kawa H, Luke MP, Yuan Z (2001) Stimulation of p300-mediated transcription by the kinase MEKK1. J Biol Chem 276:16310–16317

Senftleben U, Cao Y, Xiao G, Greten FR, Krahn G, Bonizzi G, Chen Y, Hu Y, Fong A, Sun SC, Karin M (2001) Activation by IKKalpha of a second, evolutionary conserved, NF-kappa B signaling pathway. Science 293:1495–1499

Senger K, Merika M, Agalioti T, Yie J, Escalante CR, Chen G, Aggarwal AK, Thanos D (2000) Gene repression by coactivator repulsion. Mol Cell 6:931–937

Sha WC (1998) Regulation of immune responses by NF-kappa B/Rel transcription factors. J Exp Med 187:143–146

Sheppard KA, Phelps KM, Williams AJ, Thanos D, Glass CK, Rosenfeld MG, Gerritsen ME, Collins T (1998) Nuclear integration of glucocorticoid receptor and nuclear factor-kappaB signaling by CREB-binding protein and steroid receptor coactivator-1. J Biol Chem 273:29291–29294

Sheppard KA, Rose DW, Haque ZK, Kurokawa R, McInerney E, Westin S, Thanos D, Rosenfeld MG, Glass CK, Collins T (1999) Transcriptional activation by NF-kappaB requires multiple coactivators. Mol Cell Biol 19:6367–6378

Sheridan PL, Mayall TP, Verdin E, Jones KA (1997) Histone acetyltransferases regulate HIV-1 enhancer activity in vitro. Genes Dev 11:3327–3340

Shikama N, Lyon J, La Thangue NB (1997) The p300/CBP family: integrating signals with transcription factors and chromatin. Trends Cell Biol 7:230–236

Shumilla JA, Broderick RJ, Wang Y, Barchowsky A (1999) Chromium(VI) inhibits the transcriptional activity of nuclear factor-kappaB by decreasing the interaction of p65 with cAMP-responsive element-binding protein-binding protein. J Biol Chem 274:36207–36212

Sif S, Stukenberg PT, Kirschner MW, Kingston RE (1998) Mitotic inactivation of a human SWI/SNF chromatin remodeling complex. Genes Dev 12:2842–2851

Silverman N, Maniatis T (2001) NF-kappaB signaling pathways in mammalian and insect innate immunity. Genes Dev 15:2321–2342

Sizemore N, Leung S, Stark GR (1999) Activation of phosphatidylinositol 3-kinase in response to interleukin-1 leads to phosphorylation and activation of the NF-kappaB p65/RelA subunit. Mol Cell Biol 19:4798–4805

Snowden AW, Perkins ND (1998) Cell cycle regulation of the transcriptional coactivators p300 and CREB binding protein. Biochem Pharmacol 55:1947–1954

Spencer E, Jiang J, Chen ZJ (1999) Signal-induced ubiquitination of IkappaBalpha by the F-box protein Slimb/beta-TrCP. Genes Dev 13:284–294

Stein B, Yang MX (1995) Repression of the interleukin-6 promoter by estrogen receptor is mediated by NF-kappa B and C/EBP beta. Mol Cell Biol 15:4971–4979

Stein B, Baldwin AS, Ballard DW, Greene WC, Angel P, Herrlich P (1993) Cross-coupling of the NF-kappa B p65 and Fos/Jun transcription factors produces potentiated biological function. EMBO J 12:3879–3891

Stenoien DL, Mancini MG, Patel K, Allegretto EA, Smith CL, Mancini MA (2000) Subnuclear trafficking of estrogen receptor-alpha and steroid receptor coactivator-1. Mol Endocrinol 14:518–534

Stenoien DL, Patel K, Mancini MG, Dutertre M, Smith CL, O'Malley BW, Mancini MA (2001) FRAP reveals that mobility of oestrogen receptor-alpha is ligand- and proteasome-dependent. Nat Cell Biol 3:15–23

Stewart S, Crabtree GR (2000) Transcription. Regulation of the regulators. Nature 408:46–47

Stöcklin E, Wissler M, Gouilleux F, Groner B (1996) Functional interactions between Stat5 and the glucocorticoid receptor. Nature 383:726–728

Strahl BD, Allis CD (2000) The language of covalent histone modifications. Nature 403:41–45

Tai DI, Tsai SL, Chen YM, Chuang YL, Peng CY, Sheen IS, Yeh CT, Chang KS, Huang SN, Kuo GC, Liaw YF (2000) Activation of nuclear factor kappaB in hepatitis C virus infection: implications for pathogenesis and hepatocarcinogenesis [see comments]. Hepatology 31:656–664

Tak PP, Firestein GS (2001) NF-kappaB: a key role in inflammatory diseases. J Clin Invest 107:7–11

Takeda K, Takeuchi O, Tsujimura T, Itami S, Adachi O, Kawai T, Sanjo H, Yoshikawa K, Terada N, Akira S (1999) Limb and skin abnormalities in mice lacking IKKalpha. Science 284:313–316

Tetsuka T, Uranishi H, Imai H, Ono T, Sonta S, Takahashi N, Asamitsu K, Okamoto T (2000) Inhibition of nuclear factor-kappaB-mediated transcription by association with the amino-terminal enhancer of split, a Groucho-related protein lacking WD40 repeats. J Biol Chem 275:4383–4390

Thomson S, Clayton AL, Hazzalin CA, Rose S, Barratt MJ, Mahadevan LC (1999a) The nucleosomal response associated with immediate-early gene induction is mediated via alternative MAP kinase cascades: MSK1 as a potential histone H3/HMG-14 kinase. EMBO J 18:4779–4793

Thomson S, Mahadevan LC, Clayton AL (1999b) MAP kinase-mediated signalling to nucleosomes and immediate-early gene induction. Semin Cell Dev Biol 10:205–214

Tojima Y, Fujimoto A, Delhase M, Chen Y, Hatakeyama S, Nakayama K, Kaneko Y, Nimura Y, Motoyama N, Ikeda K, Karin M, Nakanishi M (2000) NAK is an IkappaB kinase-activating kinase. Nature 404:778–782

Torchia J, Rose DW, Inostroza J, Kamei Y, Westin S, Glass CK, Rosenfeld MG (1997) The transcriptional co-activator p/CIP binds CBP and mediates nuclear-receptor function. Nature 387:677–684

Torchia J, Glass C, Rosenfeld MG (1998) Co-activators and co-repressors in the integration of transcriptional responses. Curr Opin Cell Biol 10:373–383

Unlap MT, Jope RS (1997) Dexamethasone attenuates NF-kappa B DNA binding activity without inducing IkappaB levels in rat brain in vivo. Brain Res Mol Brain Res 45:83–89

Urnov FD, Wolffe AP (2001) Chromatin remodeling and transcriptional activation: the cast (in order of appearance). Oncogene 20:2991–3006

Utley RT, Ikeda K, Grant PA, Cote J, Steger DJ, Eberharter A, John S, Workman JL (1998) Transcriptional activators direct histone acetyltransferase complexes to nucleosomes. Nature 394:498–502

Valentine SA, Chen G, Shandala T, Fernandez J, Mische S, Saint R, Courey AJ (1998) Dorsal-mediated repression requires the formation of a multiprotein repression complex at the ventral silencer. Mol Cell Biol 18:6584–6594

Van de Stolpe A, Caldenhoven E, Raaijmakers J, Van der Saag P, Koenderman L (1993) Glucocorticoid-mediated repression of intercellular adhesion molecule-1 expression in human monocytic and bronchial epithelial cell lines. Am J Respir Cell Mol Biol 8:340–347

Van de Stolpe A, Caldenhoven E, Stade BG, Koenderman L, Raaijmakers JA, Johnson JP, van der Saag PT (1994) 12-O-tetradecanoylphorbol-13-acetate- and tumor necrosis factor alpha-mediated induction of intercellular adhesion molecule-1 is inhibited by dexamethasone. Functional analysis of the human intercellular adhesion molecular-1 promoter. J Biol Chem 269:6185–6192

Van Lint C, Emiliani S, Ott M, Verdin E (1996) Transcriptional activation and chromatin remodeling of the HIV-1 promoter in response to histone acetylation. EMBO J 15:1112–1120

Vanden Berghe W, Plaisance S, Boone E, De Bosscher K, Schmitz ML, Fiers W, Haegeman G (1998) p38 and extracellular signal-regulated kinase mitogen-activated protein kinase pathways are required for nuclear factor-kappaB p65 transactivation mediated by tumor necrosis factor. J Biol Chem 273:3285–3290

Vanden Berghe W, De Bosscher K, Boone E, Plaisance S, Haegeman G (1999a) The nuclear factor-kappaB engages CBP/p300 and histone acetyltransferase activity for transcriptional activation of the interleukin-6 gene promoter. J Biol Chem 274:32091–3208

Vanden Berghe W, Francesconi E, De Bosscher K, Rèsche-Rigon M, Haegeman G (1999b) Dissociated glucocorticoids with anti-inflammatory potential repress interleukin-6 gene expression by an NF-kappa-B-dependent Mechanism. Mol Pharmacol 56:797–806

Vanden Berghe W, Vermeulen L, De Wilde G, De Bosscher K, Boone E, Haegeman G (2000) Signal transduction by tumor necrosis factor and gene regulation of the inflammatory cytokine interleukin-6. Biochem Pharmacol 60:1185–1195

Vane JR, Botting RM (1998) Anti-inflammatory drugs and their mechanism of action. Inflamm Res 47:S78–S87

Vane JR, Botting RM (2000) The future of NSAID therapy: selective COX-2 inhibitors. Int J Clin Pract 54:7–9

Vane JR, Flower RJ, Botting RM (1990) History of aspirin and its mechanism of action. Stroke 21:IV12–IV23

Vayssière BM, Dupont A, Petit F, Garcia T, Marchandeau C, Gronemeyer H, Resche-Rigon M (1997) Synthetic glucocorticoids that dissociate transactivation and AP-1 transrepression exhibit anti-inflammatory activity in vivo. Mol Endocrinol 11:1245–1255

Versaw WK, Blank V, Andrews NM, Bresnick EH (1998) Mitogen-activated protein kinases enhance long-range activation by the beta-globin locus control region. Proc Natl Acad Sci USA 95:8756–8760

Vuillard L, Nicholson J, Hay RT (1999) A complex containing betaTrCP recruits Cdc34 to catalyse ubiquitination of IkappaBalpha. FEBS Lett 455:311–314

Waddick KG, Uckun FM (1999) Innovative treatment programs against cancer: II. Nuclear factor-kappaB (NF-kappaB) as a molecular target. Biochem Pharmacol 57:9–17

Wadgaonkar R, Phelps KM, Haque Z, Williams AJ, Silverman ES, Collins T (1999) CREB-binding protein is a nuclear integrator of nuclear factor-kappaB and p53 signaling. J Biol Chem 274:1879–1882

Wan Y, Coxe KK, Thackray VG, Housley PR, Nordeen SK (2001) Separable features of the ligand-binding domain determine the differential subcellular localization and ligand-binding specificity of glucocorticoid receptor and progesterone receptor. Mol Endocrinol 15:17–31

Wang D, Baldwin ASJ (1998) Activation of nuclear factor-kappaB-dependent transcription by tumor necrosis factor-alfa is mediated through phosphorylation of RelA/p65 on serine 529. J Biol Chem 273:29411–29416

Wang LS, Chow KC, Wu CW (1999) Expression and up-regulation of interleukin-6 in oesophageal carcinoma cells by n-sodium butyrate. Br J Cancer 80:1617–1622

Wang D, Westerheide S, Hanson J, Baldwin AJ (2000) TNFalpha-induced phosphorylation of RelA/p65 on Ser529 is Controlled by casein kinase II. J Biol Chem 275:32592–32597

Webster-J Jewell-Cm Bodwell-Je Munck A, Sar M, Cidlowski-J (1997) Mouse glucocorticoid receptor phosphorylation status influences multiple functions of the receptor protein. J Biol Chem 272:9287–9293

Weinmann AS, Plevy SE, Smale ST (1999) Rapid and selective remodeling of a positioned nucleosome during the induction of IL-12 p40 transcription. Immunity 11:665–675

Weinmann AS, Mitchell DM, Sanjabi S, Bradley MN, Hoffmann A, Liou HC, Smale ST (2001) Nucleosome remodeling at the IL-12 p40 promoter is a TLR-dependent, rel-independent event. Nat Immunol 2:51–57

Wells WA (2001) Histones rule! The FASEB conference on chromatin and transcription. 7–12 July 2001. J Cell Biol 154:906–907

Wen Y, Yan DH, Spohn B, Deng J, Lin SY, Hung MC (2000) Tumor suppression and sensitization to tumor necrosis factor alpha-induced apoptosis by an interferon-inducible protein, p202, in breast cancer cells. Cancer Res 60:42–46

West MJ, Lowe AD, Karn J (2001) Activation of human immunodeficiency virus transcription in T cells revisited: NF-kappaB p65 stimulates transcriptional elongation. J Virol 75:8524–8537

White DW, Gilmore TD (1998) Bcl-2 and CrmA have different effects on transformation, apoptosis and the stability of IkB-a in chicken spleen cells transformed by temperature-sensitive v-rel oncoproteins. Oncogene 13:891–899

Whitmarsh AJ, Davis RJ (2000) A central control for cell growth. Nature 403:255–256

Wissink S, van Heerde EC, Schmitz ML, Kalkhoven E, van der Burg B, Baeuerle PA, van der Saag PT (1997) Distinct domains of the RelA NF-kappaB subunit are required for negative cross-talk and direct interaction with the glucocorticoid receptor. J Biol Chem 272:22278–22284

Wissink S, van Heerde EC, vand der Burg B, van der Saag PT (1998) A dual mechanism mediates repression of NF-kappaB activity by glucocorticoids. Mol Endocrinol 12:355–363

Witt O, Sand K, Pekrun A (2000) Butyrate-induced erythroid differentiation of human K562 leukemia cells involves inhibition of ERK and activation of p38 MAP kinase pathways. Blood 95:2391–2396

Wolffe AP (2001) Chromatin remodeling: why it is important in cancer. Oncogene 20:2988–2990

Wu WH, Hampsey M (1999) Transcription: common cofactors and cooperative recruitment. Curr Biol 9:R606–R609

Xu L, Lavinsky RM, Dasen JS, Flynn SE, McInerney EM, Mullen TM, Heinzel T, Szeto D, Korzus E, Kurokawa R, Aggarwal AK, Rose DW, Glass CK, Rosenfeld MG (1998) Signal-specific co-activator domain requirements for Pit-1 activation. Nature 395:301–306

Yamamoto Y, Gaynor RB (2001) Therapeutic potential of inhibition of the NF-kappaB pathway in the treatment of inflammation and cancer. J Clin Invest 107:135–142

Yamit-Hezi A, Dikstein R (1998) TAF(II)105 mediates activation of anti-apoptotic genes by NF-kappa B. EMBO J 17:5161–5169

Yan F, Polk DB (1999) Aminosalicylic acid inhibits IkappaB kinase alpha phosphorylation of IkappaBalpha in mouse intestinal epithelial cells. J Biol Chem 274:36631–36636

Yaron A, Hatzubai A, Davis M, Lavon I, Amit S, Manning-Am Andersen-Js Mann M, Mercurio F, Ben-Neriah Y (1998) Identification of the receptor component of the IkappaBalpha-ubiquitin ligase. Nature 396:590–594

Yin MJ, Yamamoto Y, Gaynor RB (1998) The anti-inflammatory agents aspirin and salicylate inhibit the activity of IkappaB kinase-beta. Nature 396:77–80

Young PR (1998) Pharmacological modulation of cytokine action and production through signaling pathways. Cytokine Growth Factor Rev 9:239–257

Yuan LW, Gambee JE (2000) Phosphorylation of p300 at serine 89 by protein kinase C. J Biol Chem 275:40946–40951

Zanger K, Cohen LE, Hashimoto K, Radovick S, Wondisford FE (1999) A novel mechanism for cyclic adenosine 3′,5′-monophosphate regulation of gene expression by CREB-binding protein. Mol Endocrinol 13:268–275

Zhang GJ, Adachi I (1999) Serum interleukin-6 levels correlate to tumor progression and prognosis in metastatic breast carcinoma. Anticancer Res 19:1427–432

Zhang Y, Reinberg D (2001) Transcription regulation by histone methylation: interplay between different covalent modifications of the core histone tails. Genes Dev 15:2343–2360

Zhong H, SuYang H, Erdjument-Bromage H, Tempst P, Ghosh S (1997) The transcriptional activity of NF-kappaB is regulated by the IkappaB-associated PKAc subunit through a cyclic AMP-independent mechanism. Cell 89:413–424

Zhong H, Voll RE, Ghosh S (1998) Phoshorylation of NF-kappaB p65 by PKA stimulates transcriptional activity by promoting a novel bivalent interaction with the coactivator CBP/p300. Mol Cell 1:661–671

Zhu W, Downey JS, Gu J, Di Padova F, Gram H, Han J (2000) Regulation of TNF expression by multiple mitogen-activated protein kinase pathways. J Immunol 164:6349–6358

15 DNA-Dependent Cofactor Selectivity of the Glucocorticoid Receptor

A. Dostert, T. Heinzel

15.1 Introduction 279
15.2 Ligand-Dependent GR Interaction with Known Cofactors 284
15.3 Generation of GR-DNA Complexes 286
15.4 Discussion 289
References 291

15.1 Introduction

The glucocorticoid receptor (GR) is one of the best-characterized steroid receptors and essential for the regulation of multiple physiological processes. Due to their multiple inhibitory effects on the immune system, glucocorticoids are especially suited for treatment of inflammations and autoimmune diseases, but prolonged therapy results in severe metabolic side effects due to alterations in glucose and lipid metabolism. Further side effects include osteoporosis, atrophy of the skin, myopathy, and psychosis. In the absence of its ligand, the GR is localized in the cytoplasm as part of a multiprotein complex composed of GR, heat-shock proteins, and immunophilins. Binding of glucocorticoids to the receptor induces release of GR from this complex and subsequent translocation to the nucleus. This leads to binding of receptor dimers to specific DNA motifs in the regulatory regions of target genes and to activation of their transcription (Beato et al. 1995; Mangelsdorf et al. 1995). The function of dimeric receptors is also influenced considerably by neighboring transcription factors (Leers et al. 1994; Burcin et al. 1997). The GR shows further types of action, which do not require the binding of receptor dimers to DNA-recognition sequences (Heck et al. 1994; Reichardt et al. 1998). Binding of the ligand to the receptor leads in most of these cases to repressed activity of genes which are activated through other transcription factors, for example during the immunogenic reaction. This transcriptional crosstalk with transcription factors like AP-1, NF-κB or Stat5 occurs without direct binding to DNA; presumably on the basis of protein-protein interactions with the DNA-bound transcription factors (Jonat et al. 1990; Lucibello et al. 1990; Yang-Yen et al. 1990; Schüle and Evans 1991; König et al. 1992; Stöcklin et al. 1996; Heck et al. 1997). In another crosstalk mechanism (the proliferin gene), the GR presumably binds as a monomer along with other transcription factors to a "half" recognition sequence in a "composite element" (Diamond et al. 1990). Furthermore, negative response elements (nGREs; nTREs) have been described for glucocorticoid and also thyroid hormone receptors (Drouin et al. 1993; Saatcioglu et al. 1994; Lefstin and Yamamoto 1998; Awad et al. 1999). The molecular mechanism of all of these forms of gene repression via GR remains unclear.

Multiple receptors for non-steroidal hormones do not bind to heat-shock proteins (in contrast to steroid hormone receptors) but reside in the nucleus, even in the absence of their specific ligands. They have the ability to repress genes in the absence of hormone in an active manner (silencing) (Baniahmad et al. 1992). Hormone binding to non-steroidal nuclear receptors also leads to a conformational change and usually to activation. Numerous results indicate that the activation and repression of target genes via nuclear hormone receptors requires additional coactivators or corepressors, which interact with the receptors in a ligand-dependent or -sensitive manner. Coactivators include the proteins of the SRC-1/p160 family and CBP/p300, which not only possess histone acetylase activity but also exist in a multiprotein complex together with the histone acetylase p/CAF (Onate et al. 1995; Chakravarti et al. 1996; Kamei et al. 1996; Voegel et al. 1996; Yang et al. 1996; Chen et al. 1997; Torchia et al. 1997). Hyperacetylation of histones H3 and H4 through coactivator complexes leads to decondensation of the chromatin structure. It is, therefore, an important aspect in the molecular mechanism of gene activation.

Many nuclear receptors interact in the absence of ligand (e.g., thyroid hormone receptor or retinoic acid receptor) or after binding of antagonists (e.g., progesterone and estrogen receptor) with corepressors and thus inhibit transcription. The related proteins N-CoR and SMRT, as well as SunCoR, Alien and SMRTER have been identified as corepressors (Chen and Evans 1995; Hörlein et al. 1995; Zamir et al. 1997; Dressel et al. 1999; Tsai et al. 1999). Corepressors can be part of multiprotein complexes which also contain histone deacetylases (Alland et al. 1997; Heinzel et al. 1997; Nagy et al. 1997). The exchange of corepressor and coactivator complexes with opposite chromatin-modifying enzyme activity is therefore a possible mechanism for the ligand induced switch of nuclear receptors between repression and activation (Pazin and Kadonaga 1997; Xu et al. 1999; Glass and Rosenfeld 2000).

Coactivators and corepressors contain compact sequence motifs which mediate the binding to nuclear hormone receptors. In coactivators these motifs have the consensus sequence LxxLL, which interacts as an α-helix with a hydrophobic groove of the ligand binding domain (LBD) surface (Heery et al. 1997; Torchia et al. 1997). This surface consists of helices 3, 5, and 12. Corepressors contain an extended α-helical interaction motif with the consensus sequence LxxxIxxxI/L (Hu and Lazar 1999; Perissi et al. 1999). The positioning of helix 12 determines whether coactivators or corepressors can bind (Moras and Gronemeyer 1998; Glass and Rosenfeld 2000). The type of ligand (agonist or antagonist) and the response element both influence the three-dimensional structure of the LBD and therefore the position of helix 12. The conformational change of helix 12 is also responsible for the hormone-induced dissociation of corepressors (Baniahmad et al. 1995, 1997).

Negative response elements of GR (nGREs) have been described for numerous genes. Well-characterized nGREs include the binding elements in the promoters of the pro-opiomelanocortin (POMC) (Drouin et al. 1993), osteocalcin (Strömstedt et al. 1991), α-fetoprotein (Poliard et al. 1990; Turcotte et al. 1990), and prolactin genes (Sakai et al. 1988; Subramaniam et al. 1997). POMC is the common precursor of a variety of important endocrine peptides including adrenocorticotropic hormone (ACTH). Transcription of the POMC gene is positively regulated by CRF (corticotropin releasing factor) through cyclic adenosine monophosphate (cAMP)-responsive regions. Glucocorticoids inhibit transcription of the POMC gene in the anterior pituitary and regulate the

POMC	C T G C C A G G A A G G T C A	C G T C C A	A G G C T C A
Prolactin	C T C A A A C T A G T C T C C	A G A T C T	C A C C A T C
α-Fetoprotein	C C C C A A A G A G C T C T G	T G T C C T	T G A A C A T
CRH	A A A A A T T T T T G T C A A	T G G A C A	A G T C A T A
Osteocalcin	C C A G A G G G T A T A A A C	A G T G C T	G G A G G C T
Interleukin 1β	T G C C C C A G C C A A G A A	A G G T C A	A T T T T C T
Proliferin (plfG)	T G G C T A C T C A C A G T A	T G A T T T	G T T T T A G
A	1 1 2 2 4 4 2 - 3 3 1 3 1 3 4	3 - 2 1 - 3	3 1 2 1 2 2 2
C	4 2 3 4 2 1 2 - 2 1 2 1 2 2 2	1 - - 2 6 -	1 - 1 3 1 3 1
G	- 2 2 1 - 1 2 4 - 2 2 1 2 - 1	- 7 2 1 - -	2 4 1 1 1 - 1
T	2 2 - - 1 1 1 3 2 1 2 2 2 2 -	3 - 3 3 1 4	1 2 3 2 3 2 3

Fig. 1. Sequence alignment of negative GR response elements (*nGREs*). The sequences of different nGREs derived from footprint data are compared. The conserved half site is highlighted in *gray*, the non-conserved half side is *boxed*

negative-feedback circuit at the level of the pituitary. The nGRE of the POMC gene is localized in the proximal region of the promoter (Drouin et al. 1989). The complex which forms with the nGRE may contain three GR molecules instead of a homodimer (Drouin et al. 1993).

Osteocalcin is the major non-collagenous bone protein. The corresponding gene is only expressed post-proliferatively during differentiation of normal osteoblasts, when mineralization of the extracellular matrix is occurring (Hauschka et al. 1989). The osteocalcin gene is transcriptionally regulated by a number of hormones and vitamins. It has been shown that the active metabolite of vitamin D induces osteocalcin gene transcription while glucocorticoids repress both basal-level transcription and vitamin D-induced levels (Morrison et al. 1989). Thus, the repressive glucocorticoid effect is dominant over the positive effect exerted by vitamin D. In vitro analysis has identified a glucocorticoid receptor-binding element in the human osteocalcin promoter that overlaps the TATA box (Strömstedt et al. 1991). The occupation of the dual binding sites by GR presumably prevents triiodothyronine-binding protein (TBP) from binding to the TATA box and disrupts the assembly of an active preinitiation complex.

Studies using a dimerization-defective GR mutant have shown that repression of the POMC-, prolactin- and α-fetoprotein genes requires

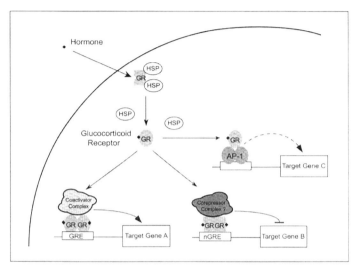

Fig. 2. Mechanisms of gene regulation by glucocorticoid receptor. In the absence of ligand, the receptor is localized in the cytoplasm as part of a multiprotein complex composed of GR, heat-shock proteins, and immunophilins. Hormone binding to the receptor induces dissociation of the complex and subsequent translocation to the nucleus. The ligand-activated receptor dimer activates gene expression by binding to specific DNA sequences (glucocorticoid response element, *GRE*) in the promoter regions of glucocorticoid-regulated genes. In contrast to the regulation of these classical GREs, repression of negatively regulated target genes is mediated by negative GREs (*nGRE*), composite GREs (not shown) or transrepression of transcription factors like AP-1 or NF-κB. This could be due to the recruitment of corepressor complexes

direct DNA binding of GR and is not based on transrepression of other transcription factors (Reichardt et al. 1998; H. Reichardt and G. Schütz, personal communication). A dimerization-defective GR which was generated by a mutation within the D loop (A485 T) of the DNA-binding domain (DBD) fails to bind DNA but can still repress activator protein 1 (AP)-regulated genes (Heck et al. 1994). Thus, transactivation can be dissociated from transrepression in cell culture. Reichardt et al. have generated mice harboring this point mutation (Reichardt et al. 1998). Those GR$^{dim/dim}$ mice are viable in contrast to GR$^{-/-}$ mice, which die shortly after birth due to defects in lung maturation. The most striking conclusion of these experiments is that GRE-mediated gene activation is not essential for development and survival. GR$^{dim/dim}$ mice are also defective in repression of genes that are regulated through nGREs. Thus, expression of POMC and prolactin is elevated in those mice. On the other hand, GR-mediated interference with AP-1 activity and repression of AP-1-regulated genes were not affected.

A comparison of GR response elements shows, that GREs and nGREs have different consensus sequences (Beato et al. 1989). The classic GREs consist of palindromic recognition sequences with the consensus GGTACA nnn TGTTCT, whereas only the 3' response element is highly conserved in nGREs (Fig. 1). In the traditional view, DBDs serve only to localize modulatory domains to the vicinity of regulated genes. However, it has been suggested that not only ligands but also the different response elements can act as an allosteric effector for the nuclear receptor (Lefstin and Yamamoto 1998). According to this model, response elements contain important information which is interpreted by the receptor. We speculate that even in the presence of an agonistic ligand, the response element could induce a receptor conformation which is permissive to the binding of corepressors (Fig. 2).

In order to test this hypothesis, we developed an assay system (AB-CDE) in which GR bound to different biotinylated GREs and nGREs is used as an affinity matrix for specific cofactor complexes. Therefore, those potential cofactors can be enriched in a context which most likely resembles the structure of GR on promoters within the cell.

15.2 Ligand-Dependent GR Interaction with Known Cofactors

To investigate whether GR binds to biotinylated response elements in a ligand-dependent manner, a good source of GR was required. The production of active GR can be problematic because of its hydrophobicity and tendency to precipitation. We decided to use two different sources: GR from mouse brain whole-cell extracts and human GR recombinantly expressed in Sf9 insect cells. Mouse brain whole-cell extract was incubated with different ligands and in vitro translated SRC-1 coactivator protein (Fig. 3A). The co-immunoprecipitation (CoIP) with anti-GR antibody shows a dexamethasone-dependent bind-

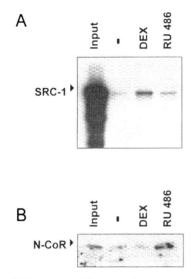

Fig. 3A, B. CoIP of GR and established cofactors of nuclear receptors. **A** Mouse brain whole-cell extract was incubated without ligand (–), in the presence of the GR agonist dexamethasone (*DEX*), or in the presence of the GR antagonist RU486 with in vitro translated, ^{35}S-labeled SRC-1. After CoIP with an anti GR antibody (Santa Cruz, M20) and sodium dodecyl sulfate (SDS) gel electrophoresis, the bound SRC-1 protein was detected by autoradiography. **B** Mouse brain whole-cell extract was incubated without ligand (–), in the presence of the GR agonist dexamethasone (*DEX*) or in the presence of the GR antagonist RU486 with an anti GR antibody (Santa Cruz, M20). After SDS gel electrophoresis the bound endogenous N-CoR was detected in Western blot

ing of SRC-1 to GR. The remaining binding activity in the absence of ligand and in the presence of the GR antagonist RU486 is presumably due to endogenous corticosteroids existing in the cell extract. Some steroid receptors such as the estrogen and progesterone receptors can interact with N-CoR in the antagonist conformation and therefore act as repressors of transcription. Since GR has a high degree of homology with the progesterone receptor, we investigated whether GR and N-CoR can interact after binding of an antagonistic ligand to GR (Fig. 3B). This result shows that GR can exist in a conformation compatible with corepressor binding at least in solution. The induction of such a conformation through allosteric effects, therefore, also seems plausible.

15.3 Generation of GR-DNA Complexes

15.3.1 ABCD Assay

The detection of DNA-dependent cofactor binding requires the generation of GR-DNA complexes which can serve as an affinity matrix. The avidin biotin complex DNA (ABCD) assay was originally developed for the investigation of DNA binding properties of thyroid hormone, retinoic acid, and retinoid X receptors (TR, RAR, and RXR) (Glass et al. 1988) and used successfully for the detection of coactivator and corepressor binding (Hörlein et al. 1995; Kurokawa et al. 1995; Kamei et al. 1996). The ABCD assay is based on the immobilization of protein DNA complexes via binding of a biotinylated oligonucleotide to a streptavidin matrix (Fig. 4, I-IV). Whole-cell extract from baculovirus-infected SF-9-cells containing recombinant human GR is incubated with buffer H (100 mM KCl; 20 mM Hepes pH 7,8–7,9; 20% glycerol; 1 mM DTT; 0,1% NP40), biotinylated oligo, herring sperm DNA (competitor) and dexamethasone 10^{-5} M for 1 h on ice. After the addition of equilibrated streptavidin agarose beads, incubation continues for an additional 30 min at 4°C on a rotator. Then the beads are washed repeatedly with buffer H-containing 50 mM KCl. The washed beads with the oligo-bound hGR are optionally incubated with HeLa- or 293T-whole-cell extract for 1 h at 4°C and washed again prior to boiling with sample-buffer and loading to a SDS-Gel.

To establish assay conditions, we used biotinylated binding elements which correspond to the classical GRE of the promoter of the tyrosine aminotransferase gene (TAT) and the nGRE of the α-fetoprotein promoter (AFP) (Fig. 5). The elements where synthesized both as a monomer and as a dimer. Samples without any binding element and a binding element for the cAMP-response element binding (CREB) protein (CRE) respectively served as a negative control. In this experiment, mouse brain whole-cell extract was used as a source for glucocorticoid receptor. Up to 20% of the receptor input is bound specifically to GREs and nGREs. However, the affinity for the AFP element is significantly lower than for the classical TAT GRE. The use of the dimerized AFP element (AFP-D) leads to enhanced binding. In subsequent experiments we were able to immobilize up to 50% of the input material using baculovirus-expressed GR (data not shown).

DNA-Dependent Cofactor Selectivity of the GR

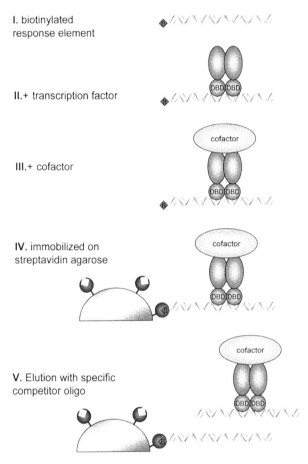

Fig. 4. ABCDE assay. A biotinylated response element (nGRE) is incubated with a cell extract containing GR (and potential cofactors) which bind to the recognition sequence. The complex is immobilized on streptavidin agarose and specifically eluted with a non-biotinylated competitor oligonucleotide

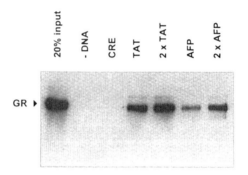

Fig. 5. GR binding to biotinylated oligonucleotides. The indicated biotinylated binding elements were incubated with mouse brain whole-cell extract in the presence of dexamethasone. Subsequent to the immobilization with streptavidin agarose and washing, the bound protein was eluted with sample buffer. GR was detected by Western blot (Santa Cruz M20 antibody)

15.3.2 ABCDE Assay

Since preliminary experiments showed that numerous proteins from different cell extracts can bind independently of the DNA element to streptavidin agarose and therefore interfere with the detection of specific cofactors, further assay development became necessary. As the binding of GR to the oligonucleotide is completely abolished at buffer concentrations higher than 200 mM KCl, we decided to elute the bound proteins with KCl in buffer H instead of the usual denaturing SDS sample buffer. Significant elution occurred at final concentrations around 250 mM (data not shown). Elution (ABCDE) of specifically bound proteins with non-biotinylated competitor oligonucleotide (GRE derived from the MMTV promoter) in a tenfold excess proved a more promising approach (Fig. 4, step V). In these experiments the elution of GR is detectable not only in Western blot but also with silver staining because of the very low background of non-specifically bound proteins (Fig. 6). Therefore, the detection of specific cofactors in the one- or two-dimensional gel electrophoresis appears to be realistic.

Having established favorable assay conditions, we used GR-DNA complexes as an affinity matrix in order to bind potential cofactors from HeLa- or 293T-cell-derived whole-cell extracts (Fig. 7). We could show that the elution with competitor DNA results in interpretable patterns of bands. The silver staining shows specific protein bands which occur

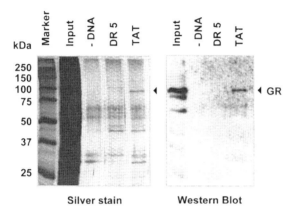

Fig. 6. Elution of glucocorticoid receptor using specific non-biotinylated competitor oligo nucleotides. The indicated biotinylated binding elements were incubated with baculovirus-expressed GR in the presence of dexamethasone (DEX). Subsequent immobilization with streptavidin agarose and washing, the bound proteins were eluted via competition with a tenfold excess of non-biotinylated mouse mammary tumor virus (MMTV)-GRE oligonucleotide. After SDS gel electrophoresis, a part of the gel was silver stained and GR was detected in parallel in Western blot (Santa Cruz P20 antibody)

only in the presence of GR and a GR-binding element, but not in negative controls (either without DNA or with a DR5 oligo which does not bind GR).

15.4 Discussion

Detection of DNA-dependent cofactors requires production of GR-DNA complexes which serve as an affinity matrix. Gelshift experiments are not suited for that purpose because it is known that GR does not create a specific gelshift unless an anti-GR antibody is added (supershift). Furthermore, only small amounts of DNA-protein complexes can be obtained. The ABCD assay was originally developed to investigate the DNA-binding properties of thyroid, retinoic acid, and retinoid X receptor (Glass et al. 1988) and successfully used for the detection of coactivators and corepressors (Hörlein et al. 1995; Kurokawa et al.

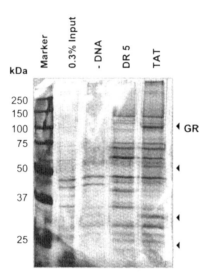

Fig. 7. Elution of GR-associated proteins. The indicated biotinylated binding elements were incubated with baculovirus-expressed GR in absence of dexamethasone. Subsequent immobilization with streptavidin agarose and washing, the GR-DNA complex was used as an affinity matrix for the binding of potential cofactors from HeLa whole-cell extracts. After further washing, the bound proteins were eluted with a tenfold excess of the MMTV-GRE oligonucleotide. Protein bands were detected by silver staining. The bands which occur only together with GR on a specific binding element (*TAT*) but not in the control elements are marked with *triangles*

1995; Kamei et al. 1996). In order to achieve less background and to use the assay for detection of unknown cofactors via mass spectrometry, we developed the ABCDE assay (avidin biotin complex DNA elution). It provides a useful tool for the investigation and comparison of different nGREs and enables us to test the hypothesis that the conformation of the DNA-binding domain could have an influence on the transcriptional regulatory surfaces in the native context.

The retinoic acid receptor provides a well-characterized example which illustrates how different response elements can change receptor activity. RAR like TR and vitamin D receptor VDR usually binds to DNA as a heterodimer with RXR. A typical response element consists of two repeats of sequences related to AGGTCA, with variable spacing between the two motifs. Differentially spaced half-site DNA elements produce distinct transcriptional responses from RAR. Similarly, differences in the response elements may cause the GR to activate transcription in the context of a GRE and repress transcription in the context of an nGRE. Hence, DNA can act as an allosteric modulator whose binding alters the transcription factor's affinity for interacting proteins, such as coactivators or corepressors.

Even though it was shown that the different domains of nuclear receptors can act independently, substantial evidence indicates that they are functionally integrated with the DBD of the native receptor. This would require a specific functional connection of the DNA-binding and transcriptional-regulatory surfaces which transmits information from the response element. It is conceivable that the conformation of the LBD is altered by different spatial arrangements of the homodimerized DBD via the hinge region and helix 1 of the LBD. This could provide an effective mechanism to differentially regulate a large number of GR-dependent genes at any given concentration of endogenous corticosteroids.

Acknowledgements. We would like to thank ChengKon Shih and Susan Goldrick for recombinantly expressed GR.

References

Alland L, Muhle R, Hou H Jr, Potes J, Chin L, Schreiber-Agus N, DePinho RA (1997) Role for N-CoR and histone deacetylase in Sin3-mediated transcriptional repression. Nature 387:49–55

Awad TA, Bigler J, Ulmer JE, Hu YJ, Moore JM, Lutz M, Neiman PE, Collins SJ, Renkawitz R, Lobanenkov VV, Filippova GN (1999) Negative transcriptional regulation mediated by thyroid hormone response element 144 requires binding of the multivalent factor CTCF to a novel target DNA sequence. J Biol Chem 274:27092–27098

Baniahmad A, Kohne AC, Renkawitz R (1992) A transferable silencing domain is present in the thyroid hormone receptor, in the v-erbA oncogene product and in the retinoic acid receptor. EMBO J 11:1015–1023

Baniahmad A, Leng X, Burris TP, Tsai SY, Tsai MJ, O'Malley BW (1995) The tau 4 activation domain of the thyroid hormone receptor is required for release of a putative corepressor(s) necessary for transcriptional silencing. Mol Cell Biol 15:76–86

Baniahmad A, Thormeyer D, Renkawitz R (1997) tau4/tau c/AF-2 of the thyroid hormone receptor relieves silencing of the retinoic acid receptor silencer core independent of both tau4 activation function and full dissociation of corepressors. Mol Cell Biol 17:4259–4271

Beato M, Chalepakis G, Schauer M, Slater EP (1989) DNA regulatory elements for steroid hormones. J Steroid Biochem 32:737–747

Beato M, Herrlich P, Schutz G (1995) Steroid hormone receptors: many actors in search of a plot. Cell 83:851–857

Burcin M, Arnold R, Lutz M, Kaiser B, Runge D, Lottspeich F, Filippova GN, Lobanenkov VV, Renkawitz R (1997) Negative protein 1, which is required for function of the chicken lysozyme gene silencer in conjunction with hormone receptors, is identical to the multivalent zinc finger repressor CTCF. Mol Cell Biol 17:1281–1288

Chakravarti D, LaMorte VJ, Nelson MC, Nakajima T, Schulman IG, Juguilon H, Montminy M, Evans RM (1996) Role of CBP/P300 in nuclear receptor signalling. Nature 383:99–103

Chen JD, Evans RM (1995) A transcriptional co-repressor that interacts with nuclear hormone receptors. Nature 377:454–457

Chen H, Lin RJ, Schiltz RL, Chakravarti D, Nash A, Nagy L, Privalsky ML, Nakatani Y, Evans RM (1997) Nuclear receptor coactivator ACTR is a novel histone acetyltransferase and forms a multimeric activation complex with P/CAF and CBP/p300. Cell 90:569–580

Diamond MI, Miner JN, Yoshinaga SK, Yamamoto KR (1990) Transcription factor interactions: selectors of positive or negative regulation from a single DNA element. Science 249:1266–1272

Dressel U, Thormeyer D, Altincicek B, Paululat A, Eggert M, Schneider S, Tenbaum SP, Renkawitz R, Baniahmad A (1999) Alien, a highly conserved protein with characteristics of a corepressor for members of the nuclear hormone receptor superfamily. Mol Cell Biol 19:3383–3394

Drouin J, Trifiro MA, Plante RK, Nemer M, Eriksson P, Wrange O (1989) Glucocorticoid receptor binding to a specific DNA sequence is required for hormone-dependent repression of pro-opiomelanocortin gene transcription. Mol Cell Biol 9:5305–5314

Drouin J, Sun YL, Chamberland M, Gauthier Y, De Lean A, Nemer M, Schmidt TJ (1993) Novel glucocorticoid receptor complex with DNA element of the hormone-repressed POMC gene. EMBO J 12:145–156

Glass CK, Rosenfeld MG (2000) The coregulator exchange in transcriptional functions of nuclear receptors. Genes Dev 14:121–141

Glass CK, Holloway JM, Devary OV, Rosenfeld MG (1988) The thyroid hormone receptor binds with opposite transcriptional effects to a common sequence motif in thyroid hormone and estrogen response elements. Cell 54:313–323

Hauschka PV, Lian JB, Cole DE, Gundberg CM (1989) Osteocalcin and matrix Gla protein: vitamin K-dependent proteins in bone. Physiol Rev 69:990–1047

Heck S, Kullmann M, Gast A, Ponta H, Rahmsdorf HJ, Herrlich P, Cato AC (1994) A distinct modulating domain in glucocorticoid receptor monomers in the repression of activity of the transcription factor AP-1. EMBO J 13:4087–4095

Heck S, Bender K, Kullmann M, Gottlicher M, Herrlich P, Cato AC (1997) I kappaB alpha-independent downregulation of NF-kappaB activity by glucocorticoid receptor. EMBO J 16:4698–4707

Heery DM, Kalkhoven E, Hoare S, Parker MG (1997) A signature motif in transcriptional co-activators mediates binding to nuclear receptors. Nature 387:733–736

Heinzel T, Lavinsky RM, Mullen TM, Soderstrom M, Laherty CD, Torchia J, Yang WM, Brard G, Ngo SD, Davie JR, Seto E, Eisenman RN, Rose DW, Glass CK, Rosenfeld MG (1997) A complex containing N-CoR, mSin3 and histone deacetylase mediates transcriptional repression. Nature 387:43–48

Hörlein AJ, Naar AM, Heinzel T, Torchia J, Gloss B, Kurokawa R, Ryan A, Kamei Y, Soderstrom M, Glass CK, et al. (1995) Ligand-independent repression by the thyroid hormone receptor mediated by a nuclear receptor co-repressor. Nature 377:397–404

Hu X, Lazar MA (1999) The CoRNR motif controls the recruitment of corepressors by nuclear hormone receptors. Nature 402:93–96

Jonat C, Rahmsdorf HJ, Park KK, Cato AC, Gebel S, Ponta H, Herrlich P (1990) Antitumor promotion and antiinflammation: down-modulation of AP-1 (Fos/Jun) activity by glucocorticoid hormone. Cell 62:1189–1204

Kamei Y, Xu L, Heinzel T, Torchia J, Kurokawa R, Gloss B, Lin SC, Heyman RA, Rose DW, Glass CK, Rosenfeld MG (1996) A CBP integrator complex mediates transcriptional activation and AP-1 inhibition by nuclear receptors. Cell 85:403–414

König H, Ponta H, Rahmsdorf HJ, Herrlich P (1992) Interference between pathway-specific transcription factors: glucocorticoids antagonize phorbol ester-induced AP-1 activity without altering AP-1 site occupation in vivo. EMBO J 11:2241–2246

Kurokawa R, Soderstrom M, Horlein A, Halachmi S, Brown M, Rosenfeld MG, Glass CK (1995) Polarity-specific activities of retinoic acid receptors determined by a co-repressor. Nature 377:451–454

Leers J, Steiner C, Renkawitz R, Muller M (1994) A thyroid hormone receptor-dependent glucocorticoid induction. Mol Endocrinol 8:440–447

Lefstin JA, Yamamoto KR (1998) Allosteric effects of DNA on transcriptional regulators. Nature 392:885–888

Lucibello FC, Slater EP, Jooss KU, Beato M, Muller R (1990) Mutual transrepression of Fos and the glucocorticoid receptor: involvement of a functional domain in Fos which is absent in FosB. EMBO J 9:2827–2834

Mangelsdorf DJ, Thummel C, Beato M, Herrlich P, Schutz G, Umesono K, Blumberg B, Kastner P, Mark M, Chambon P, et al (1995) The nuclear receptor superfamily: the second decade. Cell 83:835–839

Moras D, Gronemeyer H (1998) The nuclear receptor ligand-binding domain: structure and function. Curr Opin Cell Biol 10:384–391

Morrison NA, Shine J, Fragonas JC, Verkest V, McMenemy ML, Eisman JA (1989) 1,25-dihydroxyvitamin D-responsive element and glucocorticoid repression in the osteocalcin gene. Science 246:1158–1161

Nagy L, Kao HY, Chakravarti D, Lin RJ, Hassig CA, Ayer DE, Schreiber SL, Evans RM (1997) Nuclear receptor repression mediated by a complex containing SMRT, mSin3 A, and histone deacetylase. Cell 89:373–380

Onate SA, Tsai SY, Tsai MJ, O'Malley BW (1995) Sequence and characterization of a coactivator for the steroid hormone receptor superfamily. Science 270:1354–1357

Pazin MJ, Kadonaga JT (1997) What's up and down with histone deacetylation and transcription? Cell 89:325–328

Perissi V, Staszewski LM, McInerney EM, Kurokawa R, Krones A, Rose DW, Lambert MH, Milburn MV, Glass CK, Rosenfeld MG (1999) Molecular determinants of nuclear receptor-corepressor interaction. Genes Dev 13:3198–3208

Poliard A, Bakkali L, Poiret M, Foiret D, Danan JL (1990) Regulation of the rat alpha-fetoprotein gene expression in liver. Both the promoter region and an enhancer element are liver-specific and negatively modulated by dexamethasone. J Biol Chem 265:2137–2141

Reichardt HM, Kaestner KH, Tuckermann J, Kretz O, Wessely O, Bock R, Gass P, Schmid W, Herrlich P, Angel P, Schutz G (1998) DNA binding of the glucocorticoid receptor is not essential for survival. Cell 93:531–541

Saatcioglu F, Claret FX, Karin M (1994) Negative transcriptional regulation by nuclear receptors. Semin Cancer Biol 5:347–359

Sakai DD, Helms S, Carlstedt-Duke J, Gustafsson JA, Rottman FM, Yamamoto KR (1988) Hormone-mediated repression: a negative glucocorticoid response element from the bovine prolactin gene. Genes Dev 2:1144–1154

Schüle R, Evans RM (1991) Cross-coupling of signal transduction pathways: zinc finger meets leucine zipper. Trends Genet 7:377–381

Stöcklin E, Wissler M, Gouilleux F, Groner B (1996) Functional interactions between Stat5 and the glucocorticoid receptor. Nature 383:726–728

Strömstedt PE, Poellinger L, Gustafsson JA, Carlstedt-Duke J (1991) The glucocorticoid receptor binds to a sequence overlapping the TATA box of the human osteocalcin promoter: a potential mechanism for negative regulation. Mol Cell Biol 11:3379–3383

Subramaniam N, Cairns W, Okret S (1997) Studies on the mechanism of glucocorticoid-mediated repression from a negative glucocorticoid response element from the bovine prolactin gene. DNA Cell Biol 16:153–163

Torchia J, Rose DW, Inostroza J, Kamei Y, Westin S, Glass CK, Rosenfeld MG (1997) The transcriptional co-activator p/CIP binds CBP and mediates nuclear-receptor function. Nature 387:677–684

Tsai CC, Kao HY, Yao TP, McKeown M, Evans RM (1999) SMRTER, a Drosophila nuclear receptor coregulator, reveals that EcR-mediated repression is critical for development. Mol Cell 4:175–186

Turcotte B, Meyer ME, Bocquel MT, Belanger L, Chambon P (1990) Repression of the alpha-fetoprotein gene promoter by progesterone and chimeric receptors in the presence of hormones and antihormones. Mol Cell Biol 10:5002–5006

Voegel JJ, Heine MJ, Zechel C, Chambon P, Gronemeyer H (1996) TIF2, a 160 kDa transcriptional mediator for the ligand-dependent activation function AF-2 of nuclear receptors. EMBO J 15:3667–3675

Xu L, Glass CK, Rosenfeld MG (1999) Coactivator and corepressor complexes in nuclear receptor function. Curr Opin Genet Dev 9:140–147

Yang XJ, Ogryzko VV, Nishikawa J, Howard BH, Nakatani Y (1996) A p300/CBP-associated factor that competes with the adenoviral oncoprotein E1 A. Nature 382:319–324

Yang-Yen HF, Chambard JC, Sun YL, Smeal T, Schmidt TJ, Drouin J, Karin M (1990) Transcriptional interference between c-Jun and the glucocorticoid receptor: mutual inhibition of DNA binding due to direct protein-protein interaction. Cell 62:1205–1215

Zamir I, Dawson J, Lavinsky RM, Glass CK, Rosenfeld MG, Lazar MA (1997) Cloning and characterization of a corepressor and potential component of the nuclear hormone receptor repression complex. Proc Natl Acad Sci USA 94:14400–14405

16 The Anti-inflammatory Action of Glucocorticoid Hormones

P. Herrlich, M. Göttlicher

16.1	Introduction	297
16.2	Dimerization-Defective GR Mediates GC-Induced Anti-inflammation and Inhibition of the Immune Response	298
16.3	Mechanism of GR Interference with NF-κB Activity	299
16.4	The Precedent: The Promoter of Metallothionein IIA	300
16.5	Yeast Cells Can Support GR-Dependent or AP-1-Dependent Transcription, But Not the Cross-talk Between GR and AP-1	301
16.6	A Putative Cofactor of Cross-talk	302
16.7	Conclusions	302
References		303

16.1 Introduction

Inflammation and immune response are elaborate reactions of mammalian organisms to environmental attacks. As in any stimulated process, the existence of brakes that limit the response, is decisive for survival of the organism. Imagine a bacterial infection: the uncontrolled response, in form of cytokine release by macrophages, e.g. tumour necrosis factor (TNF)-α, would lead to septic shock and death. Excessive TNF-α levels must therefore be avoided by the induction of a brake mechanism. In the chain of reactions between contact with an agent [e.g. lipopolysaccharide (LPS)] and the expression of a program of genes (e.g. TNF-α) numerous inhibitory steps are therefore built in. A particularly important regulatory loop involves the release of glucocorticoids (GCs) which cause downmodulation of proinflammatory transcription factors. This downmodulation is transcriptional, is a function of the glucocorticoid receptor and it is not absolute. In the immune system, partial reduction of signalling or of cytokine synthesis suffices to reach levels below threshold (see, e.g. Viola and Lanzavecchia 1996).

The glucocorticoid receptor (GR) can act in two different modes: as transcription factor activating or repressing genes that carry GR-binding elements [GC response elements (GREs) and negative (n)GREs], and as modulator acting on other transcription factors such as AP-1, NF-κB, NF-AT and Stat5 (cross-talk, see reviews by Beato et al. 1995; Herrlich 2001). This brief report is concerned with the physiological role of the second mode of action, and the mechanism of the cross-talk action.

16.2 Dimerization-Defective GR Mediates GC-Induced Anti-inflammation and Inhibition of the Immune Response

Both GRE-dependent transcription and the action on nGREs require the association of two or more GR monomers, usually in form of head-to-tail/tail-to-head dimers (reviews by Evans 1988; Beato et al. 1995). A major determinant of dimerization is located in the GR D-loop. GRE-dependent transcription, in addition, involves assembly of co-activators at the GR transactivation domains. The cross-talk function of the GR, however, does neither require the presence of the transactivation domains nor the ability of the GR to dimerize (Heck et al. 1994; Stöcklin et al. 1996; and unpublished results). Also, DNA contact of the GR is dispensable (König et al. 1992; Nissen and Yamamoto 2000; Herrlich 2001). Based on the finding that the D-loop can be mutated without affecting GR cross-talk with AP-1 or NF-κB, a mutant mouse was generated carrying a D-loop point mutation (GR^{dim}; Reichardt et al. 1998). The mouse is viable, cannot or only very poorly activates GRE-dependent genes, but is fully competent in modulating AP-1 and NF-κB activity (Reichardt et al. 1998, 2001; Tuckermann et al. 1999).

The GR^{dim} mouse served to analyse the GC-induced inhibition of inflammatory and immune reactions. Indeed, phorbol ester-induced inflammation, the acute phase response and lipopolysaccharide-induced TNF-α release were efficiently counteracted by GC treatment. The synthesis and release of macrophage and T-cell cytokines were inhibited to the same degree as in wild-type mice. So was the transcription of cyclooxygenase-2 (Reichardt et al. 2001). The only GC effect on the immune system yet found to be deficient in GR^{dim} is thymocyte and peripheral T-cell apoptosis.

16.3 Mechanism of GR Interference with NF-κB Activity

The inflammatory and immune responses are by and large the result of NF-κB and AP-1 activities. NF-κB resides in the cytoplasm prior to activation. It is complexed with IκB, which masks the nuclear uptake sequence (reviewed by Bäuerle and Baltimore 1996). Upon activation of IκB kinase, the inhibitor is phosphorylated, ubiquitylated and degraded through the proteasome. NF-κB is thus released from inhibition and transported into the nucleus. As a result of NF-κB activity, IκB, itself an NF-κB target gene, is resynthesized and limits further nuclear uptake of NF-κB. IκB may, in addition, cause the move of nuclear NF-κB back into the cytoplasm.

Two modes of GC-dependent inhibition of NF-κB activity have been proposed. The GR may directly act on the NF-κB protein complex at e.g. a cytokine promoter (Caldenhoven et al. 1995; Heck et al. 1997; Nissen and Yamamoto 2000); GC may enhance the synthesis of the most ubiquitous member of the IκB family, IκBα (Auphan et al. 1995; Scheinman et al. 1995). It now appears that both mechanisms coexist. In certain cell types, e.g. mouse embryo fibroblasts, but not in CD4$^+$ T lymphocytes, we found induction of IκBα transcription and protein synthesis (Reichardt et al. 2001). This induction required the presence of a GR with intact D-loop, suggesting, as the most likely interpretation, that the IκBα promoter carries a GRE. At the same time, the TNF-α or anti-CD3-induced and NF-κB-dependent transcription of IκBα was inhibited. This inhibition was less dramatic than that of a classical (κB)$_4$-promoter/reporter (Reichardt et al. 2001), suggesting either that sequence differences in the κB elements caused differences of GC-dependent inhibition, or that IκBα transcription resulted from the combined action on a GRE as well as on the κB element.

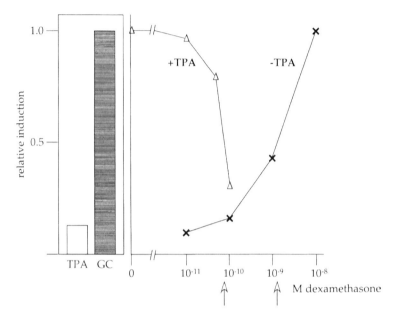

Fig. 1. Dissociation of GC actions on phorbol ester or GRE-dependent transcription from the human metallothionein IIA promoter. A promoter-reporter construct carrying the metallothionein IIA upstream sequences from −619 to beyond TATA was transiently transfected into HeLa cells. The columns indicate the relative induction by either phorbol ester (*TPA*; 100 ng/ml) or dexamethasone (*GC*; 10^{-8} M) alone. In the dose-response diagram the response to TPA was set at 1.0. Increasing concentrations of dexamethasone inhibited TPA-induced transcription until the GRE response took over beyond 10^{-10} M dexamethasone. The response curve then merges with that by GC alone. The latter is shown. The *arrows* indicate the dexamethasone concentrations for half-maximal inhibition and half-maximal induction, respectively

16.4 The Precedent: The Promoter of Metallothionein IIA

The promoter of the metallothionein IIA gene carries both a GRE at positions −247/−270 (Karin et al. 1987) and an AP-1 binding site at positions −98/−104 (Karin and Richards 1982; Angel et al. 1987). The endogenous gene is inducible by dexamethasone, so is the reporter carrying a long promoter fragment (Karin et al. 1984). The promoter is also addressed by Cadmium ions or, less strongly than by dexamethasone, by phorbol ester. If the promoter is shortened removing the GRE, the residual promoter is no longer hormone responsive, but its activity can be stimulated by phorbol ester. Phorbol ester activated transcription from this truncated promoter is strongly inhibited by dexamethasone, indicating that the AP-1 element is subjected to cross-talk inhibition (unpublished data). Also, with the endogenous promoter or long promoter-reporter constructs, the cross-talk inhibition can be detected at low hormone concentrations (e.g., half-inhibition of phorbol ester induced transcription at below 10^{-10} M dexamethasone), while the GRE-dependent transcription takes over at higher hormone levels (half-maximal at about 10^{-9} M; see Fig. 1; compare with the dose-response curves reported by Jonat et al. 1990). Thus, presumably the overall transcription of the gene represents a compromise between two modulating stimuli of opposite direction. A possible similarity to the regulation of IκBα transcription is obvious.

16.5 Yeast Cells Can Support GR-Dependent or AP-1-Dependent Transcription, But Not the Cross-talk Between GR and AP-1

We can thus conclude that the direct modulation of transcription factors such as AP-1 and NF-κB is established by GR in a molecular mode different from that required for the transcription from GRE promoters. The cross-talk mode occurs at the promoter element and no interference with the composition of the multiprotein initiation complex at the κB site has yet been detected (Nissen and Yamamoto 2000). If GR and AP-1 or NF-κB were the only participants in the decision between transcription and block of transcription, we should be able to reconstitute cross-talk in yeast cells. Indeed, hormone- and GR-dependent transcription (Picard et al. 1990) as well as Jun- and Fos-dependent transcription (Lech et al. 1988; Struhl 1988) of appropriate promoters in yeast has been reported. We established in yeast Jun-/Fos-dependent transcription from an AP-1 element promoter and could activate GR by triamcinolone, but we were unable to obtain hormone-dependent inhibition of the reporter. Although a negative result, it suggests that an additional component is lacking for which yeast cells contain no homologue.

16.6 A Putative Cofactor of Cross-talk

We undertook a yeast two-hybrid screen expressing HeLa cDNA fusions in order to fish for putative GR-interacting proteins (S. Schneider et al., unpublished data). To avoid the isolation of activation domain partners, we chose a GR construct mutated in both AF-1 and AF-2, which is totally defective in gene activation. The activation-defective GR can perfectly exert the cross-talk function. The two-hybrid screen yielded repeatedly, in a hormone-dependent fashion, one-and-the-same clone coding for a domain that interacted with the GR mutant. The completed cDNA specifies a protein of 476 amino acids that comprises, in its C-terminal half, three LIM domains. By glutathione S-transferase (GST)-pull-down experiments the LIM domains exhibited interesting and specific properties: the LIM-domain protein indeed interacted with GR, also when expressed in HeLa cells. The interaction was mediated through the most C-terminal LIM domain 3 (C. Heilbock and S. Schneider, unpublished data). Interaction with Jun was not pronounced and thus uncertain. With Fos, however, the other prominent subunit of AP-1, strong interaction occurred with LIM domain 1. The LIM-domain protein also interacted with NF-κB p65.

Using antisense experiments and expression of dominant negative mutants (deletions of individual LIM domains), we could strongly reduce the cross-talk between GR and AP-1 (S. Schneider et al., unpublished data). The LIM domain protein is thus a promising candidate factor participating in the modulating interaction between these two transcription factors. Although the LIM domain protein carried a weak "co-repressor" function in fusion experiments with the Gal DBD, the inhibition of AP-1 by GC was resistant to trichostatin or valproic acid, inhibitors of histone deacetylase activity. This suggests a mechanism that does not depend on the recruitment of histone deacetylase.

16.7 Conclusions

The work with GR^{dim} mice indicates that much of the anti-inflammatory and immune-response-suppressive actions of GC are the result of the cross-talk function of the GR. The interference with AP-1 or NF-κB appears to occur after formation of the pre-initiation complex. It is likely that novel factors participate in the inhibitory reaction. We suggest that a LIM-domain protein with binding affinity for GR, Fos and p65 fulfils this function. It remains to be seen what type of function arrests transcription at this "late" stage, after the formation of the pre-initiation complex.

References

Angel P, Imagawa M, Chiu R, Stein B, Imbra RJ, Rahmsdorf H-J, Jonat C, Herrlich P, Karin M (1987) Phorbol ester-inducible genes contain a common cis element recognized by a TPA-modulated trans-acting factor. Cell 49:729–739

Auphan N, DiDonato JA, Rosette C, Helmberg A, Karin M (1995) Immunosuppression by glucocorticoids: inhibition of NF-kappa B activity through induction of I kappa B synthesis Science 270:286–290

Baeuerle PA, Baltimore D (1996) NF-kappa B: ten years after. Cell 87:13–20

Beato M, Herrlich P, Schütz G (1995) Steroid hormone receptors: many actors in search of a plot. Cell 83:851–857

Caldenhoven E, Liden J, Wissink S, Van de Stolpe A, Raaijmakers J, Koenderman L, Okret S, Gustafsson JA, Van der Saag PT (1995) Negative cross-talk between RelA and the glucocorticoid receptor: a possible mechanism for the antiinflammatory action of glucocorticoids. Mol Endocrinol 9:401–412

Evans RM (1988) The steroid and thyroid hormone receptor superfamily. Science 240:889–895

Heck S, Kullmann M, Gast A, Ponta H, Rahmsdorf HJ, Herrlich P, Cato AC (1994) A distinct modulating domain in glucocorticoid receptor monomers in the repression of activity of the transcription factor AP-1. EMBO J 13:4087–4095

Heck S, Bender K, Kullmann M, Gottlicher M, Herrlich P, Cato AC (1997) I kappaB alpha-independent downregulation of NF-kappaB activity by glucocorticoid receptor. EMBO J 16:4698–4707

Herrlich P (2001) Cross-talk between glucocorticoid receptor and AP-1. Oncogene 20:2465–2475

Jonat C, Rahmsdorf HJ, Park KK, Cato AC, Gebel S, Ponta H, Herrlich P (1990) Antitumor promotion and antiinflammation: down-modulation of AP-1 (Fos/Jun) activity by glucocorticoid hormone. Cell 62:1189–1204

Karin M, Richards RI (1982) Human metallothionein genes – primary structure of the metallothionein-II gene and a related processed gene. Nature 299:797–802

Karin M, Haslinger A, Holtgreve H, Richards RI, Krauter P, Westphal HM, Beato M (1984) Characterization of DNA sequences through which cadmium and glucocorticoid hormones induce human metallothionein-IIA gene. Nature 308:513–519

Karin M, Haslinger A, Heguy A, Dietlin T, Cooke T (1987) Metal-responsive elements act as positive modulators of human metallothionein-IIA enhancer activity. Mol Cell Biol 7:606–613

König H, Ponta H, Rahmsdorf H-J, Herrlich P (1992) Interference between pathway-specific transcription factors: glucocorticoids antagonize phorbol ester-induced AP-1 activity without altering AP-1 site occupation in vivo. EMBO J 11:2241–2246

Lech K, Anderson K, Brent R (1988) DNA-bound Fos proteins activate transcription in yeast. Cell 52:179–184

Nissen RM, Yamamoto KR (2000) The glucocorticoid receptor inhibits NFkappaB by interfering with serine-2 phosphorylation of the RNA polymerase II carboxy-terminal domain. Genes Dev 14:2314–2329

Picard D, Khursheed B, Garabedian M J, Fortin M G, Lindquist S, Yamamoto K R (1990) Reduced levels of hsp90 compromise steroid receptor action in vivo. Nature 348:166–168

Reichardt HM, Kaestner KH, Tuckermann J, Kretz O, Wessely O, Bock R, Gass P, Schmid W, Herrlich P, Angel P, Schütz G (1998) DNA binding of the glucocorticoid receptor is not essential for survival Cell 93:531–541

Reichardt HM, Tuckermann JP, Gottlicher M, Vujic M, Weih F, Angel P, Herrlich P, Schutz G (2001) Repression of inflammatory responses in the absence of DNA binding by the glucocorticoid receptor. EMBO J 20:7168–7173

Scheinman RI, Cogswell PC, Lofquist AK, Baldwin AS Jr (1995) Role of transcriptional activation of I kappa B alpha in mediation of immunosuppression by glucocorticoids Science 270:283–286

Stöcklin E, Wissler M, Gouilleux F, Groner B (1996) Functional interactions between Stat5 and the glucocorticoid receptor. Nature 383:726–728

Struhl K (1988) The JUN oncoprotein, a vertebrate transcription factor, activates transcription in yeast. Nature 332:649–650

Tuckermann JP, Reichardt HM, Arribas R, Richter KH, Schütz G, Angel P (1999) The DNA binding-independent function of the glucocorticoid receptor mediates repression of AP-1-dependent genes in skin. J Cell Biol 147:1365–1370

Viola A, Lanzavecchia A (1996) T cell activation determined by T cell receptor number and tunable thresholds. Science 273:104–106

17 Analysis of Glucocorticoid Receptor Function in the Mouse by Gene Targeting

C. Kellendonk, F. Tronche, H.M. Reichardt, A. Bauer, E. Greiner, W. Schmid, G. Schütz

17.1	Introduction	305
17.2	Mice Deficient in the Glucocorticoid Receptor Demonstrate an Important Role for GR	308
17.3	The DNA-Binding Deficient GR Reveals the Importance of GR Crosstalk with other Transcription Factors	309
17.4	Cell-/Tissue-Specific Mutations of the GR Gene	312
17.5	Conclusions	316
References		317

17.1 Introduction

Glucocorticoids, produced in the adrenal cortex, play an important role in a variety of organ systems during development and in many physiological and pathological processes (Miller and Blake Tyrrel 1995). The glucocorticoid receptor (GR) mediates the effects of glucocorticoids by positively or negatively influencing gene activity. Glucocorticoid-controlled functions range from the regulation of metabolism, the stress response and control of innate and acquired immunity to modulation of behaviour.

A very important and well-studied function of glucocorticoids is the regulation of carbohydrate and lipid metabolism. Glucocorticoids induce the synthesis of gluconeogenic enzymes which, in consequence, lead to increased glucose synthesis. Studies in cells and animals have shown that the genes encoding gluconeogenic enzymes are direct targets of the glucocorticoid receptor. Glucocorticoids promote maturation of the lung and are highly beneficial in the treatment of the acute respiratory distress syndrome in newborn infants. Glucocorticoids are able to induce apoptosis of thymocytes, but the physiological role of this activity is still not yet defined. More recently, a role of glucocorticoids and its receptor in maturation of erythroid progenitors has been defined. The levels of glucocorticoids show a diurnal rhythm reaching peak levels at the onset of the active phase of the organism, which for rodents is the evening. Finally, glucocorticoids are important anti-inflammatory and immunosuppressive agents, an activity which has been useful for the treatment of many inflammatory disorders (Miller and Blake Tyrrel 1995).

Levels of glucocorticoids are tightly regulated by a feedback mechanism involving the hypothalamus and the pituitary gland, the hypothalamic-pituitary-adrenal (HPA) axis. Increased levels of glucocorticoid hormones in serum result in repression of corticotropin-releasing hormone (CRH) and adrenocorticotropic hormone (ACTH) at the level of transcription as well as secretion. Corticosteroids are released in response to stressful changes. Stress is considered a protective mechanism that prepares the organism to react to threatening stimuli in an appropriate way. These stimuli activate the HPA axis which in conjunction with other physiological responses coordinates the behavioural responses of the organism. Chronic changes in control of the HPA axis may have pathological consequences, since it is now well established that depression- and anxiety-related disorders are associated with dysregulation of the HPA axis.

The advances in mouse molecular genetics have allowed enormous progress in the analysis of the function of steroid receptors (Gu et al. 1994; Sauer and Henderson 1988). Inactivation of the genes for these receptors in mice has given important information on the role these molecules play in development and physiology. Often these analyses have resulted in unexpected findings such as the role of the glucocorticoid receptor in growth control (see below). As powerful as the conventional knock-out technology is, it does not allow for the generation of spatially and temporally restricted and function-selective mutations. It is therefore highly desirable to develop mouse mutants which have more refined mutations in the gene of interest, thus limiting the effect of the

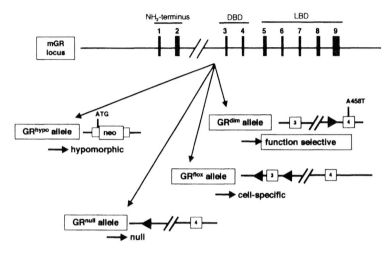

Fig. 1. Cell- and tissue-specific mutations of the glucocorticoid receptor (*GR*) gene. The four alleles of the GR gene are shown in relation to the structure of the GR locus

mutation to a specific organ or a specific cell type. Exploiting the Cre/loxP recombination system, it has become possible to generate cell- and organ-specific mutations as well as function-selective alterations (Gu et al. 1994). Furthermore, it might be of great interest to selectively alter a particular function, for example crippling the activating functions of the receptor, but retaining its inhibitory activity as we have achieved in the case of GR. Figure 1 summarizes the various types of mutations ranging from null and hypomorphic mutations to cell/tissue-specific and function-selective mutations, which have been generated with the help of the Cre/loxP recombination system.

Two closely related corticosteroid receptors have been identified, the mineralocorticoid receptor (also called the type-I receptor) and the glucocorticoid receptor (type-II receptor) (Funder et al. 1988). Glucocorticoids are able to activate both receptors, whereas mineralocorticoids are specific for the mineralocorticoid receptor (MR). Therefore, in tissues which express both receptors, e.g. the brain, glucocorticoids can affect gene expression by signalling through both receptors. Mineralocorticoid-responsive cells, such as the principal cells of the collecting duct in kidney, express the enzyme 11-beta-hydroxysteroiddehydrogenase, which is present in some mineralocorticoid target cells and which metabolizes cortisol to cortisone. This prereceptor specificity mechanism prevents binding of glucocorticoids to the MR and thus ensures mineralocorticoid-specific effects.

17.2 Mice Deficient in the Glucocorticoid Receptor Demonstrate an Important Role for GR

Mice without functional GR die shortly after birth due to atelectasis of the lungs, demonstrating an essential function of the receptor in survival (Cole et al. 1995). The specific role of the receptor in lung development is not understood. It is, however, interesting that the DNA-binding activity of the receptor is not required (see below). The basis for development of lung atelectasis is unknown and the primary target genes controlled by the receptor in lung remain to be established. Due to the early lethality, these mice have been analysed with regard to receptor function during development and gene expression during embryogenesis. It could be demonstrated that the GR is required in vivo for rapid expansion of erythroid progenitors under hypoxic conditions (Bauer et al. 1999). In Table 1 a summary of the phenotypic consequences of the GR null mutation is listed. Two inactivating mutations have been generated: a hypomorphic allele which resulted from insertion of a neomycin resistance cassette right after the AUG of the receptor coding sequence

Table 1. Alterations of physiological functions found in mice carrying a targeted disruption of the glucocorticoid receptor gene

Lung	Perinatal death due to respiratory failure in newborn mice
Liver	Reduced expression of genes encoding gluconeogenic enzymes
Adrenals	Hypertrophy and hyperplasia of the cortex with increased expression of steroidogenic enzymes; impaired chromaffin cell differentiation in the medulla and absence of epinephrine synthesis
HPA axis	Elevated serum levels for ACTH and CORT; increased expression of POMC and CRH
Bone marrow	Impaired proliferation of erythroid precursor cells
Thymus	Loss of glucocorticoid-dependent apoptosis in thymocytes

and a null allele which has been generated by excision of exon 3 (F. Tronche, unpublished data). Due to alternative splicing, a truncated mRNA leading to the synthesis of a shortened protein is generated in the former case. The allele with deletion of exon 3 mutation represents a true null and shows a fully penetrant and more severe phenotype in comparison to the GR hypo mice.

17.3 The DNA-Binding Deficient GR Reveals the Importance of GR Crosstalk with other Transcription Factors

Transcription regulation by the glucocorticoid receptor is achieved by DNA binding-dependent as well as -independent mechanisms. It is well established that binding of GR to a glucocorticoid response element (GRE) is crucial for the activation of transcription (Beato et al. 1995). Binding to so-called negative GREs (nGRE), which have been found in several genes that are negatively regulated by glucocorticoids, has also been described (Drouin et al. 1993). This hypothesis is strongly confirmed by the analysis of the dimerization-defective mutant. To study activities of the receptor which might be independent of DNA-binding, we used a knock-in approach with the Cre/loxP system in order to establish a point mutation in the D-loop, one of the dimerization domains of the receptor (Reichardt et al. 1998). This mutation selectively abolishes homodimerization of the receptor and thereby impairs binding of GR to DNA, while regulation of gene activity by protein–protein interaction remains intact. The dimerization-defective mutant called GR^{dim} represent a very useful tool to study the mechanisms of glucocorticoid action in vivo. As shown below, this enabled us to define positive as well as negative activity of the glucocorticoid receptor resulting from DNA-binding-independent mechanism. A summary of the results obtained is given in Fig. 2. The mutation at A458T is located in the D-loop, which forms an important dimerization interface. Surprisingly, homozygous mutant mice with this point mutation were fully viable, indicating that DNA binding of the receptor is not required to guarantee survival of the animals. These findings suggest that DNA binding is not required for prevention of the atelectatic lung phenotype. Using cells derived from these mutant animals as well as by performing studies in vivo, we could

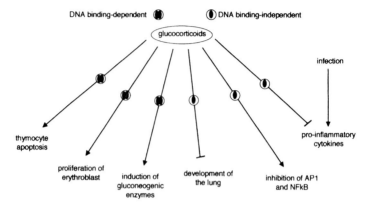

Fig. 2. The GR controls physiological responses to glucocorticoids by two different modes of action. Various functions of the GR are shown. The mode of activity – DNA binding-dependent and DNA binding-independent – is indicated

unequivocally show that important anti-inflammatory activities of the receptor are maintained in this animal model (Reichardt et al. 2001). Inflammatory responses such as the induced ear oedema following 12-O-tetradecanoylphorbol-13-acetate (TPA) treatment or phorbol myristate acetate (PMA)-induced local inflammation of the skin could be repressed by glucocorticoids in the mice with a DNA-binding-defective glucocorticoid receptor. Inhibition of cytokine synthesis following treatment with lipopolysaccharide (LPS) of thymocytes and macrophages derived from GR^{dim} could be achieved by glucocorticoids. We thus conclude that glucocorticoids repress the pro-inflammatory effect of NF-κB activity even when the receptor is impaired for DNA-binding. Since DNA-binding is impaired in GR^{dim} mice, we consider these animals to represent a very useful model to determine the contribution of protein–protein interactions of GR in anti-inflammatory responses in vivo. It could, for example, be demonstrated that DNA-binding-independent functions mediate repression of AP-1-dependent genes in skin (Tuckermann et al. 1999). Many of the side-effects of glucocorticoids like osteoporosis and steroid-induced diabetes are thought to be due to DNA-binding-dependent activities of the receptor. Therefore, these findings are encouraging since they indicate that selective ligands will

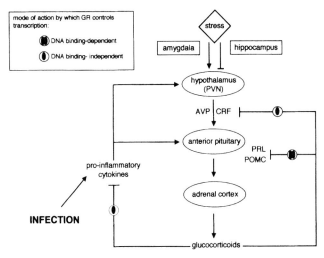

Fig. 3. Different modes of action of GR are used in control of the hypothalamic-pituitary-adrenal axis. DNA binding-dependent and -independent activities of GR are shown

be found which have full anti-inflammatory activity but lack or show diminished side-effects.

The HPA axis represents a neuroendocrine regulatory cascade which controls synthesis and release of glucocorticoids from the adrenal cortex. This cascade is based on a negative feedback circuit whereby glucocorticoids repress the synthesis and secretion of CRH at the level of the paraventricular nucleus of the hypothalamus and synthesis and secretion of ACTH in the pituitary (Miller and Blake Tyrrel 1995). The availability of the GR^{dim} mice allowed for the characterization of the mechanisms involved in this feedback loop. Using immunohistochemical analysis in the hypothalamus, we could demonstrate that CRH levels do not differ in wild-type mice from those carrying the dimerization-defective allele, strongly arguing for GR-mediated control of CRH synthesis at the level of protein–protein interaction (Reichardt et al. 1998). The mechanism of this control, however, is not known. Expression of the gene encoding pro-opiomelanocortin (POMC), the precursor of ACTH, is elevated by almost an order of magnitude demonstrating loss of negative control of POMC transcription. These findings illustrate the different modes of activities of GR in HPA-axis control (Fig. 3) and strongly support the hypothesis of negative GREs to which multimers of the receptor have to bind in order to exert this feedback control.

17.4 Cell-/Tissue-Specific Mutations of the GR Gene

Since mice without a functional GR are not viable and die shortly after birth, we are forced to develop cell-/tissue-specific mutations in order to analyse selective functions of GR. The generation of somatic mutations will not only bypass the lethality of mutant mice, but will also allow us to analyse functions of the GR in a particular cell or tissue and to estimate to which extent this function contributes to the physiology of the organism. To analyse the functions of GR in various cells/tissues, exon 3 of the receptor gene was flanked by two loxP sites using homologous recombination in embryonic stem cells (Tronche et al. 1998). To generate a specific mutation, the Cre recombinase was expressed after transgenesis in the desired cells or tissue. The penetrance of the mutation is dependent on the expression pattern of the recombinase. In order to minimize mosaic expression and to guarantee the selectivity of expression, we have used yeast and bacterial artificial chromosomes to guarantee the desired expression selectivity. Figure 4 gives an overview of the various cell-/tissue-specific GR mutations we have generated exploiting the Cre/loxP recombination system.

17.4.1 Corticosteroid Receptor Function in the Brain

To evaluate the role of GR and MR in the nervous system, deletion of the respective genes was achieved by appropriately expressing the Cre recombinase. Using the nestin promoter to drive the Cre recombinase specifically in neuronal precursor cells, the deletion of the receptor locus in the entire nervous system was achieved (Tronche et al. 1999). Lack of the GR in the nervous system is not lethal, but profoundly alters the HPA-axis equilibrium. Inactivation of the GR in the nervous system allows the discrimination between functions of the receptor in the hypothalamus and the anterior lobe of the pituitary, since the nestin promoter does not lead to expression of the Cre recombinase in the pituitary. The

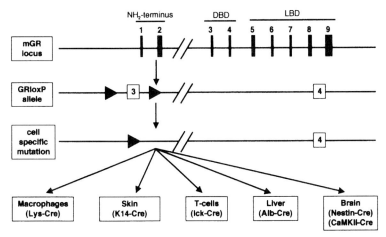

Fig. 4. Summary of cell- and tissue-specific mutations of the GR gene. Specific mutations in the GR locus were obtained by expressing the Cre recombinase in macrophages/granulocytes in the context of the lysozyme gene, in thymocytes under control of the lck gene, in hepatocytes under control of the albumin gene regulatory sequences, and in brain with regulatory sequences of the nestin and CaMKII gene. A Cre fusion gene with the ligand-binding domain of the oestradiol promoter was expressed with the cytokeratin 14 promoter in skin

missing central feedback in the paraventricular nucleus releases the inhibition of CRH production which, as a consequence, stimulates ACTH synthesis in the anterior pituitary. Testing the selectivity of the mutation we find POMC mRNA as well as ACTH immunoreactivity increased in corticotroph cells of GR^{NesCre} mice. These findings do also suggest that GR-mediated repression of POMC gene activity is not sufficient to counteract the increased levels of CRH due to central lack of GR.

Inactivation of the GR in the nervous system reduces anxiety-related behaviour. These mice are less anxious in the elevated zero maze and the dark light box, two tasks which exploit the behavioural conflict between exploring and avoiding an aversive compartment. The reduced anxiety-related responses reveal an important role of GR signalling in emotional behaviour (Tronche et al. 1999). To further define the role of MR and GR in functions of the brain, we have inactivated both receptors in identical regions. We have generated mice lacking either receptor as a consequence of expressing the Cre recombinase under the control of the CaMKIIα gene. To guarantee expression in those neurons in which CaMKIIα is active, we have generated transgenic mice expressing the recombinase in the context of a bacterial artificial chromosome (BAC), since we wanted to guarantee faithful and predictable expression of the Cre and thus the generation of the mutation (Casanova et al. 2001). Mice with loss of the GR and MR will now be used for a detailed analysis of their role during memory formation as well as in neuronal physiology. The function of these receptors in neurogenesis as well as neuronal viability will be assessed in detail. In particular, these mice will be useful to define GR and MR target genes by expression screens. We believe that detailed characterization of these mice carrying highly defined mutations in the nervous system will provide a better understanding of the role of these receptors in the development of psychiatric diseases such as depression and anxiety disorders. Furthermore, in addition to a role of the receptors in the hippocampus, it has become well established that the glucocorticoid signalling system modulates brain functions leading to drug addiction. In this context, the observation of loss of sensitization after cocaine treatment in GR-deficient mice is extremely stimulating (F. Tronche, unpublished data). An important role for GR in control of circadian timing was recently established, since glucocorticoids are able to phase shift peripheral oscillators in animals (Balsalobre et al. 2000).

17.4.2 The Glucocorticoid Receptor Has an Important Function in the Control of Postnatal Body Growth

To study the role of GR in liver functions of the adult we wished to inactivate the GR selectively in parenchymal cells of the liver. This was achieved by choosing gene regulatory sequences of the albumin gene for expression of the Cre recombinase (Kellendonk et al. 2000). Mice with a hepatocyte-specific alteration displayed, after 4 weeks, a severe growth deficit. Since no alterations in the serum levels of growth hormone and glucocorticoids could be found in these animals, we reasoned that the growth deficit resides in growth hormone signalling. Growth hormone is known to affect growth through stimulation of insulin-like growth factor (IGF)-1 by activating the activity of the transcription factor Stat5 (MacGillivray 1995). Interestingly, when we determined the mRNA levels of IGF-1 and growth hormone-regulated genes in the liver, we found that the level was drastically reduced in the mutant. The level of Stat5α and -β, which are thought to be crucial for mediation of the growth hormone signalling, was not altered nor was the phosphorylation status. Mice with the dimerization-defective mutation are of normal size and the expression of IGF-I and other growth hormone-regulated genes is unaltered. These results demonstrate that GR function in hepatocytes is crucial for body growth. The different response in GR^{dim} mice in comparison to mice with a hepatocyte-specific mutation provides strong evidence that the growth-promoting activity of the receptor does not require binding to a glucocorticoid-responsive element. It rather appears to function as a coactivating molecule thus synergizing with Stat5 activity, supporting previous in vivo experiments, which demonstrated a requirement of GR for Stat5 activation by prolactin (Stöcklin et al. 1996).

17.4.3 Anti-inflammatory Activities of Glucocorticoids Are Maintained in the Dimerization-Defective GR Mutant

To better understand the role of glucocorticoids and GR in inflammation, we are specifically inactivating receptor function in important cells of the immune system, such as thymocytes, monocytes/macrophages, dendritic cells, mast cells and eosinophils. Using the Cre/loxP system, specific mutations have and will continuously be generated to define the precise role of the receptor in innate and acquired immunity. The importance of the receptor was convincingly demonstrated in mice which lack the receptor in monocytes/macrophages and in mice which overexpress the glucocorticoid receptor (Reichardt et al. 2000; F. Tronche and A. Bauer, unpublished data). When mice without a functional receptor in macrophages were challenged by treatment with a chosen dose of LPS it was observed that all mutant mice challenged died within 36 h, whereas wild-type mice survived this challenge. In contrast, mice with two additional copies of the receptor showed enhanced resistance to the endotoxic shock. To achieve a perfect expression pattern of the receptor in transgenic mice additional copies of the receptor were introduced using yeast artificial chromosomes (YACs) (Schedl et al. 1993). Mice with the additional copies of the receptor when challenged reveal attenuated response to restraint stress and a strongly increased resistance to LPS-dependent endotoxic shock. The reduced inflammatory responses in these mice demonstrate that increased activity of the receptor due to overexpression leads to strongly increased resistance towards endotoxic shock. These results are interpreted to indicate the importance of glucocorticoid receptor-dependent signalling for controlling an overwhelming inflammatory response.

17.5 Conclusions

Exploiting the Cre/loxP recombination system in the mouse, a series of alleles of GR have been generated. These mutants have been instrumental in the molecular genetic analysis of GR functions in vivo. They allow us to define four different modes of actions of the receptor as illustrated in Fig. 5. The receptor is able to activate gene transcription by binding

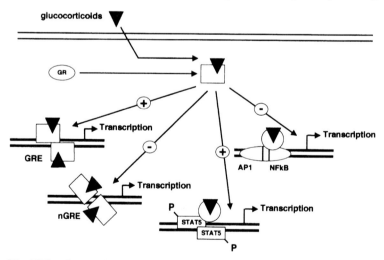

Fig. 5 The glucocorticoid receptor modulates transcription by different modes of action. The GR is able to influence transcription by binding to a GRE or nGRE. It is assumed that binding of GR to a nGRE leads to an altered GR conformation. The GR can also activate or inhibit transcription by protein–protein interaction as shown on the *right*

as a dimer to a GRE in the control region of a regulated gene. The growth-promoting activity of the receptor does not require binding of the receptor to a GRE, but is mediated by interaction of the receptor with Stat5. It therefore functions rather as a coactivator in Stat5 signalling. These results strongly support the observations which have been made in vitro. The receptor inhibits expression of genes such as the POMC gene and the prolactin gene in the anterior pituitary by binding to nGREs in the control regions of these genes. This hypothesis is strongly supported by our observations of elevated levels of the mRNAs encoding these proteins in mice with the dimerization defective and thus DNA-binding-deficient GR. The receptor is also able to modulate target gene activity by a DNA-binding-independent mechanism. It thus is able to enhance the transcriptional activity of the Stat5 proteins and inhibit the activity of AP-1- and NF-κB-dependent transcription. It will be a fascinating challenge to define the multiple activities of GR in molecular detail.

References

Balsalobre A, Brown SA, Marcacci L, Tronche F, Kellendonk C, Reichardt HM, Schütz G, Schibler U (2000) Glucocorticoid hormones can reset circadian time in peripheral tissues but not in the suprachiasmatic nucleus. Science 289:2344–2347

Bauer A, Tronche F, Wessely O, Kellendonk C, Reichardt HM, Steinlein P, Schütz G, Beug H (1999) The glucocorticoid receptor is required for stress erythropoiesis. Genes Dev 13:2996–3002

Beato M, Herrlich P, Schütz G (1995) Steroid hormone receptors: many actors in search of a plot. Cell 83:851–857

Casanova E, Fehsenfeld S, Mantamadiotis T, Lemberger T, Greiner E, Stewart AF, Schütz G (2001) A CamKIIalpha iCre BAC allows brain-specific gene inactivation. Genesis 31:37–42

Cole PJ, Blendy JA, Monaghan AP, Kriegelstein K, Schmid W, Aguzzi A, Fantuzzi G, Hummler E, Unsicker K, Schütz G (1995) Targeted disruption of the glucocorticoid receptor gene blocks adrenergic chromaffin cell development and severely retards lung maturation. Genes Dev 9:1608–1621

Drouin J, Sun YL, Chamberland M, Gauthier Y, De Lean A, Nemer M, Schmidt TJ (1993) Novel glucocorticoid receptor complex with DNA element of the hormone-repressed POMC gene. EMBO J 12:145–156

Funder JW, Pearce PT, Smith R, Smith AI (1988) Mineralocorticoid action: target tissue specificity is enzyme, not receptor, mediated. Science 242:583–585

Gu H, Marth JD, Orban PC, Mossmann H, Rajewski K (1994) Deletion of a DNA polymerase beta gene segment in T cells using cell type-specific gene targeting. Science 265:103–106

Kellendonk C, Opherk C, Anlag K, Schütz G, Tronche F (2000) Hepatocyte-specific expression of Cre recombinase. Genesis 26:151–153

MacGillivray MH (1995) In: Felig P, Baxter JD, Frohman LA (eds) Endocrinology and metabolism. McGraw-Hill, New York, pp 1619–1673

Miller WL, Blake Tyrrel J (1995) The adrenal cortex. In: Felig P, Baxter JD, Frohman IA (eds) Endocrinology and metabolism. McGraw-Hill, New York, pp 555–711

Reichardt HM, Kaestner KH, Tuckermann J, Kretz O, Wessely O, Bock R, Gass P, Schmid W, Herrlich P, Angel P, Schutz G (1998) DNA binding of the glucocorticoid receptor is not essential for survival. Cell 93:531–541

Reichardt HM, Umland T, Schütz G (2000) Mice with an increased glucocorticoid receptor gene dosage show enhanced resistance to stress and endotoxic shock. Mol Cell Biol 20:9009–9017

Reichardt HM, Tuckermann JP, Göttlicher M, Vujic M, Weih F, Angel P, Herrlich P, Schütz G (2001) Repression of inflammatory responses in the absence of DNA-binding by the glucocorticoid receptor. EMBO J 20:7168–7173

Sauer B, Henderson N (1988) Site-specific DNA recombination in mammalian cells by the Cre recombinase of bacteriophage P1. Proc Natl Acad Sci USA 85:5166–5170

Schedl A, Montoliu L, Kelsey G, Schütz G (1993) A yeast artificial chromosome covering the tyrosinase gene confers copy number-dependent expression in transgenic mice. Nature 362:258–261

Stöcklin E, Wissler M, Goullieux F, Groner B (1996) Functional interactions between Stat5 and the glucocorticoid receptor. Nature 383:726–728

Tronche F, Kellendonk C, Reichardt HM, Schütz G (1998) Genetic dissection of glucocorticoid receptor function in mice. Curr Opin Genet Dev 8:532–538

Tronche F, Kellendonk C, Kretz O, Gass P, Anlag K, Orban PC, Bock R, Klein R, Schütz G (1999) Disruption of the glucocorticoid receptor gene in the nervous system results in reduced anxiety. Nat Genet 23:99–103

Tuckermann JP, Reichardt HM, Arribas R, Richter KH, Schütz G, Angel P (1999) The DNA-binding-independent function of the glucocorticoid receptor mediates repression of AP-1-dependent genes in skin. J Cell Biol 147:1–6

18 Glucocorticoid-Inducible Genes That Regulate T-Cell Function

P.R. Mittelstadt, J. Galon, D. Franchimont, J.J. O'Shea, J.D. Ashwell

18.1	Introduction	319
18.2	Proposed Mechanisms of Glucocorticoid Receptor Activity	320
18.3	GILZ	325
18.4	IL-7 Receptor	328
18.5	Conclusion	332
References		332

18.1 Introduction

Upon antigenic challenge, the immune system must detect and respond to subtle perturbations in local microenvironments. While many important transactions entail cell-to-cell contact, coordination between multiple cells and cell types is carried out through antigen-nonspecific soluble mediators, broadly classified as cytokines and hormones. Messages delivered by the cytokines are for the most part pro-inflammatory and/or immune activating, whereas those delivered by the glucocorticoid hormones are thought to be largely inhibitory. One of the major inhibitory effects of glucocorticoids is the suppression of the synthesis of a large number of cytokines (IL-1, IL-2, IL-3, IL-4, IL-5, IL-6, IL-8, IL-10, IL-13, GM-CSF, TNF-α, and IFN-γ; Kunicka et al. 1993). When viewed individually, however, a number of effects of glucocorticoids appear to break this "rule," with glucocorticoids acting in pro-inflammatory or immunostimulatory ways. For example, glucocorticoids can increase the expression of receptors for some of the pro-inflammatory cytokines that it suppresses (IL-1, IL-6, GM-CSF, and IFN-γ; Almawi et al. 1996). A more complicated example is the effect of glucocorticoids on lymphocyte viability. Whereas thymocytes, and to a lesser extent mature lymphocytes, are induced to undergo apoptosis in response to glucocorticoid receptor (GR) occupancy (immunosuppressive), glucocorticoids inhibit apoptosis elicited by T-cell receptor (TCR)-mediated activation (immunostimulatory). This phenomenon is due to the inhibition of activation-induced Fas ligand (FasL) upregulation (Yang et al. 1995), which in normal circumstances binds Fas and elicits a death response. This chapter deals with new information about two glucocorticoid-regulated genes that appear to have negative and positive effects on immune function: GILZ and the receptor for interleukin (IL)-7 (IL-7Rα).

18.2 Proposed Mechanisms of Glucocorticoid Receptor Activity

The GR is a member of the nuclear hormone receptor superfamily of receptors, which have a common structural organization: an N-terminal constitutive activation domain (τ_1), a central DNA-binding domain (DBD) that contains two zinc-finger motifs in which dimerization-mediating residues are embedded, a second, ligand-dependent, activation domain (τ_2), and a C-terminal ligand-binding domain (LBD). The unliganded GR resides in the cytoplasm as a complex with accessory proteins including hsp90. Ligand binding causes the GR to separate from its associated proteins and migrate to the nucleus, where it binds DNA and performs its most-recognized function: regulation of gene transcription. To do so, the GR must recruit additional intermediary cofactors that control chromatin assembly and accessibility. These cofactors participate in histone-DNA contact (the SWI/SNF complex), histone-acetyl transferase recruitment (GRIP1), or histone deacetylase recruitment (NcoR and SMRT). The histone-acetyl transferases (SRC-1, p300/CBP) and histone deactylases (HDACs) themselves interact with nuclear receptors and can be considered co-regulators (Collingwood et al. 1999; Xu et al. 1999; Adcock 2001; Jenkins et al. 2001). Despite the apparent simplicity of the notion of the GR as a single-component signaling apparatus, the mechanisms by which the GR controls biologically important events are more complex.

The biological activities of the GR were originally attributed to its capacity to act as a transcriptional activator (Yamamoto and Alberts 1976). The liganded GR initiates transcription by binding to classical glucocorticoid response elements (GREs), consisting of two inverted GRE half sites separated by three residues (GGTACAnnnTGTTCT) (Yamamoto 1985; Newton 2000). Cooperative binding to both half sites induces dimerization between two GRs, a property that is required for induction of transcription through the GRE. Genes positively regulated by classical GREs participate in gluconeogenesis in the liver [tyrosine amino transferase (TAT); Jantzen et al. 1987] and cell growth (metallothionein (MT); Karin et al. 1984). The GR can also repress transcription: composite GREs, found adjacent to binding sites for other transcription factors such as activator protein (AP)-1, can confer either positive or negative regulation by the liganded GR, depending on context (Diamond et al. 1990). Recently a composite GRE with the nuclear factor of activated T cells (NFAT)-binding element has been identified in the granulocyte-macrophage (GM)-CSF promoter (Smith et al. 2001). Exclusively negative GREs (nGREs), differing in sequence from the classic GRE but requiring the binding of GR dimers, have been identified in the promoters of the pro-opiomelanocortin (POMC) (Drouin et al. 1989), IL-1β (Zhang et al. 1997), prolactin (Sakai et al. 1988), and proliferin (Mordacq and Linzer 1989) genes.

Because of the potent anti-inflammatory and immunosuppressive effects of therapeutically administered glucocorticoids, the repressive activity of the GR has been extensively studied. The GR has been shown to inhibit signaling pathways leading to activation of the inflammatory transcription factors. At the most membrane-proximal level, glucocorticoids have been shown to inhibit tyrosine phosphorylation of the TCR and associated molecules, such as the p23 form of the TCR ζ chain, the ZAP70 kinase, and the transmembrane adapter molecule linker for activation of T cells (LAT) (Van Laethem et al. 2001). This required de novo gene synthesis, and was proposed to be due to disruption of lipid rafts containing the TCR and proximal signaling molecules. Glucocorticoids have been implicated in downstream signaling as well, having been shown to directly block Jun kinase activation in HeLa cells (Gonzalez et al. 2000) and fetal hepatocytes (Ventura et al. 1999) stimulated with tumor necrosis factor (TNF)-α. Other evidence suggests that the GR can inhibit transcription by competing for coactivators shared by the GR and activation-induced transcription factors, such as CBP/p300 and SRC-1 (Kamei et al. 1996; Sheppard et al. 1998). While competition for coactivators can be demonstrated under some circumstances, the model does not account for instances in which coactivator-dependent transcription factors, such as other nuclear hormone receptors, are not sensitive to GR suppression (De Bosscher et al. 1997, 2001). Indirect models whereby the GR induces secondary genes encoding inhibitors have been proposed. Glucocorticoids can induce production of IκBα, a molecule that prevents NF-κB activation by sequestering NF-κB dimers in the cytoplasm (Auphan et al. 1995; Scheinman et al. 1995), although the physiologic importance of this has been questioned (Heck et al. 1997; De Bosscher et al. 2000).

An alternative inhibitory mechanism, termed "direct interference" or transcriptional "cross-talk," was suggested by three independent studies in 1990 on glucocorticoid repression of the phorbol myristate acetate (PMA)-induced collagenase gene promoter (Jonat et al. 1990; Schule et al. 1990; Yang-Yen et al. 1990). These reports concluded that the GR and the transcription factor activator protein (AP)-1 mutually inhibit each other's activities through direct protein–protein contact. The GR could interfere with the c-Fos and c-Jun subunits of AP-1 without detectably contacting DNA. Rather, the GR was "tethered" to AP-1, which remained bound to its DNA target (Herrlich 2001). Along with AP-1, NF-κB has been identified as a target for this mode of suppression (Ray and Prefontaine 1994). AP-1 and NF-κB are at the endpoints of many receptor-initiated signaling cascades, and participate in induction of most inflammatory cytokines as well as other types of genes the participate in the inflammatory response, such as those encoding adhesion molecules (Wisdom 1999; Karin and Ben-Neriah 2000). This model has been extended to other activating transcription factors, including Spi-1 (Gauthier et al. 1993), CREB (Imai et al. 1993), GATA-1 (Chang et al. 1993) and OCT-1 (Kutoh et al. 1992).

To facilitate rapid responses to changes in cytokine and hormone levels, the GR and many activating transcriptional regulators are constitutively expressed. Thus, direct interference can take place in the absence of new protein synthesis (Jonat et al. 1990). Indeed, glucocorticoid repression of collagenase mRNA induction by PMA took place in the presence of the protein synthesis inhibitor cycloheximide (CHX), and suppression of AP-1-dependent transcription occurred at glucocorticoid concentrations an order of magnitude lower than those needed to activate GRE-dependent transcription. The activation and repression functions of the GR can be physically separated by genetic manipulation. Some GR mutants are transcriptionally inactive but competent to mediate interference. In one of these, a deletion of the N-terminal half of the protein, including the constitutive transactivation domain τ_1, caused a 90% loss transcriptional activity, while a significant (36%) portion of its repressive activity against AP-1 was retained (Schule et al. 1990; Yang et al. 1990). A series of GR point mutants had a similar phenotype: selected amino acid substitutions in τ_1 and in the second Zn^{2+}-binding domain disrupted its ability to transactivate the GRE but not to repress the AP-1 element (Heck et al. 1994). A counterintuitive finding is that the repression function of the GR against AP-1 remains intact when its DBD is substituted with DBDs from other nuclear receptors (Schule et al. 1990; Heck et al. 1994).

The role of the GR in vivo has been explored through the use of mice in which exons encoding the GR have been deleted or modified by targeted homologous recombination. In the first such mouse, a neo gene was inserted into the second exon of the GR, which contains $\tau 1$ (Cole et al. 1995). In homozygous day-19 fetuses, the distribution of thymocyte subpopulations, as defined by CD4 and CD8 expression, were normal; however, the thymocytes did not undergo apoptosis upon treatment with Dex. It was concluded that glucocorticoids and glucocorticoid-mediated apoptosis play no role in normal T-cell development (Purton et al. 2000).

Most mice homozygous for the exon 2 disruption died shortly after birth, due to a defect in surfactant production. However, 20% of the homozygous mutant fetuses grew to adulthood, implying either GR-independent surfactant production or incomplete deletion of the GR (Kellendonk et al. 1999). Indeed, a truncated 40-kDa GR product, possibly initiated near the beginning of the third exon (residue 407, encoding a peptide of 42.3 kDa), is expressed in thymocytes and lung cells from these mice (Cole et al. 2001). This fragment, which would contain the DBD, LBD, and τ_2, exhibited near-normal hormone-binding capacity. An alternate initiation site at residue 493 would generate a product containing only the LBD and τ_2. However, the smaller expected size of this product, 33.4 kDa, makes the upstream site more plausible. Using reverse transcriptase polymerase chain reaction (RT-PCR) analysis, we found that the GR-specific mRNA of fibroblasts derived from homozygous exon 2-mutant fetuses contained exon 3 and thus the upstream (residue 407) initiation site (unpublished observations). The levels of glucocorticoids in the plasma of the exon 2-mutant mice were elevated (Kellendonk et al. 1999), indicating that feedback inhibition of the hypothalamic-pituitary-adrenal (HPA) system, which depends on inhibition of the pro-opiomelanocortin (POMC) gene by GR dimers acting at a negative GRE (see above), was disrupted, (Cole et al. 1995). Given the partial repressive activity retained by a GR 5' truncation mutant lacking most of its N-terminal activation domain (see also above), it is possible that inefficient transcriptional interference mediated by the fragment lacking exon 2 could be complemented by the increased levels of circulating glucocorticoids. In unpublished experiments, deletion of exon 3 of the GR led to more complete penetrance, with no live homozygous births and otherwise more severe phenotypes in the fetuses (Tronche et al. 1998). In the exon 3-mutant mice, the exon 2-encoded fragment, containing the constitutive activation domain, could be expressed, complicating interpretation of these mice. The formal elimination of GR-related activity may require concurrent disruption of both exons 2 and 3; a task complicated by the large distance (>50 kb) between exons 2 and 3. Thus, it is possible that glucocorticoid effects on the T-cell repertoire could remain intact in these mice.

The importance of the transcription-activating function of the GR was addressed through the creation of a mouse whose GR was replaced with a dimerization-deficient single point mutant (GR^{dim}) (Heck et al. 1994), which cannot transactivate a classical GRE but can repress AP-1 in vitro (Reichardt et al. 1998). Whereas two genes inducible by Dex in the skin, Hsp-27 and PGX-3, were not induced by Dex in Gr^{dim} mice, genes encoding metalloproteases, when induced by phorbol esters, were repressed normally by Dex (Tuckermann et al. 1999). Like the exon 2-disrupted mice, Gr^{dim} mice had elevated levels of circulating glucocorticoids due to lack of GR dimer-dependent repression of POMC transcription. Gr^{dim} thymocytes again were refractory to Dex-induced apoptosis, and in mature mice T cells appeared to have developed normally as well. It was concluded that the interfering function of the GR was responsible for the majority of the developmental effects of glucocorticoids.

18.3 GILZ

A screen for glucocorticoid-inducible genes that regulate thymocyte susceptibility to glucocorticoid-induced apoptosis identified a novel gene, which was named GILZ (glucocorticoid-induced leucine zipper) (D'Adamio et al. 1997). Expression of GILZ is restricted to resting B cells, in which it is reduced in response to stimulation with antigen (Glynne et al. 2000), thymocytes, and T cells. In the latter two it is present at low basal levels and is strongly induced by glucocorticoids. Rather than inducing apoptosis, overexpression of GILZ in a T-cell hybridoma prevented activation-induced cell death, presumably by inhibition of FasL protein upregulation (D'Adamio et al. 1997). We have previously shown that induction of FasL mRNA in T cells does not take place in the absence of new protein synthesis, an observation consistent with the notion that this gene responds to transcriptional regulators that must be synthesized de novo (Mittelstadt and Ashwell 1998). In fact, Egr-2 and Egr-3, activation-inducible and NFAT-dependent transcription factors, were found to bind to a well-defined region in the 5′ FasL promoter (the FLRE, or Fas-ligand regulatory element), and this interaction was required for FasL mRNA induction. To determine if GILZ interferes with FasL transcription, we used a luciferase reporter construct driven by the 511-bp fragment of the FasL promoter. When expressed in the Jurkat T lymphoma, this reporter is induced approximately 14-fold by stimulation with PMA and ionomycin, a response that was markedly reduced (approximately 85%) by co-expressed GILZ (Mittelstadt and Ashwell 2001). Moreover, glucocorticoid induction of a reporter driven by the FLRE alone was also inhibited by GILZ, suggesting that the Egrs were a target of this protein. Indeed, reporters driven by the respective promoters for these genes were also inhibited by GILZ.

Because Egr-2 and Egr-3 require NFAT for their transcriptional induction, the induction of NFAT-driven reporters in activated T cells was addressed. NFAT is a complex transcription factor typically composed of one cytosolic component (NFATp, NFATc, and NFAT4 are expressed in T lymphocytes) and the nuclear-resident transcription factor AP-1 (Macian et al. 2001). In fact, an isolated NFAT element as well as an AP-1-responsive reporter was blocked by GILZ. To determine the possible molecular mechanism of inhibition, in vitro binding studies were performed. Beads coated with glutathione-S-transferase (GST)-GILZ fusion proteins were offered the in vitro translated and metabolically labeled AP-1 components, c-Jun and c-Fos. Both were pulled down by GST-GILZ, but not GST alone. In contrast, neither NFATp nor NFAT4 bound the GST-GILZ fusion protein. Furthermore, the GST-GILZ fusion protein prevented AP-1 isolated from cells treated with phorbol ester from binding to an AP-1 DNA element in vitro, as determined by an electromobility shift assay (Mittelstadt and Ashwell 2001). To bind to cognate DNA target sites, c-Fos and c-Jun must heterodimerize (or c-Jun homodimerize), an interaction that requires their respective leucine zipper regions. Surprisingly, although GILZ also contains a leucine zipper that is required for its own homodimerization, this region was dispensable for the binding between GILZ and either c-Fos or c-Jun. Instead, the N-terminal 60 residues of GILZ, a region that contains no well-characterized structural features, interacted with c-Fos and c-Jun. Consistent with the binding results, the ability of GILZ to repress AP-1-dependent transcription was not significantly affected by removal of the leucine zipper (Mittelstadt and Ashwell 2001). Thus, GILZ, like the GR, can interfere with AP-1 as a monomer. GILZ did interact, via its leucine zipper, with TSC-22, a homologous small leucine zipper-containing protein also thought to serve as a transcriptional repressor. Identified on the basis of its inducibility by TGFβ, TSC-22 is also inducible by glucocorticoids (Kester et al. 1999). The leucine zipper of GILZ may serve to mediate cross regulation by similar small regulatory proteins.

In another study, the transcription factor NF-κB was identified as a target for repression by GILZ (Ayroldi et al. 2001). In this case, GILZ interacted with the NF-κB components p52 and p65. The ability of GILZ to interfere with the p52 and p65 components of NF-κB, neither of which have leucine zippers, is consistent with the leucine zipper-independent interaction between GILZ and AP-1.

The available evidence suggests that in T cells GILZ can suppress two prominent cytokine-regulating transcription factors, AP-1 and NF-κB, and thus may be general repressor of cytokine expression. Accordingly, GILZ was tested for its effect on the AP-1- and NF-κB-dependent IL-2 promoter in Jurkat cells – it inhibited the induction of this promoter quite efficiently (Mittelstadt and Ashwell 2001). Using T-cell hybridomas stably transfected with GILZ, Aryoldi et al. showed a similar repression of activation-induced upregulation of IL-2 and IL-2 receptor proteins (Ayroldi et al. 2001). If GILZ were indeed an important mediator of immunosuppression by glucocorticoids, one would predict that it would be sensitive to inhibition of protein synthesis. If direct interference between the GR and preexisting transcription factors (such as AP-1 and NF-κB) is the major mechanism, however, then de novo protein synthesis would not be required. To distinguish between these possibilities, we asked if inhibition of activation-induced IL-2 mRNA upregulation in normal T cells by glucocorticoids requires new protein synthesis. The results were remarkably clear. Using human peripheral blood mononuclear cells (PBMC) treated for 3 h with PMA and ionomycin, cycloheximide prevented Dex-mediated inhibition of IL-2 mRNA expression (Mittelstadt and Ashwell 2001). The results we obtained with normal T cells, without the use of overexpressed exogenous proteins, indicate that a newly synthesized factor (or factors) plays a major role in transcriptional interference by glucocorticoids, and suggest that direct interference with pre-existing transcription factors by the GR is not a major mechanism in T cells.

If GILZ does play a important role in mediating the effects of glucocorticoids in thymocytes and T cells, it is an interesting speculation that GILZ would be induced by glucocorticoids in T cells from Gr^{dim} mice (if in fact T cells function normally in these animals). A straightforward way for this to occur would be for monomeric GR to induce GILZ. Precedents for monomeric GR-inducing gene transcription do exist. Dex-mediated induction of mouse α-amylase-2 gene transcription mapped to a GRE half site, to which the GR DBD bound as a monomer (Slater et al. 1993). GR^{dim}, as well as a transactivation-defective GR with a point mutation in the DBD (LS-7; Helmberg et al. 1995), were able to induce the p21 gene and cause growth arrest (Rogatsky et al. 1999). Finally, without contacting DNA, the GR cooperates with Stat5 in regulating the milk-protein gene β-casein (Stoecklin et al. 1997; Rosen et al. 1998). This feature enables lactating Gr^{dim} females to produce milk normally (Reichardt et al. 2001). Alternative indirect models for the GR to induce gene transcription utilizing its repressive function can be envisioned. For example, the GR^{dim} could block synthesis of a labile transcriptional repressor. Consistent with this possibility, we found that induction of GILZ mRNA by Dex in normal PBMCs was not repressed, but further increased, by cycloheximide (unpublished observations).

The role of GILZ as an intermediate between the GR and its cytokine targets suggests an alternative site for the genetic defects previously attributed to the GR, which have been implicated in glucocorticoid resistance in conditions such as Crohn's disease (Franchimont et al. 1999). As a lymphoid-specific mediator of glucocorticoid suppression, GILZ presents a potential target for therapeutic intervention. Efforts have been made to generate synthetic glucocorticoid derivatives capable of repression while being unable to induce transcription, with the intent of avoiding the consequences of long-term activation of glucocorticoid-induced genes (Herrlich 2001). It may be possible to design glucocorticoid derivatives to selectively block GILZ upregulation while allowing the GR to carry out interference in non-lymphoid cells normally. At any rate, GILZ-mediated interference operates on both inhibiting and the enhancing modes of glucocorticoids, depending on the gene being affected. The benefit of an indirect mechanism for interference could be to increase the sensitivity of T cells to glucocorticoids. A logical extension of this notion is that GILZ serves as a dampening mechanism, initially delaying, and then prolonging, the effects of glucocorticoids.

18.4 IL-7 Receptor

The unexpected involvement of glucocorticoids in regulating the IL-7 receptor came to light because of a screen for glucocorticoid-inducible genes in T cells. cDNA from human peripheral blood mononuclear cells cultured in the presence or absence of Dex for 16 h was used to probe a cDNA microarray of almost 10,000 genes (D. Franchimont et al., submitted). Interestingly, the most strongly induced cDNA corresponded to GILZ. The fifth most strongly induced was the gene encoding the IL-7 receptor α chain (IL-7Rα). The central role that IL-7 plays in T-cell development and homeostasis prompted us to further examine the effect of glucocorticoids and IL-7 on T-cell biology.

The size of the T-cell compartment is tightly controlled, and can be maintained despite major perturbations that can be brought about by infection or by therapeutic or experimental manipulation (Van Parijs and Abbas 1998). It has been proposed that the absolute number of T cells depends on limiting "resources," or more precisely on limiting ligand(s), that drive T-cells proliferation and/or survival (Freitas and

Rocha 2000). One such well-documented class of ligands includes the products of the MHC locus complexed to antigenic peptides. The form of MHC plus peptide required for T-cell development and survival varies according to T-cell type and stage of differentiation. Whereas both naïve CD8+ (Tanchot et al. 1997; Ernst et al. 1999) and CD4+ (Kirberg et al. 1997; Rooke et al. 1997) T cells appear to depend on the same positively-selecting peptides and MHC molecules that were encountered during their thymic development, memory T cells (Zhang et al. 1998; Muraliet al. 1999), and in some studies naïve CD4+ T cells (Clarke et al. 2000; Dorfman et al. 2000), do not require an MHC-based survival signal. Given the varying dependence of T cells on TCR-derived signals for survival, it is not surprising that additional antigen-independent signals participate in T-cell homeostasis, and there is considerable evidence that IL-7 is a critical factor in this regard (Fry and Mackall 2001). During thymocyte development, the receptor for IL-7 (IL-7R) is expressed on CD4−CD8− (double negative, or DN), is down-regulated on CD4+CD8+ (double positive, DP) thymocytes, and is one again expressed on single positive thymocytes as well as peripheral T cells. IL-7 itself is expressed in locations appropriate for a role in T-cell homeostasis, such as epithelial cells of the thymus, lymph node, and intestine, as well as dendritic cells in lymph nodes. The IL-7R is composed of an α chain, which serves as the ligand binding component, and the common γ chain (γc), which is shared by the receptors for IL-2, -4, and -15, and which, through association with Jak3, transduces cytokine-triggered signals. The unique role that IL-7 plays in T-cell development is indicated by the findings that mice (Peschon et al. 1994) and humans (Puel et al. 1998) with disrupted IL-7Rα genes have greatly reduced numbers of T cells, while their B and NK cell compartments are relatively unperturbed, in contrast to the loss of the B and NK lineages that occurs when γc (DiSanto et al. 1995) or Jak3 (Nosaka et al. 1995; Russell et al. 1995) are disrupted. In the thymus, IL-7 signaling has been implicated in the DN2 (CD44+/CD25+) to DN3 (CD44−/CD25+) transition of DN cells, the stage during which rearrangement of the β chain of the TCR normally occurs (Peschon et al. 1994; von Freeden-Jeffry et al. 1995; Crompton et al. 1998; Haks et al. 1999), and in positive selection of CD4+ or CD8+ cells (Akashi et al. 1998; Brugnera et al. 2000). The poor responsiveness of the few T cells that do populate the periphery in IL-7-deficient mice (Maraskovsky et al. 1996), and the failure of normal

T cells to undergo homeostatic proliferation when transferred into T cell-depleted IL-7-deficient hosts (Schluns et al. 2000; Tan et al. 2001), indicate that IL-7 also plays a role in homeostasis of mature T cells. Memory cells, which, as mentioned above, appear to survive independently of antigen-derived signals, have been shown to depend both IL-7 and IL-15 (Zhang et al. 1998; Schluns et al. 2000).

Given the vital role IL-7 has in T-cell development and survival, it was surprising that glucocorticoids, known for their immunosuppressive activities and potency at inducing lymphocyte apoptosis, might actually upregulate the IL-7R. To explore this possibility, the IL-7Rα mRNA levels in Dex-treated PBMC were assessed. IL-7Rα mRNA increased within 2 h, and remained elevated for 18 h, in response to Dex treatment (D. Franchimont et al., submitted). The glucocorticoid antagonist RU486 blocked this induction, indicating that the GR mediates this process. To determine whether increased IL-7Rα mRNA levels correlated with increased expression of IL-7Rα protein, peripheral blood T cells and purified naïve CD4+ cord blood T cells were examined for IL-7 receptor expression by flow cytometry. Treatment with Dex overnight in both cases caused an approximately threefold increase in cell surface IL-7 receptor expression. To determine if these newly synthesized IL-7 receptors were functional, the expression of a downstream target of IL-7 signaling, the IL-2Rα protein (CD25), was assessed (Ascherman et al. 1997). Whereas treatment of naïve CD4+ cord blood T cells with Dex had no effect, treatment with IL-7 caused approximately a twofold increase in CD25 levels. Simultaneous treatment with IL-7 and Dex was synergistic, resulting in a further tenfold increase in CD25levels. This confirmed that glucocorticoid-induced IL-7 receptors are functional. Thus, the interaction between glucocorticoids and IL-7 appears to be example of an immune-enhancing activity of glucocorticoids. This finding also indicates that normally IL-7 receptor expression is limiting, and that regulation of receptor expression provides an additional mechanism to control the effects of IL-7.

Stimulation of T cells via the TCR has been reported to downregulate IL-7Rα expression (Schluns et al. 2000). This phenomenon may serve to enhance the death of activated cells by Fas ligand-mediated activation-induced cell death, and stands in contrast to the upregulation we observed in IL-7Rα in response to Dex. We asked how these stimuli modulate the T cell response to the other. Normal T cells activated with immobilized antibodies against CD3 and CD28 (the latter to provide costimulation), resulted in a sharp reduction in IL-7Rα mRNA and cell surface expression. Simultaneous stimulation with Dex prevented the activation-induced downregulation. If the increase in IL-7 levels on glucocorticoid-treated T cells is functionally meaningful, one might expect that in the presence of IL-7 glucocorticoids would increase T cell survival. Therefore, the effect of glucocorticoids and IL-7 on the spontaneous apoptosis of naïve CD4[+] cord blood T cells was determined. As expected, IL-7 alone reduced the spontaneous rate of apoptosis and Dex alone increased the rate of apoptosis. However, in the presence of IL-7, Dex actually reduced the rate of apoptosis already reduced by IL-7. This enhancement of IL-7 survival signals confirmed that the induced receptors were functional, and indicated that glucocorticoids do not disrupt the survival signals mediated by the IL-7 receptor. A corollary implication of this result is that IL-7 signaling antagonizes glucocorticoid-induced apoptosis. Thus, glucocorticoids interfere with activation-induced cell death by at least two mechanisms: inhibition of FasL upregulation and maintenance of IL-7R expression in the face of TCR-mediated activation.

Antigen recognition by T cells leads to the development of either the T helper (Th)1 [interferon (IFN)γ-secreting, primarily cellular] or Th2 (humoral) phenotype (Mosmann and Sad 1996; Romagnani 1997). In keeping with the notion that glucocorticoids and IL-7 can synergize to promote T-cell development and survival, a parallel between the two systems with respect to Th lineage commitment can be drawn, with both systems appearing to promote the Th2, or IL-4-secreting, phenotype. Production of IL-4 in vitro by human cord blood CD4[+] cells during secondary stimulation was found to depend on the presence of IL-7 during primary stimulation (Webb et al. 1997). Th2 skewing induced by treatment with IL-7 was associated with increased pathogenicity of the parasite Schistosoma mansoni, which is normally controlled by a Th1-directed humoral response (Wolowczuk et al. 1997, 1999). Glucocorticoids have also been noted to cause a shift toward Th2 responses. While glucocorticoids suppress both Th1 and Th2 cytokines, their effects on Th1 cytokine production are more pronounced, leading to relative enhancement of Th2 cytokines. Glucocorticoid-mediated Th2 redirection has been implicated in increasing the pathogenicity of infections that are normally resolved by Th1-directed immunity (Elenkov and Chrousos

1999; Ashwell et al. 2000; Franchimont et al. 2000; Sternberg 2001). Transient treatment of rat CD4+ T cells with Dex induced a Th2 cytokine response upon subsequent restimulation in the absence of Dex (Ramirez 1998). Our results indicate that the Th2 skewing caused by IL-7 and glucocorticoids are directly related: glucocorticoids enhance the responsiveness of T cells to IL-7 by increasing the expression of functional IL-7 receptors.

18.5 Conclusion

Glucocorticoids are appreciated to be a means of conveying local and global information related to metabolism and stress. With respect to the immune system, their effects have been largely considered to be inhibitory. Two mechanisms of glucocorticoid action reviewed here, which inhibit activation-induced cell death by the induction of a pro-survival receptor (IL-7Rα) and the GILZ-mediated suppression of FasL expression, provide examples of positive influences of glucocorticoids on the immune system. Elucidation of how these biological activities are coordinated in vivo will further understanding of the consequences of glucocorticoid dysregulation and of prolonged glucocorticoid therapy. For example, by increasing responsiveness of T cells to IL-7, glucocorticoids could facilitate the persistence of T cell clones recognizing low-affinity ligands. Together with interference with activation-induced cell death, glucocorticoids could paradoxically enhance autoimmunity. Manipulation of IL-7 signaling and GILZ-mediated interference may be a means with which to influence the immune-specific effects of glucocorticoids.

References

Adcock IM (2001) Glucocorticoid-regulated transcription factors. Pulm Pharmacol Ther 14:211

Akashi K, Kondo M, Weissman IL (1998) Two distinct pathways of positive selection for thymocytes. Proc Natl Acad Sci USA 95:2486

Almawi WY, Beyhum HN, Rahme AA, Rieder MJ (1996) Regulation of cytokine and cytokine receptor expression by glucocorticoids. J Leukoc Biol 60:563

Ascherman DP, Migone TS, Friedmann MC, Leonard WJ (1997) Interleukin-2 (IL-2)-mediated induction of the IL-2 receptor alpha chain gene. Critical role of two functionally redundant tyrosine residues in the IL-2 receptor beta chain cytoplasmic domain and suggestion that these residues mediate more than Stat5 activation. J Biol Chem 272:8704

Ashwell JD, Lu FW, Vacchio MS (2000) Glucocorticoids in T cell development and function. Annu Rev Immunol 18:309

Auphan N, DiDonato JA, Rosette C, Helmberg A, Karin M (1995) Immunosuppression by glucocorticoids: inhibition of NF-κB activity through induction of IκB synthesis. Science 270:286

Ayroldi E, Migliorati G, Bruscoli S, Marchetti C, Zollo O, Cannarile L, D'Adamio F, Riccardi C (2001) Modulation of T-cell activation by the glucocorticoid-induced leucine zipper factor via inhibition of nuclear factor kappaB. Blood 98:743

Brugnera E, Bhandoola A, Cibotti R, Yu Q, Guinter TI, Yamashita Y, Sharrow SO, Singer A (2000) Coreceptor reversal in the thymus: signaled CD4+8+ thymocytes initially terminate CD8 transcription even when differentiating into CD8+ T cells. Immunity 13:59

Chang TJ, Scher BM, Waxman S, Scher W (1993) Inhibition of mouse GATA-1 function by the glucocorticoid receptor: possible mechanism of steroid inhibition of erythroleukemia cell differentiation. Mol Endocrinol 7:528

Clarke SR, Rudensky AY (2000) Survival and homeostatic proliferation of naive peripheral CD4+ T cells in the absence of self peptide:MHC complexes. J Immunol 165:2458

Cole TJ, Blendy JA, Monaghan AP, Krieglstein K, Schmid W, Aguzzi A, Fantuzzi G, Hummler E, Unsicker K, Schutz G (1995) Targeted disruption of the glucocorticoid receptor gene blocks adrenergic chromaffin cell development and severely retards lung maturation. Genes Dev 9:1608

Cole TJ, Myles K, Purton JF, Brereton PS, Solomon NM, Godfrey DI, Funder JW (2001) GRKO mice express an aberrant dexamethasone-binding glucocorticoid receptor, but are profoundly glucocorticoid resistant. Mol Cell Endocrinol 173:193

Collingwood TN, Urnov FD, Wolffe AP (1999) Nuclear receptors: coactivators, corepressors and chromatin remodeling in the control of transcription. J Mol Endocrinol 23:255

Crompton T, Outram SV, Buckland J, Owen MJ (1998) Distinct roles of the interleukin-7 receptor alpha chain in fetal and adult thymocyte development revealed by analysis of interleukin-7 receptor alpha-deficient mice. Eur J Immunol 28:1859

D'Adamio F, Zollo O, Moraca R, Ayroldi E, Bruscoli S, Bartoli A, Cannarile L, Migliorati G, Riccardi C (1997) A new dexamethasone-induced gene of the leucine zipper family protects T lymphocytes from TCR/CD3-activated cell death. Immunity 7:803

De Bosscher K, Schmitz ML, Vanden Berghe W, Plaisance S, Fiers W, Haegeman G (1997) Glucocorticoid-mediated repression of NF-κB-dependent transcription involves direct interference with transactivation. Proc Natl Acad Sci USA 94:13504

De Bosscher K, Vanden Berghe W, Haegeman G (2001) Glucocorticoid repression of AP-1 is not mediated by competition for nuclear coactivators. Mol Endocrinol 15:219

De Bosscher K, Vanden W Berghe, Haegeman G (2000) Mechanisms of anti-inflammatory action and of immunosuppression by glucocorticoids: negative interference of activated glucocorticoid receptor with transcription factors. J Neuroimmunol 109:16

Diamond MI, Miner JN, Yoshinaga SK, Yamamoto KR (1990) Transcription factor interactions: selectors of positive or negative regulation from a single DNA element. Science 249:1266

DiSanto JP, Muller W, Guy-Grand D, Fischer A, Rajewsky K (1995) Lymphoid development in mice with a targeted deletion of the interleukin 2 receptor gamma chain. Proc Natl Acad Sci USA 92:377

Dorfman JR, Stefanova I, Yasutomo K, Germain RN (2000) CD4+ T cell survival is not directly linked to self-MHC-induced TCR signaling. Nat Immunol 1:329

Drouin J, Trifiro MA, Plante RK, Nemer M, Eriksson P, Wrange O (1989) Glucocorticoid receptor binding to a specific DNA sequence is required for hormone-dependent repression of pro-opiomelanocortin gene transcription. Mol Cell Biol 9:5305

Elenkov IJ, Chrousos GP (1999) Stress Hormones, Th1/Th2 patterns, pro/anti-inflammatory cytokines and susceptibility to disease. Trends Endocrinol Metab 10:359

Ernst B, Lee DS, Chang JM, Sprent J, Surh CD (1999) The peptide ligands mediating positive selection in the thymus control T cell survival and homeostatic proliferation in the periphery. Immunity 11:173

Franchimont D, Galon J, Gadina M, Visconti R, Zhou Y, Aringer M, Frucht DM, Chrousos GP, O'Shea JJ (2000) Inhibition of Th1 immune response by glucocorticoids: dexamethasone selectively inhibits IL-12-induced Stat4 phosphorylation in T lymphocytes. J Immunol 164:1768

Franchimont D, Louis E, Dupont P, Vrindts-Gevaert Y, Dewe W, Chrousos G, Geenen V, Belaiche J (1999) Decreased corticosensitivity in quiescent Crohn's disease: an ex vivo study using whole blood cell cultures. Dig Dis Sci 44:1208

Freitas AA, Rocha B (2000) Population biology of lymphocytes: the flight for survival. Annu Rev Immunol 18:83

Fry TJ, Mackall CL (2001) Interleukin-7: master regulator of peripheral T-cell homeostasis? Trends Immunol 22:564

Gauthier JM, Bourachot B, Doucas V, Yaniv M, Moreau F -Gachelin (1993) Functional interference between the Spi-1/PU.1 oncoprotein and steroid hormone or vitamin receptors. EMBO J 12:5089

Glynne R, Ghandour G, Rayner J, Mack DH, Goodnow CC (2000) B-lymphocyte quiescence, tolerance and activation as viewed by global gene expression profiling on microarrays. Immunol Rev 176:216

Gonzalez MV, Jimenez B, Berciano MT, Gonzalez-Sancho JM, Caelles C, Lafarga M, Munoz A (2000) Glucocorticoids antagonize AP-1 by inhibiting the Activation/phosphorylation of JNK without affecting its subcellular distribution. J Cell Biol 150:1199

Haks MC, Oosterwegel MA, Blom B, Spits HM, Kruisbeek AM (1999) Cell-fate decisions in early T cell development: regulation by cytokine receptors and the pre-TCR. Semin Immunol 11:23

Heck S, Bender K, Kullmann M, Gottlicher M, Herrlich P, Cato AC (1997) IκBα-independent downregulation of NF-κB activity by glucocorticoid receptor. EMBO J 16:4698

Heck S, Kullmann M, Gast A, Ponta H, Rahmsdorf HJ, Herrlich P, Cato AC (1994) A distinct modulating domain in glucocorticoid receptor monomers in the repression of activity of the transcription factor AP-1. EMBO J 13:4087

Helmberg A, Auphan N, Caelles C, Karin M (1995) Glucocorticoid-induced apoptosis of human leukemic cells is caused by the repressive function of the glucocorticoid receptor. EMBO J 14:452

Herrlich P (2001) Cross-talk between glucocorticoid receptor and AP-1. Oncogene 20:2465

Imai E, Miner JN, Mitchell JA, Yamamoto KR, Granner DK (1993) Glucocorticoid receptor-cAMP response element-binding protein interaction and the response of the phosphoenolpyruvate carboxykinase gene to glucocorticoids. J Biol Chem 268:5353

Jantzen HM, Strahle U, Gloss B, Stewart F, Schmid W, Boshart M, Miksicek R, Schutz G (1987) Cooperativity of glucocorticoid response elements located far upstream of the tyrosine aminotransferase gene. Cell 49:29

Jenkins BD, Pullen CB, Darimont BD (2001) Novel glucocorticoid receptor coactivator effector mechanisms. Trends Endocrinol Metab 12:122

Jonat C, Rahmsdorf HJ, Park KK, Cato AC, Gebel S, Ponta H, Herrlich P (1990) Antitumor promotion and antiinflammation: down-modulation of AP-1 (Fos/Jun) activity by glucocorticoid hormone. Cell 62:1189

Kamei Y, Xu L, Heinzel T, Torchia J, Kurokawa R, Gloss B, Lin SC, Heyman RA, Rose DW, Glass CK, Rosenfeld MG (1996) A CBP integrator complex mediates transcriptional activation and AP-1 inhibition by nuclear receptors. Cell 85:403

Karin M, Ben-Neriah Y (2000) Phosphorylation meets ubiquitination: the control of NF-[kappa]B activity. Annu Rev Immunol 18:621

Karin M, Haslinger A, Holtgreve H, Richards RI, Krauter P, Westphal HM, Beato M (1984) Characterization of DNA sequences through which cadmium and glucocorticoid hormones induce human metallothionein-IIA gene. Nature 308:513

Kellendonk C, Tronche F, Reichardt HM, Schutz G (1999) Mutagenesis of the glucocorticoid receptor in mice. J Steroid Biochem Mol Biol 69:253

Kester HA, Blanchetot C, den Hertog J, van der Saag PT, van der Burg B (1999) Transforming growth factor-β-stimulated clone-22 is a member of a family of leucine zipper proteins that can homo- and heterodimerize and has transcriptional repressor activity. J Biol Chem 274:27439

Kirberg J, Berns A, von Boehmer H (1997) Peripheral T cell survival requires continual ligation of the T cell receptor to major histocompatibility complex-encoded molecules. J Exp Med 186:1269

Kunicka JE, Talle MA, Denhardt GH, Brown M, Prince LA, Goldstein G (1993) Immunosuppression by glucocorticoids: inhibition of production of multiple lymphokines by in vivo administration of dexamethasone. Cell Immunol 149:39

Kutoh E, Stromstedt PE, Poellinger L (1992) Functional interference between the ubiquitous and constitutive octamer transcription factor 1 (OTF-1) and the glucocorticoid receptor by direct protein-protein interaction involving the homeo subdomain of OTF-1. Mol Cell Biol 12:4960

Macian F, Lopez-Rodriguez C, Rao A (2001) Partners in transcription: NFAT and AP-1. Oncogene 20:2476

Maraskovsky E, Teepe M, Morrissey PJ, Braddy S, Miller RE, Lynch DH, Peschon JJ (1996) Impaired survival and proliferation in IL-7 receptor-deficient peripheral T cells. J Immunol 157:5315

Mittelstadt PR, Ashwell JD (1998) Cyclosporin A-sensitive transcription factor Egr-3 regulates Fas ligand expression. Mol Cell Biol 18:3744

Mittelstadt PR, Ashwell JD (2001) Inhibition of ap-1 by the glucocorticoid-inducible protein GILZ. J Biol Chem 276:29603

Mordacq JC, Linzer DI (1989) Co-localization of elements required for phorbol ester stimulation and glucocorticoid repression of proliferin gene expression. Genes Dev 3:760

Mosmann TR, Sad S (1996) The expanding universe of T-cell subsets: Th1, Th2 and more. Immunol Today 17:138

Murali-Krishna K, Lau LL, Sambhara S, Lemonnier F, Altman J, Ahmed R (1999) Persistence of memory CD8 T cells in MHC class I-deficient mice. Science 286:1377

Newton R (2000) Molecular mechanisms of glucocorticoid action: what is important? Thorax 55:603

Nosaka T, van JM Deursen, Tripp RA, Thierfelder WE, Witthuhn BA, McMickle AP, Doherty PC, Grosveld GC, Ihle JN (1995) Defective lymphoid development in mice lacking Jak3. Science 270:800

Peschon JJ, Morrissey PJ, Grabstein KH, Ramsdell FJ, Maraskovsky E, Gliniak BC, Park LS, Ziegler SF, Williams DE, Ware CB, et al (1994) Early lymphocyte expansion is severely impaired in interleukin 7 receptor-deficient mice. J Exp Med 180:1955

Puel A, Ziegler SF, Buckley RH, Leonard WJ (1998) Defective IL7R expression in T(–)B(+)NK(+) severe combined immunodeficiency. Nat Genet 20:394

Purton JF, Boyd RL, Cole TJ, Godfrey DI (2000) Intrathymic T cell development and selection proceeds normally in the absence of glucocorticoid receptor signaling. Immunity 13:179

Ramirez F (1998) Glucocorticoids induce a Th2 response in vitro. Dev Immunol 6:233

Ray A, Prefontaine KE (1994) Physical association and functional antagonism between the p65 subunit of transcription factor NF-κB and the glucocorticoid receptor. Proc Natl Acad Sci USA 91:752

Reichardt HM, Horsch K, Grone HJ, Kolbus A, Beug H, Hynes N, Schutz G (2001) Mammary gland development and lactation are controlled by different glucocorticoid receptor activities. Eur J Endocrinol 145:519

Reichardt HM, Kaestner KH, Tuckermann J, Kretz O, Wessely O, Bock R, Gass P, Schmid W, Herrlich P, Angel P, Schutz G (1998) DNA binding of the glucocorticoid receptor is not essential for survival. Cell 93:531

Rogatsky I, Hittelman AB, Pearce D, Garabedian MJ (1999) Distinct glucocorticoid receptor transcriptional regulatory surfaces mediate the cytotoxic and cytostatic effects of glucocorticoids. Mol Cell Biol 19:5036

Romagnani S (1997) The Th1/Th2 paradigm. Immunol Today 18:263

Rooke R, Waltzinger C, Benoist C, Mathis D (1997) Targeted complementation of MHC class II deficiency by intrathymic delivery of recombinant adenoviruses. Immunity 7:123

Rosen JM, Zahnow C, Kazansky A, Raught B (1998) Composite response elements mediate hormonal and developmental regulation of milk protein gene expression. Biochem Soc Symp 63:101

Russell SM, Tayebi N, Nakajima H, Riedy MC, Roberts JL, Aman MJ, Migone TS, Noguchi M, Markert ML, Buckley RH, et al (1995) Mutation of Jak3 in a patient with SCID: essential role of Jak3 in lymphoid development. Science 270:797

Sakai DD, Helms S, Carlstedt-Duke J, Gustafsson JA, Rottman FM, Yamamoto KR (1988) Hormone-mediated repression: a negative glucocorticoid response element from the bovine prolactin gene. Genes Dev 2:1144

Scheinman RI, Cogswell PC, Lofquist AK, J Baldwin AS (1995) Role of transcriptional activation of Iκ alpha in mediation of immunosuppression by glucocorticoids. Science 270:283

Schluns KS, Kieper WC, Jameson SC, Lefrancois L (2000) Interleukin-7 mediates the homeostasis of naive and memory CD8 T cells in vivo. Nat Immunol 1:426

Schule R, Rangarajan P, Kliewer S, Ransone LJ, Bolado J, Yang N, Verma IM, Evans RM (1990) Functional antagonism between oncoprotein c-Jun and the glucocorticoid receptor. Cell 62:1217

Sheppard KA, Phelps KM, Williams AJ, Thanos D, Glass CK, Rosenfeld MG, Gerritsen ME, Collins T (1998) Nuclear integration of glucocorticoid receptor and nuclear factor-kappaB signaling by CREB-binding protein and steroid receptor coactivator-1. J Biol Chem 273:29291

Slater EP, Hesse H, Muller JM, Beato M (1993) Glucocorticoid receptor binding site in the mouse alpha-amylase 2 gene mediates response to the hormone. Mol Endocrinol 7:907

Smith PJ, Cousins DJ, Jee YK, Staynov DZ, Lee TH, Lavender P (2001) Suppression of granulocyte-macrophage colony-stimulating factor expression by glucocorticoids involves inhibition of enhancer function by the glucocorticoid receptor binding to composite nf-at/activator protein-1 elements. J Immunol 167:2502

Sternberg EM (2001) Neuroendocrine regulation of autoimmune/inflammatory disease. J Endocrinol 169:429

Stoecklin E, Wissler M, Moriggl R, Groner B (1997) Specific DNA binding of Stat5, but not of glucocorticoid receptor, is required for their functional cooperation in the regulation of gene transcription. Mol Cell Biol 17:6708

Tan JT, Dudl E, LeRoy E, Murray R, Sprent J, Weinberg KI, Surh CD (2001) IL-7 is critical for homeostatic proliferation and survival of naive T cells. Proc Natl Acad Sci USA 98:8732

Tanchot C, Lemonnier FA, Perarnau B, Freitas AA, Rocha B (1997) Differential requirements for survival and proliferation of CD8 naive or memory T cells. Science 276:2057

Tronche F, Kellendonk C, Reichardt HM, Schutz G (1998) Genetic dissection of glucocorticoid receptor function in mice. Curr Opin Genet Dev 8:532

Tuckermann JP, Reichardt HM, Arribas R, Richter KH, Schutz G, Angel P (1999) The DNA binding-independent function of the glucocorticoid receptor mediates repression of AP-1-dependent genes in skin. J Cell Biol 147:1365

Van Laethem F, Baus E, Smyth LA, Andris F, Bex F, Urbain J, Kioussis D, Leo O (2001) Glucocorticoids attenuate T cell receptor signaling. J Exp Med 193:803

Van Parijs L, Abbas AK (1998) Homeostasis and self-tolerance in the immune system: turning lymphocytes off. Science 280:243

Ventura JJ, Roncero C, Fabregat I, Benito M (1999) Glucocorticoid receptor down-regulates c-Jun amino terminal kinases induced by tumor necrosis factor alpha in fetal rat hepatocyte primary cultures. Hepatology 29:849

von Freeden-Jeffry U, Vieira P, Lucian LA, McNeil T, Burdach SE, Murray R (1995) Lymphopenia in interleukin (IL)-7 gene-deleted mice identifies IL-7 as a nonredundant cytokine. J Exp Med 181:1519

Webb LM, Foxwell BM, Feldmann M (1997) Interleukin-7 activates human naive CD4+ cells and primes for interleukin-4 production. Eur J Immunol 27:633

Wisdom R (1999) AP-1: one switch for many signals. Exp Cell Res 253:180

Wolowczuk I, Delacre M, Roye O, Giannini SL, Auriault C (1997) Interleukin-7 in the skin of Schistosoma mansoni-infected mice is associated with a decrease in interferon-gamma production and leads to an aggravation of the disease. Immunology 91:35

Wolowczuk I, Nutten S, Roye O, Delacre M, Capron M, Murray RM, Trottein F, Auriault C (1999) Infection of mice lacking interleukin-7 (IL-7) reveals an unexpected role for IL-7 in the development of the parasite Schistosoma mansoni. Infect Immun 67:4183

Xu L, Glass CK, Rosenfeld MG (1999) Coactivator and corepressor complexes in nuclear receptor function. Curr Opin Genet Dev 9:140

Yamamoto KR (1985) Steroid receptor regulated transcription of specific genes and gene networks. Annu Rev Genet 19:209

Yamamoto KR, Alberts BM (1976) Steroid receptors: elements for modulation of eukaryotic transcription. Annu Rev Biochem 45:721

Yang Y, Mercep M, Ware CF, Ashwell JD (1995) Fas and activation-induced Fas ligand mediate apoptosis of T cell hybridomas: inhibition of Fas ligand expression by retinoic acid and glucocorticoids. J Exp Med 181:1673

Yang-Yen HF, Chambard JC, Sun YL, Smeal T, Schmidt TJ, Drouin J, Karin M (1990) Transcriptional interference between c-Jun and the glucocorticoid receptor: mutual inhibition of DNA binding due to direct protein-protein interaction. Cell 62:1205

Zhang G, Zhang L, Duff GW (1997) A negative regulatory region containing a glucocorticosteroid response element (nGRE) in the human interleukin-1β gene. DNA Cell Biol 16:145

Zhang X, Sun S, Hwang I, Tough DF, Sprent J (1998) Potent and selective stimulation of memory-phenotype CD8+ T cells in vivo by IL-15. Immunity 8:591

19 Structural Analysis of the GR Ligand-Binding Domain

U. Egner

19.1	Introduction	341
19.2	Comparative Modelling of the Glucocorticoid Receptor LBD	344
19.3	Comparison of LBD Structures Within the AR/PR/GR Subfamily	347
19.4	Orientation of Ligands in the LBP Within the AR/PR/GR Subfamily	354
19.5	Summary and Outlook	354
References		355

19.1 Introduction

The human glucocorticoid receptor (hGRα and hGRβ) belongs to the superfamily of nuclear receptors (NRs) which act as ligand-inducible transactivation factors (Gronemeyer and Laudet 1995; Moras and Gronemeyer 1998; Resche-Rigon and Gronemeyer 1998; McKay and Cidlowski 1999; Oakley et al. 1999). Other members of this family are the steroid (e.g. progesterone, PR; androgen, AR; mineralocorticoid, MR; oestrogen, hERα and hERβ), the thyroid (TR) and retinoic (X, RXR; acid, RAR) receptors and a variety of orphan receptors, where no ligand is known yet. Based on conserved sequence and function, all NRs can be divided into five to six domains (Bourguet et al. 2000; Egea et al. 2000). The structurally best-characterised domains are the DNA-binding domains and the ligand-binding domains (LBDs). LBDs harbour the ligand-binding pocket (LBP) as well as a ligand-dependent transactiva-

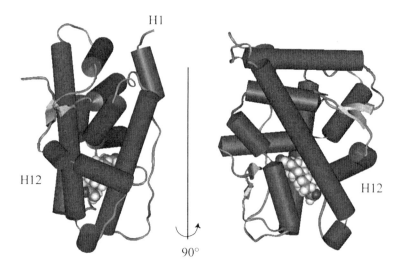

Fig. 1. Overall fold of PR LBD in complex with progesterone (Williams and Sigler 1998), α-helices are shown as *cylinders*, β-strands as *arrows*. The LBP is located in the lower part of the LBD. The cartoon was prepared with the Insight2000 software (Accelrys, San Diego, USA)

tion function. The DNA-binding domain and the LBD are both involved in receptor dimerisation. Many pharmaceutical companies aim at developing tailored ligands for the steroid LBD, to control or specifically modulate hormone-dependent processes like fertility control, osteoporosis, cancer and skin inflammatory diseases.

Presently, the crystal structures of more than 35 nuclear receptor LBD-ligand complexes are deposited in the latest version of the Protein Data Bank (Berman et al. 2000). All these structures confirm the existence of a common fold encompassing approximately 12 α-helices and up to 3 β-sheets arranged as a three-layered α-helical "sandwich" (Fig. 1) (Egea et al. 2000). Comparison of the LBD crystal structures revealed a structurally conserved "upper" part (helices H1, H4, H5, C-terminal part of H7, H8-H10) and a structurally more flexible "lower" part (N-terminal part of helix H7, H11, H12, sheets 1 and 2, loop H11/H12) where the LBP is located. Based on sequence identity and conserved function, the steroidal subfamily of NRs can be divided into two subgroups. The oestrogen receptors hERα and hERβ belong to one group while AR, PR, GR and MR are to be grouped into a second subgroup. The sequence identity within the subgroups is around 60% for the hERs and between 51% and 55% for the other subgroup. Currently, there is no experimental structure available for the GR LBD. Since the elucidation of the first crystal structures of a nuclear receptor LBD, several models for the GR LBD were developed based on related LBD structures with varying sequence homology (Wurtz et al. 1996; Lind et al. 2000; Dey et al. 2001). The crystal structures of the highly homologous AR and PR LBD provide a suitable basis to generate more accurate GR LBD models. We generated a homology model of the hGRα LBD based on the available crystal structure of the hAR LBD- and hPR LBD-R1881 and the hPR LBD-progesterone complexes. The hGRβ receptor is a splice variant of the human GR gene resulting in a protein, which does not bind the hormone any longer. This uncommon observation is most likely due to the considerably shorter C-terminus compared to the hGRα sequence (Oakley et al. 1999).

The crystal structures of the AR and PR LBD as well as the GR LBD model structures were used to characterise the LBPs in more detail (Williams and Sigler 1998; Matias et al. 2000; Sack et al. 2001). This analysis helps (1) in identifying structurally conserved areas as well as more flexible regions in the LBPs of this highly homologous subgroup and (2) in understanding how the relative positions of helices influence the size of the LBPs. The results of such an analysis are applied when ligands are fitted into a binding niche. In order to achieve a better understanding of the structural features that contribute to specificity, it is of great importance to learn how different receptors accommodate to the same ligand. The crystal structures of the AR and PR LBD in complex with the same ligand metribolone (R1881) gave us the opportunity to compare their LBP in detail and to achieve a better understanding of compound selectivity within the AR/PR/GR subgroup of steroid receptors (Matias et al. 2000). The AR and PR LBD crystal structures as well as our model structure of the hGRα LBD were used to fit progesterone, R1881 and dexamethasone into the LBP and to analyse if we can qualitatively explain their experimentally available relative binding affinities on the bases of the 3D structures.

19.2 Comparative Modelling of the Glucocorticoid LBD

19.2.1 Overall LBD Structure

A model of the hGR LBD was generated based on the coordinates of the hAR LBD-R1881 (PDB entry code 1e3g), the hPR LBD-R1881 complex (molecule A of PDB entry code 1e3k) and the hPR LBD-progesterone complex (molecule A of PDB file 1a28) and based on the sequence alignment shown in Fig. 2 using the Insight2000/Modeller software (Accelrys, San Diego, USA) applying standard parameters. All loops were kept as generated by Modeller. In the calculation, 15 models were generated and the model with the lowest energy and probability density function values was used for further analysis. The model was analysed using both PROCHECK (Laskowski et al. 1993) and Profiles-3D (as implemented in Insight2000).

In this model of the 1st run, most of the residues (95.2%) had their backbone dihedral angles located in the favoured or most-favoured regions (4.4%) of the Ramachandran plot. The side chain stereo-parameters are inside the range or better than the statistics derived from a set of crystal structures of at least 2.1 Å resolution. The analysis of the 3D-profile showed no regions with a negative score indicating areas which need further inspection. The overall probability density function (PDF) value for the GRα LBD model is very high as compared to the maximum value calculated for a protein of that size (105 "units" out of 114 possible). For comparison, the PDF value of the AR LBD structure used for model building is 95 and that of PR LBD is 101. The region Asn708–Trp713 has the lowest score (score≈0). This area is far away from the binding niche (region H9–H10) and a score value of that size can be accepted for that area. When the root-mean-square deviation (RMSD) of the GRα LBD model to the AR and PR LBD crystal structures was calculated, it was observed that the GR LBD structure was very similar to the PR LBD structures with a RMSD of 0.2 Å, while the RMSD to the AR LBD structure was much higher (1.1 Å for 211 Cα atoms). For comparison, the RMSD between the AR and PR LBD crystal structures is 1.2 Å. To test if this result was biased towards the number of templates which were used for model generation, a second calculation was performed (second run). In this run, only one structure from PR LBD and one structure from AR LBD were used (Table 1).

Structural Analysis of the GR Ligand-Binding Domain

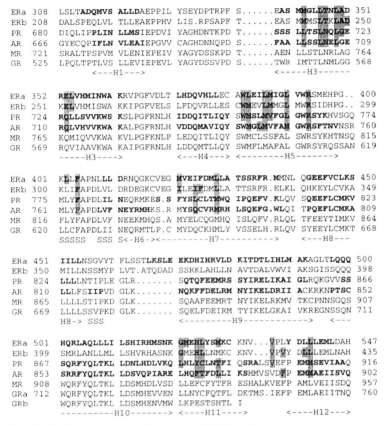

Fig. 2. Multiple sequence alignment of LBDs of human steroid receptors. Shaded in *grey* are those residues which are in contact to ligands. Residues in *bold* are located in α-helices or β-strands

As expected, the resulting second GR LBD model has a similar RMSD to both template structures (0.8 Å), the RMSD between the two GR LBD models is of similar size (0.7 Å). But is it really necessary to use both structures in the further analysis? Both models are very similar to each other, the biggest differences can be found in loop regions. Even the sizes of the LBP are very similar, the second model features a slightly broader LBP in the area of Gln642. It can be concluded that the model generated strongly depends on the number of template molecules

Table 1. Details of the generation of the hGRα LBD model structures

Model	Crystal structures used in model generation	Energy	PDF	RMSD to AR/PR [Å]
1st run	AR LBD – R1881 PR LBD progesterone PR LBD – R1881	7,365	593	1.1/0.2/0.2
2nd run	AR LBD – R1881 PR LBD progesterone	6,697	608	0.7/0.8

RMSD, root-mean-square deviation; PDF, probability density function.

used in the calculation. But in case of the hGRα model structure, both model structures are very similar and based on its slightly better stereochemistry (PROCHECK), we choose the first model for further discussions.

19.2.2 Ligand-Bound LBD Structure

Dexamethasone, progesterone and R1881 were manually fitted into the LBP to obtain a GRα LBD model with ligands bound (Figs. 3, 4). The hydrogen bonds to Arg611 and Asn564 were maintained as a guide to manually dock the ligands into the LBP. The energy of the complexes were minimised using protocols similar to those described elsewhere (Letz et al. 1999). Hydrogen bonds of the ligand 3-carbonyl oxygen to GR residues Arg611 and Glu570 were kept fixed during energy minimisation using the software Discover97.0 (Accelrys, San Diego, USA).

The LBP is lined with 20 residues, of which 14 contribute to the hydrophobic nature of the cavity. Polar residues are located in the two opposite sites of the cavity: Arg611 and Gln570 (GR numbering) inter-

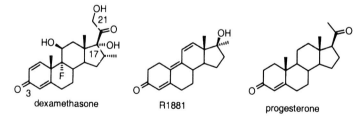

Fig. 3. Scheme of ligands

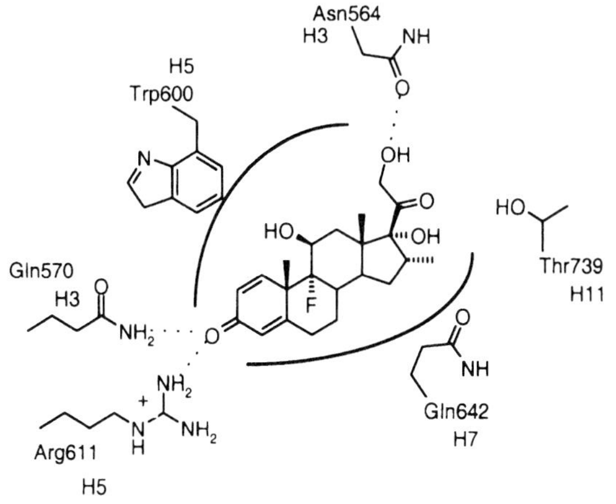

Structural Analysis of the GR Ligand-Binding Domain

Fig. 4. Schematic view of the interactions between dexamethasone and the hGRα LBD in the ligand-binding pocket. The amino acid side-chains, including the Cα atoms, are shown

acting with the 3-keto group of the steroidal A-ring and Asn564 and Thr739 forming hydrogen bonds to the positions 20 and 21 of the D-ring. In the energy minimisation studies, an alternative conformation of the Thr739 side chain was observed with the methyl group pointing to the position 20 of the ligand. The complexes with the differing Thr side chain conformers have a similar energy and it remains to be seen if this hydrogen bond is indeed observed when the crystal structure of the GRα LBD is available.

19.3 Comparison of LBD Structures Within the AR/PR/GR Subfamily

19.3.1 Overall LBD Structure

A comparison of the crystal structures of the AR and PR LBDs reveals several areas which differ between the structures:

– The conformation of helix H6 and the beginning of helix H7

And to a lesser extent:

- The bending of helix H11
- The size of the loop H11/ H12
- The relative position of helix H12 in the agonist position (Fig. 5A)

The most pronounced structural differences occur in the region of helix H6. The conformation of this region varies considerably among the structurally known steroid receptors influencing the size of the LBP in the vicinity of a steroidal D-ring (Müller-Fahrnow and Egner 1999; Egea et al. 2000). For comparison, the distance between equivalent atoms (Gly420 Cα) in the ERα LBD can be as high as 4.0 Å in, for example, the ERα LBD-oestradiol and ERα-tamoxifen crystal structures. In the known PR LBD structures the situation is similar and distances of up to 4.6 Å between equivalent Cα atoms (residue Glu791) are observed. The variation in the structures is shown in Fig. 5B where all monomers of the dimeric PR LBD structures are superimposed (complexes with progesterone and R1881 coloured in light grey). All structurally known AR LBDs crystallised as monomers, therefore, three structures were superimposed: the wild-type human AR LBD, the rat AR LBD and a (T877A) mutant rat AR LBD in complex with dihydrotestosterone (DHT) (Sack et al. 2001) (Fig. 5A, coloured in black). In the AR LBD structures, the difference in the helix H6 conformation is not as pronounced as for the ER and PR LBD structures. Conformational changes of helix H6 may also affect the N-terminal part of helix H7 (Fig. 5B): the distance between the Cα atoms of the equivalent residues Gln783 in AR and Leu797 in PR is 2.1 Å.

The bending of helix H11 varies between the AR and PR LBD structures. The position of helix H11 is closer to the LBP in the AR structures than in all PR structures, therefore decreasing the size of the LBP in the AR in that area. The size of the loop H11/H12 in the GRα

Fig. 5A, B. Superposition of the publicly known PR and AR LBD structures. **A** The PR LBD monomers of the dimeric crystal structure are coloured in *light grey*, and the individual monomeric AR LBD structures are highlighted in *black*. **B** Detailed view of the binding niches of the PR and AR LBD crystal structures (in *light grey* and *black*, respectively). The most flexible residues as well as some residues with conserved orientations are shown

Structural Analysis of the GR Ligand-Binding Domain

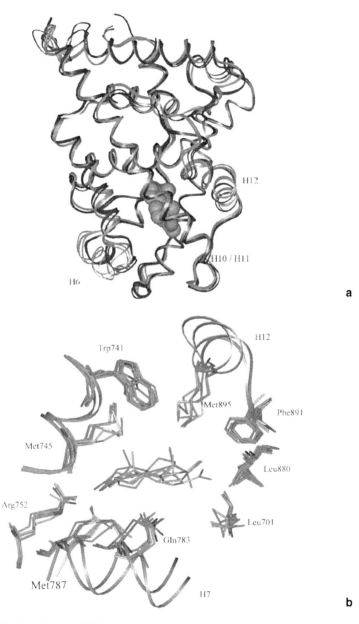

Fig. 5 a,b. Legend see p. 348

LBD model structure is shorter by one amino acid residue as compared to AR and PR. The loop is more than 10 Å away from the D-ring area of a steroidal ligand and probably does not influence the size of the binding niche. The relative position between the N-terminal parts of helix H12 varies only slightly with differences between equivalent residues (Cα atoms) of less than 1 Å.

19.3.2 The Vicinity of the LBP

Although the overall sequence identity within the AR/PR/GR subgroup is relatively high (>50%), the similarity of the vicinity of the LBPs is even higher. If only those amino acid residues are considered that line the LBP, the sequence identity is as high as 75% (residues shaded grey in Fig. 2). The amino acid residues lining the LBP which differ most within the AR/PR/GR subgroup are distributed all over the sequence, but with respect to the 3D structure, most of these residues are located in the vicinity of the steroidal D-ring. Exceptions are Ala605 and Gln642, which are located close to the steroidal B- and C-rings.

The differences in sequence in the individual receptors do not contribute alone to receptor specificity. The flexibility of amino acid sidechain conformations in the LBP as observed in the known AR and PR LBD structures influence the size as well as the local environment of ligands when bound in the LBP. Depending on the ligand bound, different conformations are observed for residues Trp741/ 600 (AR/GR numbering), Met745/604 and Met787/646 (all near the A-/B-ring of a steroidal ligand), Gln783/642 (near C-ring) and Met895/Leu753, Phe891/749, Leu880/Thr739, Leu701/Met560 near the D-ring area (Fig. 5B).

The flexibility in the helix H6 area influences the position of the N-terminal part of helix H7. The different orientations of Gln642 (GR) and Met646 (GR) as compared to the AR and PR structures lead to an individual local environment in the LBP of each receptor near the B-/C-ring of a steroidal ligand. Moreover, Gln642 is not conserved in the AR/PR/GR subgroup being either a glutamine (AR, GR) or a leucine (PR).

In the AR crystal structures different conformations of Trp600 and Met604 are observed depending on binding R1881 or dihydrotestosterone. The 19-methyl substituent of dihydrotestosterone causes a reorientation of the Trp600 side chain and as a direct consequence the Met604 side chain.

The distribution of non-conserved amino acid residues in the LBP together with the identification of flexible residues leads to the conclusion that the area around the steroidal D-ring and below the C-/D-ring are those positions in a steroidal ligand where substituents experience an individual local environment in the LBP of each receptor. Appropriate substituents at those positions in the steroidal D-ring will generate selective ligands. These conclusions, which are drawn on the basis of the 3D structures of the individual receptors, are also reflected in the experimental biochemical data on ligand-binding affinities.

19.4 Orientation of Ligands in the LBP Within the AR/PR/GR Subfamily

With the analysis of the differences in overall structure and especially with the identification of individual local environments in the LBP of the AR/PR/GR subgroup, we may now attempt to fit selected ligands into the LBP and see if we can qualitatively explain their relative binding affinities (RBAs). How do different receptors accommodate to the same ligand? How much contributes the local environment of the LBP to ligand specificity?

Since the 1960s when the first progestin was used in oral contraceptives, a lot of steroidal ligands have been synthesised and published varying in the number of double bonds and ring substituents (Raynaud et al. 1979; Doré et al. 1985). We fitted dexamethasone, progesterone and R1881, which have different D-ring substituents, in the LBP and tried to explain the preference for their substituents for AR, PR and GR on the basis of the available crystal and model structures (Table 2). Here, we would like to concentrate on those well-known ligands listed in Table 2.

The crystal structures of two closely related receptor LBDs (AR and PR LBD) in complex with the same ligand R1881 gave us the opportunity to compare their LBPs in detail and achieve a better understanding of compound selectivity (Matias et al. 2000). R1881 has high RBA values for both AR and PR, but not for GR. In the crystal structures of

Table 2. RBA values of selected ligands for AR, PR and GR (from Doré et al. 1985; Teutsch et al. 1994)

ligand		PR	AR	GR
progesterone		100	20	115
R1881		190	290	4
dexamethasone		0.1	0.1	100

the AR and PR LBD in complex with R1881, different hydrogen bonds are observed for the ligand. Hydrogen bonds are observed to the conserved arginine (Arg752 in AR, Arg766 in PR) and asparagine residue (Asn705 in AR, Asn719 in PR). The hydrogen bond to this arginine is only observed in one monomer of the PR LBD dimer. In the AR LBD, an additional hydrogen bond is observed of the 17β-OH to Thr877. The equivalent residue in PR, Cys891, is too far away for such an interaction. In addition, in AR LBD, the side chain of Leu880 makes van der Waals contacts to the ligand which are not observed in PR LBD as the equivalent residue at this position is shorter (Thr894). Due to the more planar scaffold of the R1881 ligand, it can only be fitted into the binding niche of the GRα LBD when the hydrogen bond to Asn564 is broken. The probably missing second key interaction partner for R1881 in the GR LBP may account for the strongly decreased RBA value (RBA=4).

The RBA values of progesterone in the three receptors can immediately be understood on the basis of the structures. It is not possible to accommodate progesterone in the LBP of the AR LBD. The bulkier side chain of Leu880 (AR) (Thr894 in PR), along with the substitution of a

Structural Analysis of the GR Ligand-Binding Domain

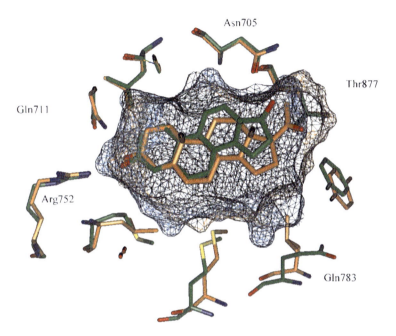

Fig. 6. Van der Waals presentation of the LBP of the AR and PR LBD crystal structures. Superposition of AR LBD-R1881 (*green, blue surface*) and PR LBD-progesterone (*orange, orange surface*). Some residues from AR are labelled

threonine into a cysteine (Thr877 in AR, Cys891 in PR) is very likely responsible for the specific recognition of a 17β-hydroxyl group as in R1881 contrary to the elongated 17β-acetyl group of progesterone. The LBP of AR in that region is much smaller than that of PR (Fig. 6). The decreased pocket volume in AR may be responsible for the low RBA values of ligands with bulkier substituents in this area. This conclusion is reflected in the experimentally observed RBA values (Doré et al. 1985). In the GRα LBD there is enough space to accommodate the 17β-acetyl group, which explains the high RBA value of progesterone in GR (RBA=115). Similarly to PR, a hydrogen bond is not necessarily observed between the carbonyl oxygen of the acetyl group and the LBD.

The interactions between dexamethasone and the GRα LBD are discussed in more detail in Sect. 19.2.2 (see also Fig. 4). The low RBA values for this ligand for PR and AR are most likely due to the presence of the 16α-methyl group, which is located near Gln642 in GR. In PR, the side chain of the equivalent residue Leu797 points into the binding niche and would be too close to the 16α-methyl group. In AR, Gln783 points away from the LBP, but the position of helix H7 is shifted in this area with respect to the AR, PR and GR LBD structures (see Sect. 19.3.1). This shift leads to a reduced volume of the AR LBP in this region and unfavourable contacts between the 16α-methyl group and the AR LBD might occur thus leading to the observed decrease in RBA values.

There is an ongoing debate in the literature as to whether Gln642 (GR) is a hydrogen bond partner to the 17α-hydroxyl group which is preferred in GR ligands (Lind et al. 2000). In the present GR LBD model, the side-chain of Gln642 is too far away for a hydrogen bond even if other energetically favourable conformations of Gln642 are taken into account. Moreover, a 16α-methyl group would hinder such an interaction.

The discussion of a selected number of ligands shows that it is possible to qualitatively explain preferences for steroidal agonists with varying substituents near the steroidal D-ring in the three receptors. Individual environments and local flexibility in the LBP have to be considered if ligands are fitted into the binding niche. Still it is still a long way to correctly predict RBA values on the basis of an experimental structure. In order to develop more predictive functions and to deepen our understanding of protein–ligand interactions, additional structures have to be solved for the AR/PR/GR subgroup of steroid receptors.

19.5 Summary and Outlook

A model of the hGRα LBD was generated based on the crystal structures of the hAR LBD and hPR LBD in complex with progesterone and R1881. The model was assessed to be of good quality, comparable to an X-ray structure of medium resolution. The crystal and the model structures of the AR, PR and GRα LBD were compared to each other to identify regions which play a role in receptor specificity. The model structure can be used for a qualitative description of the binding of steroidal agonists with varying D-ring substituents with respect to their experimental RBA values. With the knowledge gained in the analysis of the model structure, ligand complexes may contribute to the structure-based drug design of new receptor-specific ligands. It remains to be seen which new and unexpected details we learn once the crystal structure of the hGRα LBD is solved.

Acknowledgements. I would like to thank my colleagues in Structural Biology and Computational Chemistry as well as our collaboration partners Dr. Dino Moras, IGBMC Strasbourg, and Prof. Maria Armenia Carrondo, ITQB Lisbon, for fruitful discussions.

References

Berman HM, Westbrook J, Feng Z, Gillilan G, Bath TN, Weissig H, Shindyalov IN, Bourne PE (2000) The Protein Data Bank. Nucleic Acids Res 28:235–242

Bourguet W, Germain P, Gronemeyer H (2000) Nuclear receptor ligand-binding domains, three-dimensional structures, molecular interactions and pharmacological implications. Trends Pharmacol Sci 10:381–388

Dey R, Roychowdhury P, Mukherjee C (2001) Homology modelling of the ligand-binding domain of glucocorticoid receptor: binding site interactions with cortisol and corticosterone. Protein Eng 14:565–571

Doré JC, Gilbert J, Ojasoo T, Raynaud JP (1985) Correspondence analysis applied to steroid receptor binding. J Med Chem 29:54–60

Egea PF, Klaholz BP, Moras D (2000) Ligand-protein interactions in nuclear receptors of hormones. FEBS Lett 476:62–67

Gronemeyer H, Laudet V (1995) Transcription factors 3: nuclear receptors. Protein Profile 2:1173–1308

Laskowski RA, MacArthur MW, Moss DS, Thornton JM (1993) PROCHECK: a program to check the stereochemical quality of protein structures. J Appl Cryst 26:283–291

Letz M, Bringmann P, Mann M, Mueller-Fahrnow A, Reipert D, Scholz P, Wurtz JM, Egner U (1999) Investigation of the binding interactions of progesterone using muteins of the human progesterone receptor ligand binding domain designed on the basis of a three-dimensional protein model. Biochim Biophys Acta 1429:391–400

Lind U, Greenidge P, Gillner M, Koehler KF, Wright A, Carlstedt-Duke J (2000) Functional probing of the human glucocorticoid receptor steroid-interacting surface by site-directed mutagenesis. J Biol Chem 275:19041–19049

Matias PM, Donner P, Coelho R, Thomaz M, Peixoto C, Macedo S, Otto N, Joschko S, Scholz P, Wegg A, Basler S, Schafer M, Ruff M, Egner U, Carrondo MA (2000) Structural evidence for ligand specificity in the binding domain of the human Androgen receptor: implications for pathogenic gene mutations. J Biol Chem 275:26164–26171

McKay LI, Cidlowski JA (1999) Molecular control of immune/inflammatory responses: interaction between nuclear receptor factor-Kappa B and steroid-receptor-signalling pathways. Endocrine Rev 4:435–459

Moras D, Gronemeyer H (1998) The nuclear receptor ligand-binding domain: structure and function. Curr Opin Cell Biol 10:384–391

Müller-Fahrnow A, Egner U (1999) Ligand-binding domain of estrogen receptors. Curr Opin Biotechnol 10:550–556

Oakley RH, Jewell CM, Yudt MR, Bofetiado DM, Cidlowski JA (1999) The dominant negative activity of the human glucocorticoid receptor b isoform. J Biol Chem 274:27857–27866

Raynaud JP, Ojasoo T, Bouton MM, Philibert D (1997) Receptor binding as a tool in the development of new bioactive steroids. In: Ariens EJ (eds) Drud Design. Academic Press, New York, pp 169–214

Reche-Rigon M, Gronemeyer H (1998) Therapeutic potential of selective modulators of nuclear receptor action. Curr Opin Chem Biol 2:701–507

Sack JF, Kish KF, Wang C, Attar RM, Kiefer SE, An Y, Wu GY, Scheffler JE, Salvati ME, Krystek SR, Weinman R, Einspahr HM (2001) Crystallographic structures of the ligand-binding domains of the androgen receptor and its T877A mutant complexed with the natural agonist dihydrotestosterone. Proc Natl Acad Sci 98:4904–4909

Teutsch G, Goubet F, Battman T, Bonfils A, Bouchoux F, Cerede E, Gofflo D, Gaillard-Kelly M, Philibert D (1994) Non-steroidal antiandrogens: synthesis and biological profile of high-affinity ligands for the androgen receptor. J Steroid Biochem Mol Biol 48:111–119

Williams SP, Sigler PB (1998) Atomic structure of progesterone complexed with its receptor. Nature 393:392–396

Wurtz JM, Bourguet W, Renaud JP, Vivat V, Chambon P, Moras D, Gronemeyer H (1996) A canonical structure for the ligand-binding domain of nuclear receptors [published erratum appears in Nat Struct Biol 1996 Feb3(2):206]. Nat Struct Biol 3:87–94

20 SEGRAs: A Novel Class of Anti-inflammatory Compounds

H. Schäcke, H. Hennekes, A. Schottelius, S. Jaroch,
M. Lehmann, N. Schmees, H. Rehwinkel, K. Asadullah

20.1	Introduction	357
20.2	Hypothesis and Aims of the Project	362
20.3	Proof of Concept	362
20.4	Summary	368
References		369

20.1 Introduction

Glucocorticoids (GCs) have been widely used in the treatment of acute and chronic inflammatory diseases for more than 50 years. They are very effective and represent standard therapies for many inflammatory conditions with immunologic or non-immunologic backgrounds. Unfortunately, the desired anti-inflammatory and immunosuppressant effects are often accompanied by severe and/or partially non-reversible side effects (Table 1). Beside unknown individual factors, dosage and duration of therapy are major contributors to the frequency and severity of these side effects. Thus, long-term and high-dose GC therapy is problematic and partially dangerous. However, no good alternative exists so far, as other anti-inflammatory therapies usually are either less effective (at least in certain indications) or display a high side-effect potential as well. Consequently, there is a strong medical need to develop substances with an anti-inflammatory potency similar to that of

Table 1. GC-mediated side effects appearing after systemic therapies (summarized by A. Cato from Piper et al. 1991 and Lukert and Raisz 1990)

Early in therapy
  Insomnia
  Emotional lability
  Truncal obesity
  Diabetes mellitus
  Peptic ulcer
  Acne
Anticipated with sustained treatment
  Cushingoid habitus
  Hypothalamic-pituitary-adrenal suppression
  Infection diathesis
  Osteonecrosis
  Myopathy
  Impaired wound healing
Delayed effects
  Osteoporosis
  Skin atrophy
  Cataracts
  Atherosclerosis
  Growth retardation
  Fatty liver
Rare and unpredictable effects
  Psychosis
  Pseudotumor cerebri
  Glaucoma
  Epidural lipomatosis
  Pancreatitis
  Hirsutism or Virilism
  Convulsion
  Hepatomegaly
  Congestive heart failure

classical GCs but with reduced side effects. New insights into the molecular mechanisms of GC-mediated actions have revealed new opportunities to potentially find such substances with a better effect/side-effect profile.

SEGRAs: A Novel Class of Anti-inflammatory Compounds

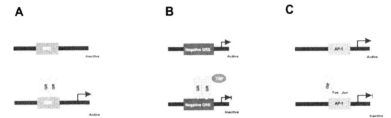

Fig. 1A–C. Molecular mechanisms of GC mediated effects. **A** Transactivation: the ligand-activated GR homodimer binds to GREs in promoter region of GC-sensitive genes, inducing gene transcription. **B** Transrepression: the ligand-activated GR homodimer binds to negative GREs in promoter region of GC-regulated genes (e.g., POMC promoter), thus inhibiting gene transcription. **C** Transrepression: ligand-activated GR monomer binds to a subunit of another transcription factor (e.g., c-Fos subunit of AP-1 or p65 subunit of NF-κB), thus inhibiting the induction of gene transcription by these factors

The effects of GCs are mediated by the specific glucocorticoid receptor (GR). After ligand binding, the cytoplasmic GR translocates into the nucleus and modulates gene transcription either by a positive (transactivation) or a negative (transrepression) mode of regulation. The positive regulation of target genes occurs by specific binding of the DNA-binding domain of the activated GR homodimer to GC response elements (GREs) in the promoter or enhancer region of responsive genes followed by an induction of gene transcription. While the transactivation is controlled by this protein–DNA interaction, the negative regulation by the GR may be exerted by protein–DNA as well as protein–protein interactions. In one case, the activated GR can bind to so-called negative GC response elements (nGREs) leading to a repression of gene transcription, via a protein–DNA interaction. Alternatively, in the more common way of transrepression, the GR may interact as a monomer via a protein–protein interaction with other transcription factors, e.g., AP-1, NF-κB, or STAT5, thus preventing activation of gene transcription (Fig. 1).

The crucial question is whether these different molecular mechanisms (transactivation/protein–DNA interaction on the one hand and transrepression/protein–protein interaction on the other hand) are responsible for distinct clinical effects and can be separated. For a few

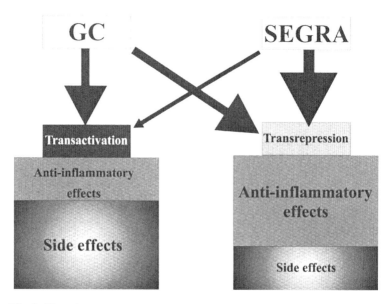

Fig. 2. Hypothesized profile for the induction of molecular mechanisms for classical GCs and dissociated GCs (*SEGRAs*)

years, there has been growing evidence that certain side effects of glucocorticoids are mainly mediated by transactivation of genes, whereas anti-inflammatory effects predominantly result from transrepression (Belvisi et al. 2001) (Fig. 2). While this hypothesis provides a good general rule for GC action, some investigations showed that the induction of gene transcription of anti-inflammatory genes such as secretory leukocyte proteinase inhibitor (Sallenave et al. 1994), lipocortin-1 (Flower and Rothwell 1994), and interleukin (IL)-1 receptor antagonist (Levine et al. 1996) is also involved in anti-inflammatory responses of GCs. Similarly, not all side effects seem to be mediated exclusively by transactivation (Table 2). Steroid diabetes and steroid glaucoma seem to be induced mainly via the transactivation mechanism. The two most important enzymes of the gluconeogenesis, an essential pathway in the development of diabetes, the phosphoenolpyruvate carboxykinase (O'Brien et al. 1995; Crosson et al. 2000) and the glucose-6-phosphatase (Yoshiuchi 1998) are both induced by GCs. Steroid-in-

Table 2. Summary of the important underlying molecular mechanisms of a number of GC side effects

---
Side effects are mediated predominantly via:
Transrepression
  ACTH suppression
  Disturbed wound healing
  Infections
  Osteoporosis
Transactivation
  Diabetes mellitus
  Glaucoma
  Muscle atrophy/myopathy
  Osteoporosis
Unknown mechanisms
  Cataract
  Peptic ulcer
  Hypertension

---

duced glaucoma is caused by an increased synthesis of extracellular matrix proteins. Prolonged GC treatment leads to a major progressive induction of the 55-kDa TIGR/MYOC gene product in human trabecular meshwork cells (Lutjen-Drecoll et al. 1998). In contrast, an impaired wound healing caused by GCs depends on their anti-inflammatory and immunosuppressive activities via transrepression (Beer et al. 2000). Finally, in osteoporosis as a GC-induced side effect, transactivation as well as transrepression mechanisms are involved, besides a number of indirect effects (Weinstein et al. 1998; Silvestrini et al. 2000).

The differentiation of GR activity mediated via transrepression or transactivation into desired and undesired actions of GCs remains, therefore, a hypothesis, which is supported by some observations. However, the molecular mechanisms of all GR actions are still not completely understood, especially in the case of unwanted effects of GCs.

20.2 Hypothesis and Aims of the Project

Based on the new insight into the molecular mechanisms of GC-mediated action, we and others (e.g., Roussel-Uclaf, Michelle Resche-Rigon) hypothesized that activation of the GR with compounds inducing a predominant induction of transrepression over transactivation should lead to the majority of the anti-inflammatory effects of GCs with less side effects. Such *se*lective GR *a*gonists (SEGRAs) would represent a novel class of anti-inflammatory compounds with a favorable effect/side-effect profile (Fig. 2). For the identification of such compounds, it had to be demonstrated first as a precondition that transrepression alone is sufficient to mediate the anti-inflammatory action. Second, compounds with a dissociated profile in vitro had to be identified, and third, their superiority had to be demonstrated in vivo.

20.3 Proof of Concept

20.3.1 Efficacy of Classical GCs in $GR^{dim/dim}$ Mice

Schütz and coworkers recently generated a transgenic mouse carrying a dimerization-incompetent GR (Reichardt et al. 1998). Due to this mutation, the GR could not bind to the DNA anymore, excluding any GC-induced transactivation of gene expression. Results obtained with these transgenic mice gave the strong suggestion that dimerization of the GR leading to transactivation of target genes is not required to ensure the fetal development and the survival, whereas in contrast, GR knockout mice are embryonally lethal. (Cole et al. 1995). Moreover, in the $GR^{dim/dim}$ mice the induction of anti-inflammatory effects by classical GC treatment was observed (Reichardt et al. 2000; G. Schütz, personal communication). Thus, there is evidence that using only one of the two major GC-mediated molecular mechanisms (transrepression) is sufficient to reach anti-inflammatory effects.

20.3.2 Identification of Selective Glucocorticoid Receptor Agonists

In 1997 we started a research program aiming at the separation of different GR mechanisms by a new class of GR ligands. We are searching for compounds which bind to the GR and preferentially induce GR–protein interactions in the nucleus. Compounds are selected on the basis of their GR binding and dissociation between transrepression and transactivation. For determination of the transrepression and transactivation capacities we use reporter gene assays as well as the determination of the effects on endogenous tyrosine aminotransferase (TAT) production in a rat hepatoma cell line (transactivation) and IL-8 secretion in promyeloic THP-1 cells (transrepression).

After screening of about 500,000 compounds and analyzing the effects of several hundred compounds with high GC-binding potential in more detail, we were able to identify numerous GC-binding compounds which show transrepression activities at least as good as dexamethasone. Further investigation of these mainly non-steroidal compounds led to the identification of three subgroups with regard to their transactivation profile: (1) agonists, (2) partial agonists, and (3) antagonists in transactivation. Moreover, some steroidal and non-steroidal compounds did show neither agonistic nor antagonistic effects on transactivation, while a transrepression was observed. The majority of compounds turned out to be agonists in transactivation. The subgroup of partial agonists in transactivation is significantly smaller. Only a few compounds were identified as being antagonists in transactivation, suggesting that this type of compound is more challenging to obtain.

One criterion for further compound selection was the degree of separation between transrepression and transactivation potencies. Interestingly, we found that prednisolone represents a slightly dissociated steroid (factor 7.6) better than dexamethasone (factor 5.4). Thus, the hurdle compounds had to take a dissociation better than found for prednisolone. Compounds with a dissociation factor higher than 10 were identified. However, one compound we investigated in more detail after topical application (see below) showed a dissociation factor comparable to that of prednisolone (factor 8), while being clearly more potent in transrepression.

Taken together, we demonstrated the existence of non-steroidal GR ligands which induce a predominant transrepression. Similar results were recently reported from other groups (Vayssiere et al. 1997; Resche-Rigon and Gronemeyer 1998; Vanden Berghe et al. 1998; Belvisi et al. 2001).

20.3.3 Dissociation of Anti-inflammatory Effects from Side Effects by Use of SEGRAs in Animal Models

We have analyzed compounds with an in vitro dissociation between transactivation and transrepression in rodents. In vivo activities were examined first in the croton oil model of ear inflammation in mice and rats. Induction of TAT in the liver of mice and rats was used as a surrogate parameter for catabolic effects of GCs. The promoter of the TAT gene contains regulatory sequences enabling an activation of transcription by GCs (Hargrove and Granner 1987; Jantzen et al. 1987; Rigaud 1991). The standard GCs, dexamethasone (strong) and prednisolone (weak), were used as controls.

As a parameter for anti-inflammatory activity we measured the inhibition of ear edema. Both classical GCs as well as dissociated GCs were quite effective. Our compounds showed anti-inflammatory activity after systemic applications in the croton oil-induced ear inflammation model of mice and rats at least comparable to that of prednisolone. They reached a similar efficacy (80%–100%) compared to prednisolone (90%) after subcutaneous treatment of mice. Slight differences with regard to the potencies were observed in mice (ED_{50} of prednisolone: ca. 5 mg/kg; compound A: ca 2 mg/kg and compound D: ca. 1 mg/kg). As a parameter for transactivation mediated side effects, we determined the induction of the TAT. If at all, a significant induction of TAT was only observed at the highest dosages for all tested novel compounds (at most a twofold increase), while for prednisolone an approximately sixfold increase of TAT activity was measured with the highest dosage of 30 mg/kg.

Thus, dissociated GCs show a good separation between inhibition of edema and TAT induction, which is in contrast to classical GCs. These in vivo data demonstrate that the translation from in vitro dissociation of transrepression and transactivation into in vivo separation of therapeutic efficacy and side effects is possible.

As a clinically more relevant parameter than the TAT induction we investigated the potential of increasing the blood glucose level in rats after systemic treatment. This is a particularly interesting parameter since it may serve as a surrogate parameter for the risk of diabetes induction, a frequent and dangerous side effect connected with the long-term use of GCs. Based on the knowledge of the underlying molecular mechanisms, dissociated GCs should not influence this parameter strongly. Furthermore, selected compounds were tested with regard to their effects on adrenocorticotropic hormone (ACTH) level. ACTH is a cleavage product of the pro-opiomelanocortin (POMC), whose transcription is negatively regulated by GCs. In the POMC promoter, the activated GRE-bound GR interferes with the binding of essential transcription factors, thus inhibiting the induction of gene transcription (Drouin et al. 1993). This negative regulation seems to be DNA-binding-dependent and should not be affected by dissociated GCs. However, ACTH secretion into the serum is underlying a DNA-binding-independent mechanism. The synthesis of the corticotropin-releasing hormone (CRH), responsible for ACTH release, is negatively regulated by the GR via a protein–protein interaction (Reichardt et al. 1998, 2000). Therefore, we did not expect an advantage of the SEGRA compounds compared with standard GCs in in vivo experiments.

In contrast to classical GCs none of the tested dissociated compounds showed a significant increase in the blood glucose level after systemic treatment. While systemic prednisolone treatment induced the blood glucose dose dependently up to approximately 200% of the control with the highest dosage of 30 mg/kg, no increase was observed with compound D and only a 20% increase with all dosages of compound A. However, as expected, with regard to ACTH suppression, the compounds were as effective as prednisolone.

The idea to develop GCs with an improved effect/side-effect profile for systemic therapies is also attractive for other companies. Very recently, results were published by Belvisi and coworkers (2001) concluding that the in vitro separation of their compound was not confirmed in vivo. The investigators focused on a lung edema model in rats to demonstrate anti-inflammatory activity. Regarding side-effect induction, effects on bones, body weight, and thymus were analyzed. The in vitro dissociated compound was anti-inflammatory active in vivo showing a potency even better than budesonide and prednisolone. Unfortunately, the compound was as effective as budesonide and prednisolone in induction of the measured side effects (serum marker of bone metabolism, femoral head histology, body weight, thymus involution), suggesting that in vitro dissociation does not always translate to an improved therapeutic ratio for GCs in vivo (Belvisi et al. 2001). Reasons for the obtained results could be that the authors did not distinguish between side effects regarding the underlying molecular mechanisms and, on the other hand, the compound might have been metabolized very rapidly to an active metabolite, which does not show such a dissociated profile. A separation of transactivation from transrepression activities in vitro should result in an in vivo separation of anti-inflammatory activity from e.g., diabetes induction measured by TAT induction in the liver or of glucose levels in serum of treated animals. Extensive pharmacokinetic characterization of the compound might help for a better understanding of the discrepancies between the obtained in vitro profile and the in vivo activity of the compound.

In addition, Coghlan and coworkers (2001) reported non-steroidal compounds that bind selective and with a high affinity to the GR. In an in vivo asthma model, one of the described compounds showed a functional profile similar to that of prednisolone, that means a good anti-inflammatory activity but no dissociation with regard to side effect induction.

Thus, the principle of identifying dissociated ligands of the GR, steroidal or non-steroidal, is still a very attractive one, and a few companies are on the way to developing such compounds with the expectation of an improved effect/side-effect profile in the therapy of chronic inflammatory diseases.

20.3.4 Introduction of a First Drug Candidate for Topical Application

GCs are the most widely used drugs in dermatology. The introduction of topical hydrocortisone and in particular of the first halogenated corticosteroid, triamcinolone acetonide, represented great advances compared to previously available therapies. However, adverse effects became apparent and the subsequent backlash of opinion against topical GCs has created confusion and prejudice against all steroid-containing preparations, in its extreme as "steroid phobia," which is still a considerable concern (see Sterry and Asadullah in this book). Importantly, not only cutaneous side effects like skin atrophy can occur during topical GC therapy but systemic side effects as well.

In order to get a first impression on the potential of SEGRAs, we therefore extensively investigated activity and side-effect induction after topical application of one of our early compounds. Interestingly, the compound was significantly active in the croton oil-induced ear inflammation model after subcutaneous application (80% inhibition of ear edema; ED_{50}: ca. 10 mg/kg). Moreover, after systemic application, if at all, the compound induced TAT only at the highest given dosage of 30 mg/kg (2.4-fold induction). It showed a slightly weaker potency (ED_{50}: ca. 0.01%) after topical administration, but it reached the full efficacy of the standard (prednisolone; ED_{50}: ca. 0.003%). However, after long-term topical treatment of rats, using 30 times the ED_{50} (effective dosage) of both the compound and prednisolone, topical (skin thickness and breaking strength) and systemic side effects (reduction in body weight gain and reduction of adrenal weight) were clearly less pronounced in compound-treated than in prednisolone-treated rats. With regard to the topical side effects, the compound induced only 50% to 60% of the prednisolone effects. In addition the compound did not affect the adrenal weight and the body weight gain was only reduced by 20% (prednisolone reduced the body weight gain by 90%). Further on, the compound induced significantly less systemic immune deterioration (loss of thymus by 55% and spleen weight by 10%) than prednisolone (loss of thymus by 85% and spleen weight by 50%).

Here we show, that a compound with a dissociated in vitro profile with regard to transactivation and transrepression is effective in the treatment of acute inflammation while it induces less side effects. The compound displays potent anti-inflammatory activity after topical and systemic application and reaches an inhibition of the inflammatory response by 60%–80%. Importantly, the induction of side effects by this compound is clearly lower than by the weak GC prednisolone.

20.4 Summary

Dissociated GCs show a separation between anti-inflammatory effects and certain side effects. This renders them as attractive compounds with better effect/side-effect profile as promising drug candidates and tool compounds for analyzing the molecular mechanisms of single side effects.

A number of the GC-mediated side effects (e.g., osteoporosis, skin atrophy) are regulated in a very complex manner and use more than one molecular mechanism of the GR. Thus, theoretical predictions about the behavior of selective GR agonists regarding these effects are very difficult to make. Investigations of SEGRA compounds in relevant animal models will be the only way to get this important information. By availability of these tool compounds we now are in the advantageous situation to test them in vivo and to learn more about the possibilities and even the limitations of the selective GR agonists.

Considering that the compounds have a non-steroidal structure, i.e., totally unrelated to steroids or other hormones at all, displaying only partially the molecular effects of GCs and are dissociated in their clinical profile, they should not be considered as GCs. Therefore, we introduced the term selective glucocorticoid receptor agonists (SEGRAs).

These SEGRAs seem to represent a useful novel therapeutic modality which may complement existing therapeutic principles for the topical and especially the systemic treatment of inflammatory diseases.

In summary, we and others are convinced that dissociated GCs are therapeutic compounds that exert many of the anti-inflammatory and immunosuppressive effects of standard GCs, while their potential to induce side effects is reduced. Whereas the in vitro dissociated profile of other compound classes (Belvisi et al. 2001) was not translated into a separation between anti-inflammatory activity and the induction of side effects in in vivo models, we could demonstrate this for the SEGRA compounds. Regarding the diversity of molecular mechanisms involved in mediating the complex side effects of GCs, it might be that only some of these unwanted effects can be reduced. However, as GCs are one of the most important anti-inflammatory therapeutics in the treatment of severe and chronic inflammatory diseases, even a partial reduction of side effect induction would be a great advantage for many patients.

Acknowledgements. The authors would like to thank Andrew Cato for providing Table 1 and Wolf-Dietrich Döcke and Ulrich Zügel for their helpful suggestions during the preparation of this manuscript.

References

Beer H-D, Fässler R, Werner S (2000) Glucocorticoid-regulated gene expression during cutaneous wound repair. Vitam Horm 59:217–239
Belvisi MG, Wicks SL, Battram CH, Bottoms SEW, Redford JE, Woodman P, Brown TJ, Webber SE, Foster ML (2000) Therapeutic benefit of a dissociated glucocorticoid and the relevance of in vitro separation of transrepression from transactivation activity. J Immunol 166:1975–1982
Coghlan MJ, Kym PR, Elmore SW, Wang AX, Luly JR, Wilcox D, Stashko M, Lin C-W, Miner J, Tyree C, Nakane M, Jacobson P, Lane BC (2001) Synthesis and characterization of non-steroidal ligands for the glucocorticoid receptor: selective quinoline derivatives with prednisolone-equivalent functional activity. J Med Chem 44:2879–2885
Cole TJ, Blendy JA, Monaghan AP, Krieglstein K, Schmidt W, Aguzzi A, Fantuzzi G, Hummler E, Unsicker K, Schütz G (1995) Targeted disruption of the glucocorticoid receptor gene blocks adrenergic chromaffin cell development and severely retards lung maturation. Genes Dev 9:1608–1621
Crosson SM, Roesler WJ (2000) Hormonal regulation of the phosphoenolpyruvate carboxykinase gene. J Biol Chem 275:5804–5809
Drouin J, Sun YL, Chamberland M, Gauthier Y, De LA, Nemer M, Schmidt TJ (1993) Novel glucocorticoid receptor complex with DNA element of the hormone-repressed POMC gene EMBO J 12:145–156
Flower RJ, Rothwell NJ (1994) Lipocortin-1: cellular mechanisms and clinical relevance. Trends Pharmacol Sci 15:71–76
Hargrove JL, Granner DK (1987) Biosynthesis and intracellular processing of tyrosine aminotransferase. In: Christen P, Metzler DE (eds) Transaminases. John Wiley & Sons, New York, pp 511–532
Jantzen H-M, Strähle U, Gloss B, Stewart F, Schmidt W, Boshart M, Miksicek R, Schütz G (1987) Cooperativity of GC response elements located far upstream of the tyrosine aminotransferase gene. Cell 49:29–38
Levine SJ; Benfield T, Shelhamer JH (1996) Corticosteroids induce intracellular interleukin-1 receptor antagonist type I expression by a human airway epithelial cell line. Am J Respir Cell Mol Biol 15:245–251
Lukert BP, Raisz LG (1990) Glucocorticoid-induced osteoporosis: pathogenesis and management. Ann Intern Med 112:352–364

Lutjen-Drecoll E, May CA, Polansky JR, Johnson DH, Bloemendal H, Nguyen TD (1998) Localization of the stress proteins alpha B-crystallin and trabecular meshwork inducible glucocorticoid response protein in normal and glaucomatous trabecular meshwork. Invest Ophthalmol Vis Sci 39:517–525

O'Brien RM, Noisin EL, Suwanichkul A, Yamasaki T, Lucas PC, Wang J-C, Powell DR, Granner DK (1995) Hepatic nuclear factor 3- and hormone-regulated expression of the phosphoenolpyruvate carboxykinase and insulin-like growth factor-binding protein-1 genes. Mol Cell Biol 15:1747–1758

Piper JM, Ray WA, Daugherty JR, Griffin MR (1991) Corticosteroid use and peptic ulcer disease: role of nonsteroidal anti-inflammatory drugs. Ann Intern Med 114:735–740

Reichardt HM, Kaestner KH, Tuckermann J, Kretz O, Wessely O, Bock R, Gass P, Schmidt W, Herrlich P, Angel P, Schütz G (1998) DNA binding of the glucocorticoid receptor is not essential for survival. Cell 93:531–541

Reichardt HM, Tronche F, Bauer A, Schütz G (2000) Molecular genetic analysis of glucocorticoid signaling using the Cre/loxP system. Biol Chem 381:961–964

Resche-Rigon M, Gronemeyer H (1998) Therapeutic potential of selective modulators of nuclear receptor action. Curr Opin Chem Biol 4:501–507

Rigaud G, Roux J, Pictet R, Grange T (1991) In vivo footprinting of rat TAT gene: dynamic interplay between the GC receptor and a liver-specific factor. Cell 67:977–986

Sallenave JM, Shulmann J, Crossley J, Jordana M, Gauldi J (1994) Regulation of secretory leukocytes proteinase inhibitor (SLPI) and elastase-specific inhibitor (ESI/elafin) in human airway epithelial cells by cytokines and neutrophilic enzymes. Am J Respir Cell Mol Biol 11:733–741

Silvestrini G, Ballanti P, Patacchioli FR, Mocetti P, Di Grezia R, Martin Wedard B, Angelucci L, Bonucci E (2000) Evaluation of apoptosis and the GC receptor in the cartilage growth plate and metaphyseal bone cells of rats after high-dose treatment with corticosterone. Bone 26:33–42

Vanden Berghe W, Francesconi E, De Bosscher K, Resche-Rigon M, Haegeman G (1998) Dissociated glucocorticoids with anti-inflammatory potential repress interleukin-6 gene expression by a nuclear factor-κB-dependent mechanism. Mol Pharm 56:797–806

Vayssiere BM, Dupont S, Choquart A, Petit F, Garcia T, Marchandeau C, Gronemeyer H, Resche-Rigon M (1997) Synthetic glucocorticoids that dissociate transactivation and AP-1 transrepression exhibit antiinflammatory activity in vivo. Mol Endocrinol 11:1245–1255

Weinstein RS, Jilka RL, Parfitt AM, Manolagas SC (1998) Inhibition of osteoblastogenesis and promotion of apoptosis of osteoblasts and osteocytes by GCs. Potential mechanisms of their deleterious effects on bone. J Clin Invest 102:274–282

Yoshiuchi I, Shingu R, Nakajima H, Hamaguchi T, Horikawa Y, Yamasaki T, Oue T, Ono A, Miyagawa JI, Namba M, Hanfusa T, Matsuzawa Y (1998) Mutation/polymorphism scanning of glucose-6-phosphatase gene promoter in noninsulin-dependent diabetes mellitus patients. J Clin Metab 83:1016–1019

Subject Index

14-3-3   185

ABCD assay   286, 289
ABCDE assay   288, 290
acetylation   233, 236, 251, 256
acute phase response   298
adhesion molecules   10
adrenocorticotropic hormone
  (ACTH)   86, 306, 311
airway hyperresponsiveness   14
allergic disorders   153
androgen   341
anisomycin   137
antagonism   133
anti-inflammatory proteins   6
antigen   153
AP-1   132, 144, 298, 317, 322,
  324, 326
apoptosis   10, 155, 299
arachidonic acid metabolism   156
asthma   1, 133
asthmatic children   87
asthmatic inflammation   14
ATF2   133

bone density   56
bone mineral density   56

c-Fos   132, 301
c-Jun N-terminal kinase (JNK)   132
calcium   160

calcium/phosphate homeostasis   59
cAMP response element-binding
  protein (CREB)-binding protein   133
casein gene expression   214
cataract   72, 75
chaperone   96, 179
chromatin   92
– remodeling   111
– structure   4
co-activator   133
cofactor   302
compression fractures   56
corepressors   281, 285
corticosteroids   1, 14, 65, 66, 69,
  77, 84, 87
– receptor   312
– resistance   15, 16
– therapy   73
see also glucocorticoids
cross-talk   133, 298
cyclooxygenase-2   299
cytokine   8, 298, 319
cytokine genes   156
cytoskeletal proteins   182

D-loop   298
degranulation   156
dendritic cells   12
dexamethasone   134, 140

dimerization-defective GR 138, 298, 315
dominant negative inhibitor 201
dominant negative mutants 302

Elk-1 133
endothelial cells 13
eosinophils 11
episcleritis 66
epithelial cells 13
Erk-1/2 phosphorylation 164
extracellular-regulated kinase (ERK)1/2 133, 164

Fas ligand 320
fluorescence loss in photobleaching (FLIP) 118
fluorescence recovery after photobleaching (FRAP) 98, 118
Fos 301

gene expression 233, 243, 246, 257
GILZ 320, 325, 328
glaucoma 66, 74
glucocorticoids 1, 25, 39, 91, 155, 214, 220, 234, 246, 297, 319, 321, 323, 330, 331
– receptor 2, 59, 91, 111, 118, 138, 144, 177, 215, 217, 246, 247, 249, 251, 257, 279, 302, 305, 308, 314
  – hGRβ 197, 198
  – monomers 298
– therapy 25, 39
– treatment 83
see also corticosteroids
GMCF 61
GR-dim 138
GR-interacting proteins 184
GR-LS7 138

haematopoietic progenitors 154
HAT 241
HDAC 237, 239, 253
Herpes simplex type1 (HSV-1) 188
histone deacetylase 302
homology model 343
hsp90 92, 93
hypothalamic-pituitary-adrenal (HPA) 306, 311, 324

IκB 299
IκB kinase 299
IgE 153, 160
IL-2 319, 326, 330
IL-7R 328, 330
immune response 297
immunostimulatory 319
immunosuppressive 320, 321, 330
inflammation 6, 181, 234, 246, 255, 297, 364, 367
inflammatory enzymes 9
inflammatory receptors 9
inhibitory 319, 322
inositol 1,4,5-trisphosphate (IP3) 160
interference 322, 328
intracellular localization 183
intranuclear movement 111

Jak-Stat pathway 220
JNK 162
JNK phosphorylation 162
Jun 301

ligand-binding pocket (LBP) 341, 343, 344, 346, 347, 348, 350, 352
LIM domains 302
lipopolysaccharide 298
luteinizing hormone/ follicle-stimulating hormone (LH/FSH) 59

Subject Index

macrophages 11
mammary epithelial cells 213, 216
mast cells 12, 153
– activation 154, 155
– development 154
– growth factors 154
– mediator release 155
– number 154
– survival 155
matrix-assisted laser desorption ionization-time of flight-base 190
mesenchymal stem cells 61
metallothionein IIA 300
mitogen-activated protein kinase (MAPK) 133, 237, 242
MKP-1 166
monitoring
– flow cytometric 27
– immune 26, 27
MSK1 242, 255
mucus secretion 14

negative response elements 280, 281
nerve growth factor 154
neutrophils 13
NF-κB 144, 181, 234, 235, 236, 239, 240, 241, 245, 247, 249, 251, 255, 257, 258, 298, 317, 322, 326
NF-AT 298
nonsteroidal anti-inflammatory drug (NSAIDs) 234, 244, 255
nuclear export 102
nuclear hormone receptors 215, 226
nuclear receptor degradation 101 ff.
nuclear receptor superfamily 131
nuclear transport 93
nucleus 92

ophthalmology 65, 66, 77
orientations 350
osmotic shock 137
osteoclastogenesis 61
osteoclasts 60

p23 94
p38MAPK 137
phorbol ester 298
phospholipase A2 160
phospholipase C 160, 162
pro-inflammatory 319
progesterone 341, 344, 346, 352
prolactin 213, 216, 219, 220
proteasome 100, 168
protein kinase C (PKC) 160

Raf-1 185
RANKL 61
RBL-2H3 157
recycling 94
repressive 321
retinoic acid receptor (RAR) 133
RU486 138

scleritis 67, 70
septic shock 297
serum response element (SRE) 133
signal transduction 159, 160
spontaneous fractures 56
Stat5 (signal transducer and activator of transcription) 214, 216, 217, 219, 298, 317
stem cell factor 154
suppression 319, 322

T cells 321, 324, 325, 328, 331
T-lymphocytes 12
thymocytes 320, 323, 324
thyroid hormone (TR) 133
TIGR gene 74
TNF-α 137, 140, 297

TPA-response elements
  (TREs)   132
transactivation   95, 131, 157, 359,
  362, 363, 364, 366, 367
transactivation domains   298
transcription   98, 301, 320, 321,
  324, 325
transcriptional coactivators   218
transcriptional cross-talk   221, 309
transcriptional synergy   215
transrepression   132, 157, 359,
  362, 363, 364, 366, 367

triamcinolone   301
two-hybrid screen   302
tyrosine phosphatase   166

ubiquitin   100
ultraviolet (UV) radiation   132
uveitis   66, 70, 77

valproic acid   302

yeast cells   301

Ernst Schering Research Foundation Workshop

Editors: Günter Stock
Monika Lessl

Vol. 1 *(1991)*: Bioscience ⇆ Society – Workshop Report
Editors: D. J. Roy, B. E. Wynne, R. W. Old

Vol. 2 *(1991)*: Round Table Discussion on Bioscience ⇆ Society
Editor: J. J. Cherfas

Vol. 3 *(1991)*: Excitatory Amino Acids and Second Messenger Systems
Editors: V. I. Teichberg, L. Turski

Vol. 4 *(1992)*: Spermatogenesis – Fertilization – Contraception
Editors: E. Nieschlag, U.-F. Habenicht

Vol. 5 *(1992)*: Sex Steroids and the Cardiovascular System
Editors: P. Ramwell, G. Rubanyi, E. Schillinger

Vol. 6 *(1993)*: Transgenic Animals as Model Systems for Human Diseases
Editors: E. F. Wagner, F. Theuring

Vol. 7 *(1993)*: Basic Mechanisms Controlling Term and Preterm Birth
Editors: K. Chwalisz, R. E. Garfield

Vol. 8 *(1994)*: Health Care 2010
Editors: C. Bezold, K. Knabner

Vol. 9 *(1994)*: Sex Steroids and Bone
Editors: R. Ziegler, J. Pfeilschifter, M. Bräutigam

Vol. 10 *(1994):* Nongenotoxic Carcinogenesis
Editors: A. Cockburn, L. Smith

Vol. 11 *(1994)*: Cell Culture in Pharmaceutical Research
Editors: N. E. Fusenig, H. Graf

Vol. 12 *(1994):* Interactions Between Adjuvants, Agrochemical and Target Organisms
Editors: P. J. Holloway, R. T. Rees, D. Stock

Vol. 13 *(1994):* Assessment of the Use of Single Cytochrome P450 Enzymes in Drug Research
Editors: M. R. Waterman, M. Hildebrand

Vol. 14 *(1995):* Apoptosis in Hormone-Dependent Cancers
Editors: M. Tenniswood, H. Michna

Vol. 15 *(1995):* Computer Aided Drug Design in Industrial Research
Editors: E. C. Herrmann, R. Franke

Vol. 16 (1995): Organ-Selective Actions of Steroid Hormones
Editors: D. T. Baird, G. Schütz, R. Krattenmacher

Vol. 17 (1996): Alzheimer's Disease
Editors: J.D. Turner, K. Beyreuther, F. Theuring

Vol. 18 (1997): The Endometrium as a Target for Contraception
Editors: H.M. Beier, M.J.K. Harper, K. Chwalisz

Vol. 19 (1997): EGF Receptor in Tumor Growth and Progression
Editors: R. B. Lichtner, R. N. Harkins

Vol. 20 (1997): Cellular Therapy
Editors: H. Wekerle, H. Graf, J.D. Turner

Vol. 21 (1997): Nitric Oxide, Cytochromes P 450,
and Sexual Steroid Hormones
Editors: J.R. Lancaster, J.F. Parkinson

Vol. 22 (1997): Impact of Molecular Biology
and New Technical Developments in Diagnostic Imaging
Editors: W. Semmler, M. Schwaiger

Vol. 23 (1998): Excitatory Amino Acids
Editors: P.H. Seeburg, I. Bresink, L. Turski

Vol. 24 (1998): Molecular Basis of Sex Hormone Receptor Function
Editors: H. Gronemeyer, U. Fuhrmann, K. Parczyk

Vol. 25 (1998): Novel Approaches to Treatment of Osteoporosis
Editors: R.G.G. Russell, T.M. Skerry, U. Kollenkirchen

Vol. 26 (1998): Recent Trends in Molecular Recognition
Editors: F. Diederich, H. Künzer

Vol. 27 (1998): Gene Therapy
Editors: R.E. Sobol, K.J. Scanlon, E. Nestaas, T. Strohmeyer

Vol. 28 (1999): Therapeutic Angiogenesis
Editors: J.A. Dormandy, W.P. Dole, G.M. Rubanyi

Vol. 29 (2000): Of Fish, Fly, Worm and Man
Editors: C. Nüsslein-Volhard, J. Krätzschmar

Vol. 30 (2000): Therapeutic Vaccination Therapy
Editors: P. Walden, W. Sterry, H. Hennekes

Vol. 31 (2000): Advances in Eicosanoid Research
Editors: C.N. Serhan, H.D. Perez

Vol. 32 (2000): The Role of Natural Products in Drug Discovery
Editors: J. Mulzer, R. Bohlmann

Vol. 33 (2001): Stem Cells from Cord Blood, In Utero Stem Cell Development, and Transplantation-Inclusive Gene Therapy
Editors: W. Holzgreve, M. Lessl

Vol. 34 (2001): Data Mining in Structural Biology
Editors: I. Schlichting, U. Egner

Vol. 35 (2002): Stem Cell Transplantation and Tissue Engineering
Editors: A. Haverich, H. Graf

Vol. 36 (2002): The Human Genome
Editors: A. Rosenthal, L. Vakalopoulou

Vol. 37 (2002): Pharmacokinetic Challenges in Drug Discovery
Editors: O. Pelkonen, A. Baumann, A. Reichel

Vol. 38 (2002): Bioinformatics and Genome Analysis
Editors: H.-W. Mewes, B. Weiss, H. Seidel

Vol. 39 (2002): Neuroinflammation – From Bench to Bedside
Editors: H. Kettenmann, G. A. Burton, U. Moenning

Vol. 40 (2002): Recent Advances in Glucocorticoid Receptor Action
Editors: A. Cato, H. Schaecke, K. Asadullah

Vol. 41 (2002): The Future of the Oocyte: Basic and Clinical Aspects
Editors: J. Eppig, C. Hegele-Hartung, M. Lessl

Supplement 1 (1994): Molecular and Cellular Endocrinology of the Testis
Editors: G. Verhoeven, U.-F. Habenicht

Supplement 2 (1997): Signal Transduction in Testicular Cells
Editors: V. Hansson, F. O. Levy, K. Taskén

Supplement 3 (1998): Testicular Function:
From Gene Expression to Genetic Manipulation
Editors: M. Stefanini, C. Boitani, M. Galdieri, R. Geremia, F. Palombi

Supplement 4 (2000): Hormone Replacement Therapy
and Osteoporosis
Editors: J. Kato, H. Minaguchi, Y. Nishino

Supplement 5 (1999): Interferon:
The Dawn of Recombinant Protein Drugs
Editors: J. Lindenmann, W.D. Schleuning

Supplement 6 (2000): Testis, Epididymis and Technologies
in the Year 2000
Editors: B. Jégou, C. Pineau, J. Saez

Supplement 7 (2001): New Concepts in Pathology and Treatment
of Autoimmune Disorders
Editors: P. Pozzilli, C. Pozzilli, J.-F. Kapp

Supplement 8 (2001): New Pharmacological Approaches
to Reproductive Health and Healthy Ageing
Editors: W.-K. Raff, M. F. Fathalla, F. Saad

Supplement 9 (2002): Testicular Tangrams
Editors: F.F.G. Rommerts, K.J. Teerds

Supplement 10 (2002): Die Architektur des Lebens
Editors: G. Stock, M. Lessl